Advances in Intelligent Systems and Computing

Volume 232

For further volumes:
http://www.springer.com/series/11156

Luis F. Castillo · Marco Cristancho
Gustavo Isaza · Andrés Pinzón
Juan Manuel Corchado Rodríguez
Editors

Advances in Computational Biology

Proceedings of the 2nd Colombian Congress on Computational Biology and Bioinformatics (CCBCOL)

Editors
Luis F. Castillo
University of Caldas
Manizales
Colombia

Marco Cristancho
Cenicafé - Centro Nacional de Investigaciones del Café en Colombia

Gustavo Isaza
University of Caldas
Manizales
Colombia

Andrés Pinzón
BIOS - Centro Bioinformática y Biologia Computacional de Colombia
Manizales
Colombia

Juan Manuel Corchado Rodríguez
Department of Computer Science
School of Science
University of Salamanca
Salamanca
Spain

ISSN 2194-5357 ISSN 2194-5365 (electronic)
ISBN 978-3-319-01567-5 ISBN 978-3-319-01568-2 (eBook)
DOI 10.1007/978-3-319-01568-2
Springer Cham Heidelberg New York Dordrecht London

Library of Congress Control Number: 2013944537

Printed on acid-free paper

Springer is part of Springer Science+Business Media (www.springer.com)

Preface

Bioinformatics and Computational Biology are areas of knowledge that have emerged due to advances that have taken place in the Biological Sciences and its integration with Information Sciences. The expansion of projects involving the study of genomes has led the way in the production of vast amounts of sequence data which needs to be organized, analyzed and stored to understand phenomena associated with living organisms related to their evolution, behavior in different ecosystems, and the development of applications that can be derived from this analysis.

In Colombia the main reference to consider for the advancement of Science is the National Development Plan 2010-2014, which is based on what the government calls, the locomotives of growth. The areas or economic activities to be included that have been selected with priority in the four years' duration are (i) new innovation based sectors, (ii) agriculture and rural development, (iii) Housing and friendly cities, (iv) energy mining development and expansion, (v) transport infrastructure.

The first locomotive is focused on the need to promote the development of emerging sectors based on innovation, which features information technology and telecommunications, and biotechnology, among others. Several strategies have been proposed to generate knowledge that can be applied to production processes and the solution of problems affecting the community. This requires the formation of human resources, project financing and organization of institutions to promote research and innovation.

Bioinformatics becomes a crucial area of development in the National policy "COMMERCIAL DEVELOPMENT OF BIOTECHNOLOGY FROM THE SUSTAINABLE USE OF BIODIVERSITY", because it can support the process of search and discovery of molecules, genes or active ingredients that are present in our biodiversity, so through biotechnology they can be industrially produced in a sustainable scheme.

We can envision Colombia's effort to strengthen this field of research, on data from the Biodiversity Information System of Colombia - SiB, analyzed in the Report on the State of Renewable Natural Resources and the Environment

2009, which describes the departments with the largest number of known species and include Quindío, Risaralda, Caldas, Cundinamarca, Valle, Antioquia and Boyacá, located in the central Andean region.

The CCBCOL'13 Congress has the following objectives: Submit progress in research in computational biology and related areas and their relations and international scope, identify strengths and weaknesses in relation to infrastructure, research training and development strategies of computational biology in Colombia, advance the establishment of agreements that allow the integration of infrastructure, cooperation and development of research projects relevant and competitive, nationally and internationally, advance the establishment of agreements that allow the integration of infrastructure, cooperation and development of research projects relevant and competitive, nationally and internationally, encourage contact between scientists from multiple disciplines (computer science, biology, mathematics, statistics, chemistry, etc.) that conduct research in computational biology and related areas in the country, and launch the Colombian Society for Computational Biology.

Another important achievement and articulated with this event, is the founding of the Colombia's Computational Biology and Bioinformatics Center (BIOS), which is headquartered in the city of Manizales (the central Colombian coffee production zone), a leading national supercomputing facility devoted to providing services to government, academia and businesses interested in Biotechnology research, Development and Bioprospecting.

This volume compiles accepted contributions for the 2nd Edition of the Colombian Computational Biology and Bioinformatics Congress CCBCOL, after a rigorous review process in which 54 papers were accepted for publication from 119 submitted contributions.

Luis F. Castillo
Marco Cristancho
Gustavo Isaza
Andrés Pinzón
CCBCOL'13 Programme Co-chairs

Organization

General Co-chairs

Luis Fernando Castillo	Universidad de Caldas, Colombia
Andrés Pinzón	BIOS Centro de Bioinformática y Biología Computacional, Colombia
Marco Cristancho	Cenicafé, Colombia

Organizing Committee

Luis Fernando Castillo	Universidad de Caldas, Colombia
Gustavo A. Isaza E.	Universidad de Caldas, Colombia
Marco Cristancho	Cenicafé, Colombia
Andrés Pinzón	BIOS Centro de Bioinformática y Biología Computacional, Colombia
Dago Bedoya	BIOS Centro Bioinformática y Biología Computacional, Colombia
Emiliano Barreto	Universidad Nacional de Colombia, Colombia
Diego Mauricio Riaño	Universidad de los Andes, Colombia

Scientific Committee

Luis Fernando Castillo	Universidad de Caldas, Colombia
Gustavo Isaza E.	Universidad de Caldas, Colombia
Carlos Alberto Ruiz Villa	Universidad de Caldas, Colombia
Oscar Julián Sanchez	Universidad de Caldas, Colombia
Lucimar Gomes Dias	Universidad de Caldas, Colombia
German LopezGartner	Universidad de Caldas, Colombia
Andrés Paolo Castaño	Universidad de Caldas, Colombia
Maria Helena Mejía	Universidad de Caldas, Colombia
Maria Mercedes Zambrano	Corpogen, Colombia
Juan Manuel Anzola	Corpogen, Colombia

Jorge Duitama	CIAT, Colombia
Adriana Muñoz	University of Maryland, USA
Patricia Vélez	Universidad del Cauca, Colombia
Mauricio Rodriguez	Centro Bioinformática y Biología Computacional, Colombia
Andrés Pinzón	Centro Bioinformática y Biología Computacional, Colombia
Alvaro Gaitán	Cenicafé, Colombia
Marco Cristancho	Cenicafé, Colombia
Carlos Ernesto Maldonado	Cenicafé, Colombia
Emiliano Barreto	Universidad Nacional de Colombia, Colombia
NestorDario Duque	Universidad Nacional de Colombia, Colombia
Luis Fernando Cadavid	Universidad Nacional de Colombia, Colombia
Luz Mary Salazar Pulido	Universidad Nacional de Colombia, Colombia
Edgar Antonio Reyes	Universidad Nacional de Colombia, Colombia
Luis Fernando Niño	Universidad Nacional de Colombia, Colombia
Silvia Restrepo	Universidad de los Andes, Colombia
Diego Mauricio Riaño	Universidad de los Andes, Colombia
Adriana Bernal	Universidad de los Andes, Colombia
Felipe García Vallejo	Universidad del Valle, Colombia
Pedro Antonio Moreno	Universidad del Valle, Colombia
Mauricio Corredor Rodríguez	Universidad de Antioquia, Colombia
Omar Triana	Universidad de Antioquia, Colombia
Carlos Muskus	Universidad de Antioquia, Colombia
Juan Manuel Corchado	Universidad de Salamanca, Spain
Juan F. de Paz	Universidad de Salamanca, Spain
Emilio Corchado	Universidad de Salamanca, Spain
Sara Rodriguez	Universidad de Salamanca, Spain
Florentino Fdez-Riverola	Universidad de Vigo, Spain
Daniel Glez-Peña	Universidad de Vigo, Spain
Miguel Reboiro Jato	Universidad de Vigo, Spain
Hugo López F.	Universidad de Vigo, Spain
Analia Lourenco	Universidad de Vigo, Spain
David Torrens	Centro Supercomputación Barcelona (BSC), Spain
Jorge Enrique Gómez	Universidad del Quindio, Colombia
Jannet Gonzalez	Universidad Javeriana, Colombia
Nelson Fernández	Universidad de Pamplona, Colombia
Manuel Alfonso Patarroyo	Fundación Instituto de Inmunología de Colombia (FIDIC), Colombia
Juan F. Alzate Restrepo	Centro Nacional de Secuenciación Genómica-CNSG, Colombia
Sarah Ayling	The Genome Analysis Centre (TGAC), United Kingdom
Leonardo Mariño	NCBI, USA

Acknowledgement

This book has been sponsored by Colciencias through the National Call 612/2013 to form a bank of eligible scientific events National and International component of social appropriation of knowledge that take place in Colombia between the second half of 2013 and the first half of 2014.

Contents

Predictive Modeling of Signaling Transduction Mediated by Tyrosine-Kinase Receptors

Ivan Mura

EAN University
Carrera 11 No. 78 - 47 Bogotá - Colombia
imura@ean.edu.co

Abstract. HER members of the tyrosine-kinase family of transmembrane receptors are initiators of signaling cascades driving crucial cellular process, such as gene transcription, cell cycle progression, apoptosis. Given their capacity of oncogenic transformation these receptors are the target of selective anticancer drugs, which in-vivo are however not as effective as anticipated by in-vitro experiments. Translating HER inhibitors into effective therapies to block the oncogenic signaling cascades will be facilitated by models that can provide reliable predictions for the evolution of the intricate HER mediated signaling networks. This work presents a process-algebra based approach to compactly specify and simulate HER signaling models. The proposed HER activation model can be easily reused as a building block in larger models of signaling.

Keywords: Tyrosine kinase receptors, Signaling pathways, Cancer therapies, Computational modeling, Stochastic simulation.

1 Introduction

The tyrosine-kinase family of transmembrane receptors includes at least 17 different classes of receptors, among which the human epidermal growth factor receptors (hereafter, HER). There are four structurally related HER receptors: HER1, HER2, HER3, and HER4. HERs play a crucial role in cell signaling, mediating cell proliferation, migration, differentiation, apoptosis, and cell motility, owing to their ability to activate important cytoplasmic signaling relaying molecules such as PI3K, Ras, Stat3, Grb2 among others [1].

HER family receptors are often over-expressed, amplified or mutated in many forms of cancer. HER1 is found to be over-expressed in more than 80% of head and neck cancers, 50% of gliomas, 10 to 15% of non-small cell lung cancers in the west [7]. Amplification and overexpression of HER2 is seen in about 25 to 30% of breast cancers [13]. Given their frequent altered expression or dysregulation in human tumors, HERs are target of selective anticancer drugs.

In spite of the effective inhibition shown in vitro, only a small percent of the patients that receive HER antagonist therapies respond to it (for instance, 20-30% in the case the HER2 inhibitor *trastuzumab* [6]). This indicates the existence of a complex set of intertwined relationships that the HER family members

L.F. Castillo et al. (eds.), *Advances in Computational Biology*,
Advances in Intelligent Systems and Computing 232,
DOI: 10.1007/978-3-319-01568-2_1,

exhibit. A drug is attempting to switch off the signaling cascade by specifically targeting one receptor type, while at the same time other HER members are compensating for the effects of the drug with sustaining signaling activation [2]. This unexpectedly complex behavior may well explain the disappointing results of various trials of candidate HER antagonists drugs and offers an excellent challenge for the deployment of Systems Biology modeling approaches [12].

In this paper we consider the modeling of the initial events of the HER signaling pathways, which are triggered by the binding of the HER receptors with their ligands, the dimerization of receptors and the phosphorylation of their intracellular domains. More specifically, inspired by the recent study reported in [2], we consider a scenario where the HER1 and HER2 receptors are sinergistically working to sustain signaling. This scenario only considers a few molecules, yet the temporal evolution of the system results in an combinatorial mesh of interacting partners. This complexity offers at least two types of challenges to the modeling tools: From a model specification point of view, such complexity makes model definition long and cumbersome (and hence error prone), whereas from the solution point of view it makes model simulation a computationally intensive task, due to the large number of possible reactions.

To tackle these issues, we shall be using an approach based on the modeling language BlenX [3], which adopts a process-algebra model specification to nicely manage the complexity inherent to the combinatorial explosion of the number of species configurations. We offer two contributions in this paper. The first one is represented by the example of application of the process-algebra modeling approach, which finds in the complexity of the HER activation models an ideal application area. The second contribution consists in the model itself, which can be used as a building block in larger models of signaling or easily adapted to define activation models for other tyrosine-kinase receptors.

This paper is organized as follows. We provide in Section 2 an introduction to the structure and function of the HER family receptors, and then we focus in Section 3 on the definition of a HER activation and phosphorylation model in BlenX. Section 4 shows the results of model validation and finally Sections 5 and 6 provide a short discussion and conclusions for the paper, respectively.

2 HER Receptors

We provide in this section a short description of the molecular details of the HER activation process. Our discussion is limited to HER1 and HER2, but it readily applies to HER3 and HER4 as well.

HER receptors are mostly located on the cell membrane, and are made up of an extracellular region or ectodomain, a single transmembrane-spanning region, and a cytoplasmic tyrosine kinase domain. HER proteins are capable of forming homodimers, heterodimers, and possibly higher-order oligomers upon activation by a subset of potential growth factor ligands. There are many growth factors that activate HER1 receptors, among which EGF, TGF-α and neuregulins. Although unliganded homodimers and heterodimers can also form, they are unlikely to be active [10]. Multiple phosphorylation sites exist in the intracellular domains of HERs. For instance, at least 5 sites appear to be relevant for downstream signal rely in HER1 [5] and at least 4 in HER2 [9]. The dimer formation is reversible; HER dimers can dissociate and reassociate regardless of their phosphorylation status.

This means that each single HER molecule can exist in many possible configurations. If each HER molecule carried 4 phosphorylation sites, the total number of possible configurations would be of the order of 2^9. Such a combinatorial number of configurations makes most modeling approaches cumbersome if not impracticable. Any modeling tool requiring the explicit encoding of all possible species configurations and of all the reactions they participate in would be impossible to use. We shall see in the next section how the system can be easily encoded with the BlenX approach, which does not require fully unfolding the set of possible configurations of the species.

3 Modeling Methods

BlenX [3] is a modeling language based on the process calculi and rule-based paradigms. It is specifically designed to account for the complexity of biological networks. The advantages of a rule-based approach become evident when the biological system exhibits a combinatorial number of possible configurations as in the case of the HER early signaling network. Given the space limitation, we just provide in this section a few clues about the BlenX modeling approach. A complete explanation with examples of application, can be found in [4].

BlenX uses a general abstraction of a biological network that separately considers biological entities and their interaction capabilities. Biological entities are encoded in BlenX as objects called *boxes*. Boxes expose interfaces called *binders*, which mimic domains of interaction, for instance for complexation and phosphorylation. The interaction capabilities of binders are determined by its type attribute, which is controlled by each box internal *process*, which updates them according to the box state.

BlenX resembles a normal programming language. A box for the EGF ligand of HER1 would be defined as follows:

```
let egf : bproc = #(egfrec,egfrec)[nil];
```

This text is declaring the box *egf* as having one binder named *egfrec*, having a type *egfrec*. This binder models the site across which EGF binds to the receptor. The text within square brackets is the box process. In the case of EGF, the process is `nil`, i.e. the null process, which does nothing and hence the type of binder *egfrec* will ever be changed. This models the fact that the complexation with HER1 is always possible for EGF molecules. A box for a HER1 molecule needs more binders: one called *h1lig* for the interaction with the ligand, one called *h1dim* for the dimerization with another HER molecule, plus at least one binder *h1ph* to model a single phosphorylation site. Moreover, it would need a non-null internal process to model the phosphorylation process so that it can happen only after ligand binding and dimerization. We declare HER1 and HER2 boxes (for the sake of conciseness, each one with just one phosphorylation site), as follows:

```
let her1 :: bproc = #(h1lig,h1lig),#(h1dim,h1dim),
    #(h1ph,free)[h1_proc];
let her2 :: bproc = #(h2dim,h2dim),#(h2ph,free),
    [h2_proc];
```

Binding interaction capabilities are called *affinities* in BlenX, and are defined for pairs of binder types. For instance, to model the reversible complexation of EGF and HER1, we declare in BlenX a tuple as follows: (egfrec,h1lig,k_{on},k_{off}), where k_{on} and k_{off} are the rate of complex formation and dissociation, respectively. The rate information is used at simulation time. BlenX uses a discrete-space discrete-time interpretation of model evolution, according to Gillespie stochastic molecular dynamics [8]. Intuitively, the rates are proportional to the speed with which the biochemical transformations occur.

The internal processes keep track of the history of the boxes and appropriately change the state of the binders. For instance, process h1_proc of box her1 will change the type of binder h1ph from free to phospho when an egf box binds h1lig and an HER molecule binds h1dim.

To model our scenario of HER1 and HER2 interaction, we just need the 3 box declarations given above, and the specification of the two internal processes h1_proc and h2_proc, a BlenX program that is as compact as 15 lines of code. Additionally, we need 4 affinities declaration, one for the HER1-EGF binding, and 3 for the hetero and homodimerizations. Overall, a very compact BlenX model accounts for all the possible configurations of the species and their reactions.

4 Results

We present in this section the results of simulation experiments aiming at validating the HER1-HER2 interaction model. We start our experiments by reproducing an experimental setting that was used in the paper [11], where the activation of HER1 in hepatocytes is considered. In that study, an in-vitro culture of HER1 cells is stimulated by a single EGF pulse, and the phosphorylation level of the receptors is measured over time by immunoprecipitation. By using the kinetics of EGF binding, dimerization and phosphorylation given in [11] to instanciate the BlenX model, we could reproduce the HER1 time course of phosphorylation over the time window [0-120] seconds, as shown in Fig.1.

Then, we used the same rates in the complete model, when also HER2 is considered. Without introducing any new model parameter, we reproduced the relative proportions of HER1-HER1 homodimers and HER1-HER2 heterodimers at equilibrium (10 minutes after EGF stimulation) in four breast cancer cell lines. Table 1 shows the very good match of simulation results with respect to

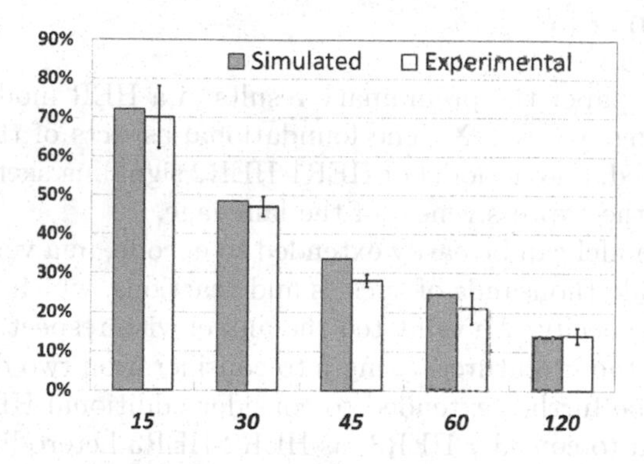

Fig. 1. Comparison of experimental and simulated HER1 phosphorylation over time. Experimental results are provided with error bars. Simulation results were estimated at 95% confidence level and are within 10% of the estimated value.

Table 1. Results of HER1-HER1 homodimer versus HER1-HER2 heterodimer ratio simulations for various breast cancer cell lines. Confidence intervals of simulation are within 10% of the estimated value.

Cell Line	Num HER1	Num HER2	Ratio Homodimers/Heterodimers	
			Experimental results	Simulation results
AU565	204560	1447688	0.071	0.072±0.0012
SKBR3	143559	1402832	0.042	0.050±0.0014
SKOV3	387771	657088	1.627	1.393±0.0035
H1650	158872	53810	15.625	14.934±0.0776

the experimental data obtained in [2]. These results validate the BlenX model and demonstrate its predictive capabilities.

5 Discussion

The BlenX model presented in this paper appears to be able to reproduce the experimentally observed behavior of HER receptors under different conditions. Main advantages of the proposed modeling approach are its compactness and easy extensibility. For instance, to pass from a pure HER1 model, i.e. one considered in the experimental setting of Kholodenko, to the HER1-HER2 interaction model, we just *added* the HER2 specifications to the existing code, without having to *change* it. This is due to the basic modeling choice of not specifying explicitly the reactions each species can participate in, but just to define its possible interactions. Specifically, the definition of interactions is made at the level of the binders, which represent the domains of molecules. Such a subtle difference obviate the necessity of enumerating the set of reactions, their reactants and products.

6 Conclusions

We showed in this paper the preliminary results of a HER model based on the process-algebra language BlenX. The foundational aspects of the modeling approach are presented, and a model of HER1-HER2 signaling sketched to provide some clues about the expressiveness of the language.

The proposed model can be easily extended to encode, in a very compact way, systems that include thousands of species and reactions, which would be otherwise impossible to specify. We validated the model with respect to experimental data coming from the literature, taking into consideration two different studies. The model could be further extended to consider additional HER members. In particular, we plan to consider HER3, as HER2-HER3 heterodimers are among the most active complexes in relaying growth factor signals.

References

1. Citri, A., Yarden, Y.: EGF-ERBB signalling: towards the systems level. Nat. Rev. Mol. Cell Biol. 7, 505–516 (2006)
2. DeFazio-Eli, L., Strommen, K., Dao-Pick, T., Parry, G., et al.: Quantitative assays for the measurement of HER1-HER2 heterodimerization and phosphorylation in cell lines and breast tumors. Breast Cancer Res. 13 (2011)
3. Dematté, L., Priami, C., Romanel, A.: Modelling and simulation of biological processes in BlenX. Perform Eval. Rev. 35, 32–39 (2008)
4. Dematté, L., Priami, C., Romanel, A.: The blenX language: A tutorial. In: Bernardo, M., Degano, P., Zavattaro, G. (eds.) SFM 2008. LNCS, vol. 5016, pp. 313–365. Springer, Heidelberg (2008)
5. Downward, J., Parker, P., Waterfield, M.D.: Autophosphorylation sites on the epidermal growth factor receptor. Nature 311, 483–485 (1984)
6. Esteva, F.J., Valero, V., Booser, D., Guerra, L.T., et al.: Unraveling resistance to trastuzumab (Herceptin): insulin-like growth factor-I receptor, a new suspect. J. Clin. Oncol. 20, 1800–1808 (2002)
7. Frederick, L., Wang, X.Y., Eley, G., James, C.D.: Diversity and frequency of epidermal growth factor receptor mutations in human gliobastomas. Cancer Res. 60, 1383–1387 (2000)
8. Gillespie, D.T.: A general method for numerically simulating the stochastic time evolution of coupled chemical reactions. J. Comp. Physics 22, 403–434 (1976)
9. Hazan, R., Margolis, B., Dombalagian, M., Ullrich, A., et al.: Identification of autophosphorylation sites of HER2/neu. Cell Growth Differ. 1, 3–7 (1990)
10. Jura, N., Endres, N.F., Engel, K., Deindl, S., et al.: Mechanism for activation of the EGF receptor catalytic domain by the juxtamembrane segment. Cell 137, 1293–1307 (2009)
11. Kholodenko, B.N., Demin, O.V., Moehren, G., Hoek, J.B.: Quantification of short term signaling by the epidermal growth factor receptor. Biol. Chem. 274, 30169–30181 (1999)
12. Kitano, H.: Systems Biology: a brief overview. Science 295, 1662–1664 (2002)
13. Slamon, D.J., Godolphin, W., Jones, L.A., Holt, J.A., et al.: Studies of the HER-2/neu proto-oncogene in human breast and ovarian cancer. Science 244, 707–712 (1989)

Bioinformatic Analysis of Two Proteins with Suspected Linkage to Pulmonary Atresia with Intact Ventricular Septum

Oscar Andrés Alzate Mejía[1] and Antonio Jesús Pérez Pulido[2]

[1] Docente Universidad Autónoma de Manizales, Colombia
[2] Docente Universidad Pablo de Olavide, Sevilla - España
oalzate@autonoma.edu.co, ajperez@upo.es

Abstract. Pulmonary atresia with intact ventricular septum (PA-IVS) is a congenital heart disease characterized by occlusion of the pulmonary valve causing complete obstruction of the outflow tract from the right ventricle to the lungs. Some authors attribute the origin of disease to genetic causes and the mutation of genes WFDC8 and WFDC9 have been proposed as related to with its pathogenesis. Based on this suspicion, a bioinformatic analysis to their gene products was made to find the relationship between the mutation and disease.

Were reviewed the annotations, domains and structures of these proteins to study their biological characteristics; to find equivalent sequences in other species the orthologous of proteins were searched, so it was made a phylogenetic analysis and were searched conserved domains using alignments. Similarly, were searched the tissues in which genes are expressed to find its relation to heart and was studied the intergenic sequence to uncover regulatory sequences associated with cardiac development.

The results suggest that the human proteins WFDC8 and WFDC9, currently related to the immune system, they are also related to extracellular matrix proteins, and they could be expressed in heart tissue and embryonic, in addition, was found that the corresponding intergenic sequence has different binding sites for factors transcription related to the development of heart and heart valves.

Keywords: Pulmonary atresia, WFDC8, WFDC9, heart tissue.

1 Introduction

Congenital heart diseases are disorders of the heart and great vessels that exist before birth. They describe structural or functional injuries of one or more of the four cardiac chambers or septum which separate them, or their respective valves.

The Pulmonary atresia with intact ventricular septum (PA-IVS) is a congenital heart defect characterized by a pulmonary valve atresia, a complete obstruction of the outflow tract from the right ventricle to the lungs. But unlike other diseases of heart septum that separates both ventricles, it is intact. The etiology of the disease begins to be known. Some familial cases suggest genetic basis (1). A report of PA-IVS was

L.F. Castillo et al. (eds.), *Advances in Computational Biology*,
Advances in Intelligent Systems and Computing 232,
DOI: 10.1007/978-3-319-01568-2_2, © Springer International Publishing Switzerland 2014

published in monozygotic twins (2); in them and through an analysis of comparative genomic hybridization (aCGH) revealed a deletion of 55 kb at the level of the chromosome 20q13.12 involving WFDC8 and WFDC9 genes in both patients.

In humans, WFDC genes are encoding small proteins that can contain one or more domains WFDC. It is knows that the domain WFDC performs part of innate immunity, which is present in some inhibitors serine protease that stop endogenous peptidases secreted by cells proinflammatory, thus avoiding damage and inflammation of the tissues. Similarly, it has been shown to inhibit proteases secreted by exogenous microorganisms showing potent antibacterial, antifungal, and antiviral properties (3). Currently does not exist in the literature references linking the proteins WFDC8 and WFDC9 with the heart or great vessels. In this work, we analyze bioinformatics in order to find the relation of the mutation and pulmonary atresia with intact ventricular septum.

To this purpose, functional annotations, domain arquitecture and 3D structures of proteins were reviewed in order to analyze their biological characteristics; to find sequences equivalents in other species (orthologs), a phylogenetic analysis was made and we searched for domains maintained by alignments. Similarly, we looked for tissues where genes are expressed to find his relationship with the heart and, finally, we studied the intergenic sequence to discover regulatory sequences related to cardiac development.

The results reveal the relationship of the WAP for WFDC8 and WFDC9 domains with extracellular matrix proteins, which constitute the architecture of the heart valves. So the same, the expression of the proteins was found in cardiac and embryonic tissue. Finally, it shows that the sequence between both genes presents different binding sites for factors of transcription which are related to the development of the heart and heart valves.

2 Materials and Methods

2.1 Review of Annotations from WFDC8 and WFDC9

To assign the biological characteristics of proteins WFDC8 and WFDC9 the database UniProt was used. The available information about the origin, attributes, annotations, ontology and sequences were obtained from this important knowledge base. Review of annotations was extended thanks to cross-references offering UniProt. The program Rasmol V2.7.4.2 was used to study the regions of interest and view the structure obtained.

2.2 Search for Orthologs and Similarity

To search of homologous sequenceswe used three methods. First we looked for significant orthologs in the results provided by the UniProt Blast. Second, an algorithm written in Perl programming language based on the location of signals of short length sequence was used (4). Finally, we searched for orthologs with NCBI Blast tool tblastn program against database EST (sequences coming from mRNA sequencing).

2.3 Comparison of Sequences

We compared the foundorthologs using the Dot Plot program UGENE V1.10.4. We obtained arrays of points by comparing the human sequence with each of their possible orthologs to observe the best positions to be evolutionarily conserved.

2.4 Multiple Alignment and Phylogeny

We perform the multiple alignment between human proteins and their orthologues using ClustalX V2.0.10. To edit and analyze the alignments also applied the program Bioedit V7.0.4.1. Another purpose with alignments was to carry out phylogenetic analysis, for this purpose, the multiple alignment data were treated with TreeView V1.6.6 to visualize the phylogenetic trees

2.5 Gene Expression Data

Gene expression data was originated from results from different experiments stored in the ArrayExpress database of EBI and the database of proteins in human, nextprot BETA.

2.6 Analysis Bioinformatics to the Intergenic Sequence

The coordinates of deletion of the work of Stefano were presented (2) in the graphical interface UCSC Genome browser (2006), to obtain the graphic that showed the deletion reported in this paper. On the other hand, were sought in the JASPAR database factors of transcription of the human species related heart embryogenesis. In addition, another application of an algorithm in Perl programming language search with weight matrices common transcription factors in obtained alignments. Finally a file in format gff added as tracks in the Genomes of UCSC, allowed to analyze the relationship between the conserved intergenic sequence and transcription factors.

2.7 Experimental Analysis to the Intergenic Sequence

Different experiments are underway at the moment to check the regulatory elements of conserved intergenic sequence. To do so, we conduct important molecular biology techniques such as PCR, electrophoresis, techniques of purification obtaining of colonies and digestion with EcoR1.

3 Results

Only the work of Stefano in 2008 (2) has related both WFDC8 and WFDC9 proteins to PA-IVS. To perform a search in the UniProt database and your links we find that domains WFDC8 and WFDC9 are associated with the extracellular matrix. To apply the tblastn to proteins WFDC we find that the orthologs of WFDC8 and WFDC9 are expressed in embryonic and cardiac tissue. On the other hand the databases of gene expression, Array Express, for WFDC8 shows results of differential expression in organs different than the heart. However, it highlights expression of genes in different cell types, including embryonic stem cells. The results stored in other databases of proteins reported the expression of WFDC8 in ESTs of fetuses. Using the coordinates of deletion of the work of De Stefano (2) was obtained the region deleted of the cases studied (Figure 1A), this region includes the WFDC9 gene, intergenic sequence with the promoter region of WFDC8 and the initial region of this gene. To observe the intergenic sequence was detailed the relationship of this sequence with JunD genes and c-Jun (Figure 1), which has been governing the heart expression (5), (6).

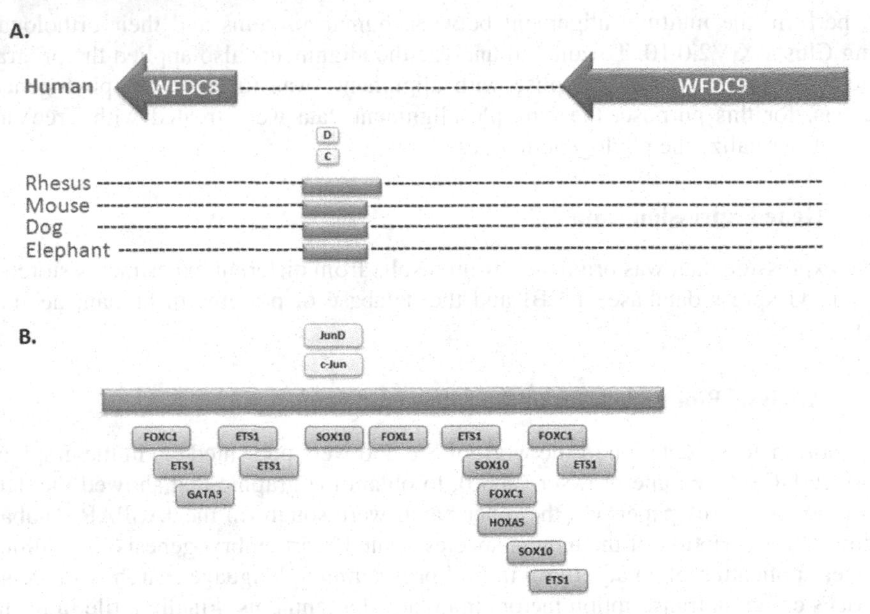

Fig. 1. Extent of mutation causative of PA-IVS (**A**). Show the WFDC9 gene, intergenic sequence and the region 5' of the WFDC8 gene. Details, in the midle of the intergenic sequence, a region preserved in 4 different species and the emergence in this region of binding sites for the transcription factors JunD and c-Jun. (**B**) Enlargement of the region preserved in mammals where we found binding sites for 15 more cardiac embryogenesis related transcription factors. Highlights the increased presence of ETS1 distributed into several blocks and is detailed in the middle of the sequence, the binding site for JunD and c-Jun shared with Sox10.

On the other hand, we found 16 factors of transcription in the human species related with the development of the heart, angiogenesis and the proteins of the extracellular matrix. So the factors of transcription of the human species were sought from Jaspar and analyzed its ontology using Ensembl database. Then and after applying an algorithm in programming language Perl was found in the conserved sequence 15 binding sites for transcription factors related to cardiac embryogenesis, with an identity of 80% compared to weight matrices. In addition, other important factors of transcription met identities a little lower, as it is the case of Sox 9 with 70% of identity against the arrays.

Finally, we analyze the important provision of these transcription factors on the intergenic sequence conserved. The factors complement the listed JunD and c-Jun (Figure 1B)

4 Discussion

Current research relates proteins WFDC8 and WFDC9 with the immune system. The bioinformatics analysis applied to these proteins allowed to clarify its relationship with the development of the heart, heart valves and PA-IVS.

According to the results of this work, the proteins WFDC8 and WFDC9 have WAP domains that bind the entire molecule to proteins of the extracellular matrix by means of its amino acids exposed. The mutation of the genes and the consequent lack of proteins can lead to the disruption of the extracellular matrix and therefore the des-funcionalización of the valves during the development of the heart.

The expression of the proteins WFDC8 and WFDC9 was confirmed with databases of gene expression in embryonic stem cells and human fetuses ESTs. Still, there are more preponderant evidence confirming its expression in heart and embryonic tissue. Their homologous sequences in rodents are quite significant for this fact.

In the present work, wehave found evidence suggesting that the mechanisms required for the structuring of the heart valves have relationship with different transcription factors. ETS1 is a transcription factor which has a wide range of biological functions including the regulation and cell growth, hematopoiesis, lymphocyte development, development and remodeling of vessels and mainly participation in the development of different organs (7). Some studies involve him significantly in the development of the heart of mammals and in the pathogenesis of some congenital heart (8).

On the other hand, the Sox proteins constitute a long family of factors involved in different processes of embryonic development. Sox 8, Sox 9 and Sox 10 exhibit a dynamic expression during embryogenesis of the heart correlated with the septation and differentiation of the connective tissue of the valve (9). Particularly, Sox 9 is required for proliferation and differentiation of cardiac valves progenitor cells, is required for the expression of collagen in the modeling of valve and its absence is related to diseases of the heart valve, including increases in calcium in the area (10).

The relationship of c-jun transcription factors, JunD, and conserved intergenic sequences also have significance to the development of the heart (5), (6). During embryogenesis, c-jun is required for the development of the cardiovascular system, but its regulation diminishes in the adult heart. For his part, JunD is dispensable for the cardiovascular development of the embryo, however is expressed in all cardiac development and is kept in the cells of the adult heart.

Finally the mutation of conserved intergenic sequence of WFDC8 and WFDC9 includes the deletion for different binding sites of ETS1, Sox 9, Sox10, c-jun, JunD, and other factors of transcription. This probably shows that the absence of sequence prevented the function of these important factors and proper development of the heart valve. Equally, the area of mutation as well as conserved sequence included the gen WFDC9, the promoter region and the initial part of WFDC8 (Figure 1), which possibly confirms the lack of expression of these two proteins.

5 Conclusions

An *in silico* analysis link to (o relates to) the domains WAP of WFDC8 and WFDC9 with extracellular matrix proteins, which could constitute the architecture of heart valves, where the mutation of genes can lead to your disfunctionalization. So the same, we showed that proteins WFDC8 and WFDC9, are expressed in human embryonic tissue and in the heart, in any of its orthologs. Finally study the intergenic sequence finding a site conserved in different species, which were important binding sites for transcription factors related to the embryonic development of heart and heart valves, suggesting that the deletion could prevent valvular embryogenesis is promoted. Currently wish to strengthen these results, and now are looking for experimentally in the intergenic sequence enhancers that promote cardiac expression. Conclusions are expected to confirm the causes of PA-IVS.

References

1. Grossfeld, P.D., Lucas, V.W., Sklansky, M.S., Kashani, I.A., Rothman, A.: Familial occurrence of pulmonary atresia with intact ventricular septum. American Journal of Medical Genetics 72 (1997)
2. De Stefano, D., Li, P., Xiang, B., Hui, P., Zambrano, E.: Pulmonary Atresia With Intact Ventricular Septum (PA-IVS) in Monozygotic Twins. American Journal of Medical Genetics 146A, 525–528 (2008)
3. Clauss, A., Lilja, H., LundWall, A.: The evolution of a genetic locus encoding small serine proteinase inhibitors. Biochemical and Biophysical Research Communications 333, 383–389 (2005)
4. Mier, P., Perez Pulido, A.: Fungal Smn and Spf30 homologues are mainly present in filamentous fungi and genomes with many introns: Implications for spinal muscular atrophy. Gene. 491, 135–141 (2012)
5. Hilfiker-Kleiner, D., Hilfiker, A., Kaminsky, K., Schaefer, A., Park, J.-K., Michel, K., Quint, A., Yaniv, M., Weitzman, J., Drexler, H.: Lack of JunD Promotes Pressure Overload–Induced Apoptosis, Hypertrophic Growth, and Angiogenesis in the Heart. Circulation 112, 1470–1477 (2005)
6. Eferl, R., Sibilia, M., Hilberg, F., Fuschbichler, A., Kufferath, I., Guertl, B., Zenz, R., Wagner, E., Zatloukal, K.: Functions of c-Jun in Liver and Heart Development. The Journal of Cell Biology 145, 1049–1061 (1999)
7. Oikawa, T., Yamada, T.: Molecular Biology of the Ets family of transcription factors. Gene. 303, 11–34 (2003)
8. Ye, M., Coldren, C., Liang, X., Mattina, T., Goldmuntz, E., Benson, W., Ivy, D., Perryman, M.B., Garret-Shinha, L.A., Grossfeld, P.: Deletion of ETS-1, a gene in the Jacobsen syndrome critical region, causes ventricular septal defects and abnormal ventricular morphology in mice. Human Molecular Genetics 19, 648–656 (2010)
9. Montero, J.A., Giron, B., Arrechedera, H., Cheng, Y.-C., Scotting, P., Chimal-Monroy, J., Garcia-Porrero, J.A., Hurle, J.M.: Expression of Sox8, Sox9 and Sox10 in the developing valves and autonomic nerves of the embryonic heart. Mechanisms of Development 118, 199–202 (2002)
10. Lincoln, J., Kist, R., Sherer, G., Yutzey, K.: Sox9 is required for precursor cell expansion and extracellular matrix organization during mouse heart valve development. Developmental Biology 305, 120–132 (2007)

Construction and Comparison of Gene Co-expression Networks Based on Immunity Microarray Data from *Arabidopsis*, Rice, Soybean, Tomato and Cassava

Luis Guillermo Leal, Camilo López, and Liliana López-Kleine

Department of Statistics, Department of Biology,
Universidad Nacional de Colombia, Bogotá, Colombia
{lgleala,celopezc,llopezk}@unal.edu.co

Abstract. A big challenge in gene expression data analyses is to reveal the coordinated expression of different genes. Gene co-expression networks (GCNs) are graphic representations where nodes symbolize genes while edges reconstruct the coordinated transcription of genes to certain external stimuli. In this paper, an enhanced novel methodology for construction and comparison of GCNs is proposed. Microarray datasets from pathogen infected plants (*Arabidopsis*, rice, soybean, tomato and cassava) were used. Initially, similarity metrics that find linear and non-linear correlations between gene expression profiles were evaluated. A similarity threshold was chosen and GCNs were constructed. Afterwards, GCNs were characterized by graph variables and a principal component analysis on these variables was applied to differentiate them. The results allowed the discovery of topologically and non-topologically similar networks among species. Potentially conserved biological processes, like those related to immunity in plants could be studied from this work.

Keywords: Gene co-expression networks, similarity metrics, principal component analysis, plants immunity.

1 Introduction

The integration of expression data offers significant insights to understand the transcriptional mechanisms of living organisms. Gene co-expression networks (GCNs) are graphic representations for this purpose [1]. They depict the coordinated transcription of genes by means of linked nodes in a graph. Accordingly, GCNs show the functional associations between co-expressed genes when external treatments are induced [2].

Commonly, the procedure to construct a GCN involves a similarity metric assessed between expression profiles of genes [2].The Absolute value of Pearson Correlation Coefficient (APCC) has been the most used similarity metric [3]. However, as gene expression profiles are not always correlated linearly, many non-linearly correlated genes are not retained for inclusion in the final GCN [4]. Subsequently, a similarity threshold is selected to find relevant pairs of genes connected

L.F. Castillo et al. (eds.), *Advances in Computational Biology*,
Advances in Intelligent Systems and Computing 232,
DOI: 10.1007/978-3-319-01568-2_3,

in the network [5]. Once GCNs are constructed, they can be analyzed in order to provide functional annotations for genes which's function is unknown [6]. Besides, GCNs have been useful in studies of translational functional genomics such as networks alignment [6] and comparisons based on graph variables [7]. During the last case, networks have mainly been described by topological variables and a principal component analysis (PCA) on these variables has been applied to characterize them. Projections resulting from PCA reflect structural similarity through closeness on the PCA space. Nevertheless, networks comparison based on topological variables has for now only allowed the discovery of similar graph motifs while valuable biological conclusions remain unrevealed [5].

Because of the importance of these applications, both the construction and comparison of GCNs demands specific and well developed strategies. In this work, we concentrate on shortcomings of current approaches and propose an enhanced novel methodology to overcome them. We compared the performance of APCC with Mutual Information Coefficient (MIC) [4] and the Normalized Mean Residue Similarity (NMRS) [8], and chose the metric that better detects linear and non-linear correlations. To characterize the GCNs, we added new non-topological variables, such as tolerance to pathogen attacks and assortativity coefficients related to functional annotations. These variables describe better the networks from a biological perspective and are less dependent on network size. To evaluate our methodology, pathogen resistance microarray datasets from *Arabidopsis thaliana*, rice [*Oryza sativa*], soybean [*Glycine max*], tomato [*Solanum lycopersicum*] and cassava [*Manihot esculenta*] were queried from public repositories and used to construct and compare a total of 59 GCNs on different datasets.

Summarizing, we have considered not only the use of appropriated similarity metrics but also the comparison of networks using multivariate methods. This strategy allowed us to find functionally similar GCNs from immunity processes in plants. Moreover, the current biological knowledge on model organisms and less studied species is widened with our results and can be a starting point for emitting biological hypothesis related to plant immunity processes.

2 Materials and Methods

Raw microarray datasets from pathogen infected plants experiments were obtained from GEO DataSets repositories and previous studies [9]. Datasets were independently pre-processed through noise reduction, quantile normalization and log2 transformation. Expression matrices were obtained independently from each dataset after removing non-differentially expressed genes [10].

A square similarity matrix was calculated for every single expression matrix. The similarities ($s_{i,j}$) between pairs of genes i and j were calculated using a metric. As mentioned previously, we compared the similarities obtained with APCC ($s_{i,j}^{APCC}$), MIC ($s_{i,j}^{MIC}$) and NMRS ($s_{i,j}^{NMRS}$) [3][4][8]. These measures take values in the same interval [0,1], where 0 indicates non-dependence between expression profiles, and 1 indicates total dependence or maximum similarity. Subsequently, a similarity threshold (τ) was selected for each similarity matrix. The τ allowed us to determine the GCN's edges according to an adjacency function [3]. We followed a method based on network's clustering coefficient for τ selection [5]. The GCNs were characterized with eigth graph variables: Clustering coefficient, centralization, heterogeneity, network density, two assortativity coefficients related to gene ontology and PFAM annotations, attack tolerance and the Kendall's tau correlation between node degree and presence of immunity domains (WRKY, TIR, NBS, LRR, kinase and LysM) [11].

A characterization matrix of 59 GCNs by eight variables was formed and a PCA was conducted on this matrix [7]. PCA was used to reduce the characterization matrix's dimensionality and the principal components (PCs) retaining more information were used to place the networks on the plane formed by them. Networks were analyzed using the PCA projections.

3 Results

3.1 Construction of GCNs

After datasets preprocessing, *Arabidopsis* and rice were the plants with more data available: 40 and 8 expression matrices were formed respectively. On the other hand, soybean, tomato and cassava are less studied plants and therefore the number of experiments was limited (5, 3 and 3 expression matrices respectively). From the expression matrices, similarity measures were evaluated. The pairwise comparisons of $s_{i,j}^{APCC}$ vs. $s_{i,j}^{MIC}$ allow to see a V-shape (Fig. 1a). Genes with linearly correlated expression profiles are placed in the upper right corner and genes with non-linearly correlated expression profiles can be found in the upper left corner. Those cases where $s_{i,j}^{APCC} < s_{i,j}^{MIC}$ represented co-expressed genes detected by MIC but conducing to a low APCC value. Pairwise comparisons of $s_{i,j}^{APCC}$ and $s_{i,j}^{NMRS}$ conform a less defined V-shape (Fig. 1b). There are also many pairs where $s_{i,j}^{APCC} < s_{i,j}^{NMRS}$ due to the fact that NMRS is a measure that identifies magnitude of proximity between profiles. Based on these results we concluded that NMRS and MI are both useful measures in detecting linear and non-linear correlations. Nevertheless, non-linear correlations are better revealed by MIC. For any $\tau > 0.5$, the number of edges from non-linearly correlated profiles will be higher if the MIC is used. Given our interest is to construct GCNs including linear and non-linear relationships between genes, we decided that MIC is the best metric among the three approaches evaluated. Afterwards, similarity thresholds were obtained for each similarity matrix and GCNs were constructed.

3.2 Comparison of GCNs by Principal Component Analysis (PCA)

GCNs were characterized by graph variables and summarized on principal components (PCs) using PCA. The first three PCs were selected and represented in figure 2. PC1, PC2 and PC3 explain 33%, 20% and 14% of the total variance respectively. The correlation circle of PC1-PC2 plane (Fig. 2b) shows that heterogeneity, density and clustering coefficient are highly correlated to PC1,

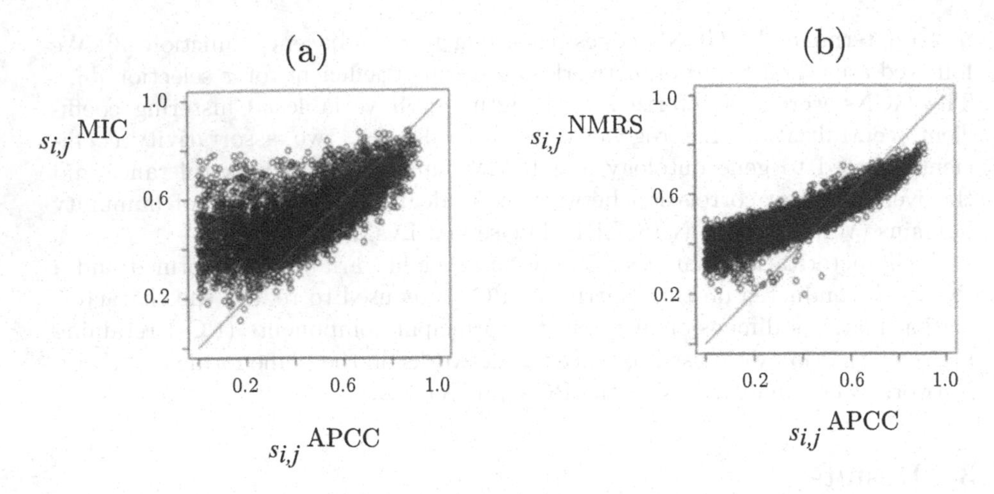

Fig. 1. Dispersion plots of similarities calculated from one *Arabidopsis* expression matrix: *(a)* Pairwise comparison of $s_{i,j}^{APCC}$ vs. $s_{i,j}^{MIC}$ *(b)* Pairwise comparison of $s_{i,j}^{APCC}$ vs. $s_{i,j}^{NMRS}$

while assortativity coefficients are better correlated to PC2. In this way, PC1 is associated mainly to the topological information from GCNs and PC2 is associated to non-topological information. From the correlation circle of PC1-PC3 plane (Fig. 2d) we found a high correlation between tolerance to attacks and the dependence immunity-degree along PC3. These variables were not explained by PC1 or PC2, consequently PC3 is associated mainly with the robustness of the immunity processes.

Positions of GCNs on the planes can be explained with the contribution of each variable to the PCs. On quadrant II of PC1-PC2 plane (Fig. 2a), we find GCNs influenced by high clustering coefficients, centralization and density. On quadrant III, the separation of GCNs is not strong but there are networks with high assortativity coefficients. In relation to PC1-PC3 plane (Fig. 2b), GCNs with high tolerance to attacks and dependence immunity-degree are placed on quadrants I and II.

We found that some closer pairs of GCNs are greatly equivalent and that immunity processes represented in these networks could share some similarities. In a few cases, networks related to the same pathogen were separated because not only different pathogen mutants but also mutant plants were compared. These results indicate that our methodology of GCNs construction and comparison is able to detect gene co-expressions characteristic of each experiment retaining complexity of the processes analyzed.

4 Discussion

Results showed that the methodology is capable of detecting co-expressed genes whose expression profiles are not linearly correlated which enhances the

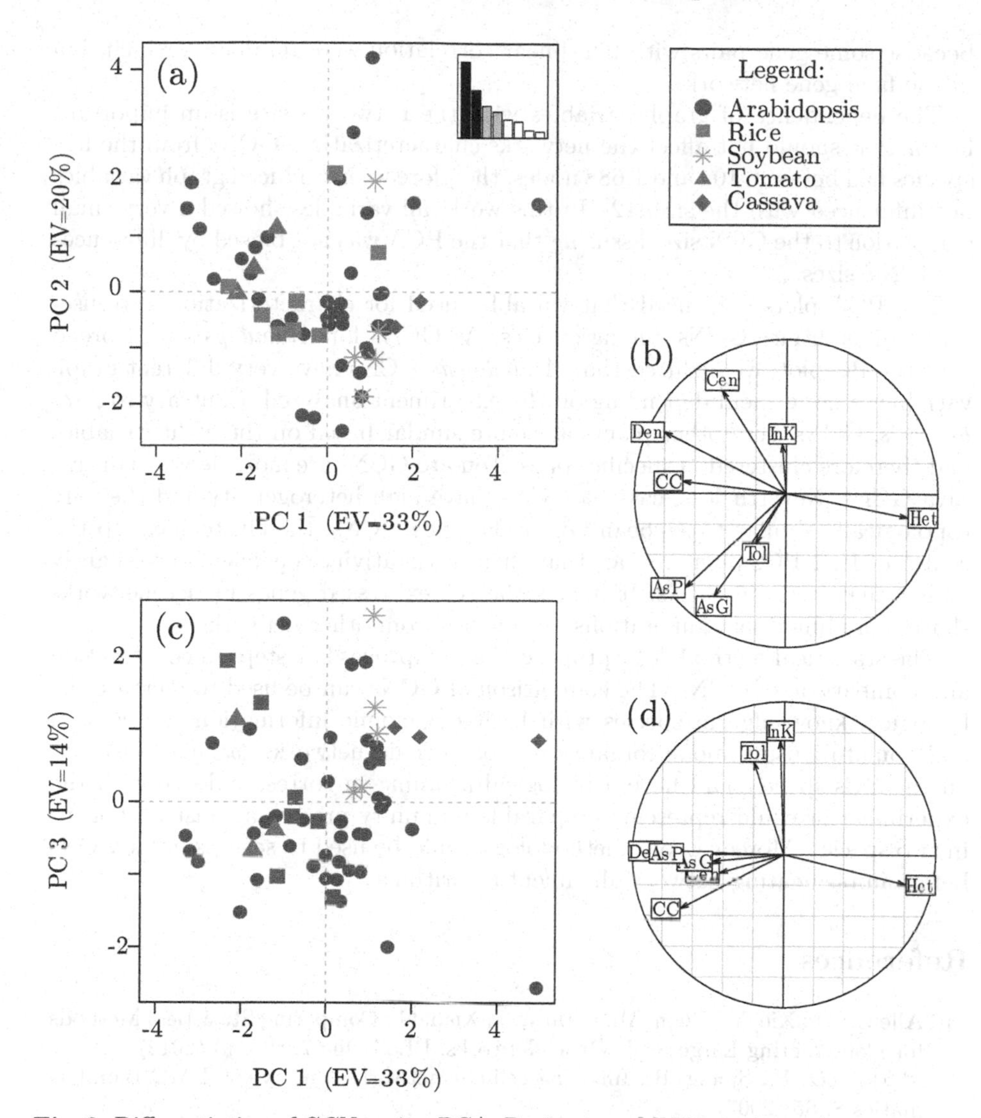

Fig. 2. Differentiation of GCNs using PCA. Projection of GCNs and correlation circles on planes *(a)(b)* PC1-PC2 and *(c)(d)* PC1-PC3. Axes labels show the percentage of explained variance (EV) by each principal component. Bar plot of eigenvalues is showed. Variables are represented with labels: Clustering coefficient (CC), centralization (Cen), heterogeneity (Het), density (Den), assortativity coefficient from GO (AsG), assortativity coefficient from PFAM (AsP), tolerance to attacks (Tol) and correlation between node degree and presence of immunity domains (InK).

knowledge on biological processes represented on the final networks. The V-Shape in figure 1a is consistent with previous works [4]. This result was especially important when τ was chosen in the gene pairwise similarity matrix, because some gene pairs with non-linear correlation were retained for inclusion in the final gene network.

The dependence of graph variables with the network's size is an important factor that should not affect the networks characterization. GCNs from the five species had between 104 and 1,688 nodes, therefore we introduced graph variables not influenced with the size [12]. In this work, all variables showed a very small correlation to the GCN size, assuring that the PCA was not biased by differences in GCNs' sizes.

The PCA plots evidenced that variables used for characterization were useful to differentiate GCNs among species. As GCNs for *Arabidopsis* are spread over the PC plot, we deduced that *Arabidopsis* GCNs have very different graph variables among them depending on the experiment analyzed. Contrary to *Arabidopsis*, GCNs from other plants are more similar based on the eight variables and therefore clustered in specific zones. Tomato GCNs are more dense than cassava GCNs. Nevertheless, cassava GCNs, have high heterogeneity and they are topologically similar to soybean GCNs. For rice GCNs, the cluster near to the center of PC1-PC2 plane reveals that their assortativity coefficients are slightly higher than average GCNs. It means that co-expressed genes in rice networks share more functional annotations than genes from other networks.

The statistical methodology proposed here improves key steps in construction and comparison of GCNs. The comparison of GCNs can be used to increase the biological knowledge in species with limited genomic information (i.e cassava and tomato), using model organisms. For specific networks, a small distance on the PCs space could have a biological meaning in correspondence with the experiment or could represent comparable immunity process against pathogens in two species. Moreover, our methodology could be used to select similar GCNs before implementing network alignment algorithms.

References

1. Allen, J.D., Xie, Y., Chen, M., Girard, L., Xiao, G.: Comparing Statistical Methods for Constructing Large Scale Gene Networks. PLoS One 7, e29348 (2012)
2. Markowetz, F., Spang, R.: Inferring cellular networks – a review. BMC Bioinformatics 8, S5 (2007)
3. Zhang, B., Horvath, S.: A general framework for weighted gene co-expression network analysis. Stat. Appl. Genet. Mol. Biol. 4, Article17 (2005)
4. Daub, C.O., Steuer, R., Selbig, J., Kloska, S.: Estimating mutual information using B-spline functions–an improved similarity measure for analysing gene expression data. BMC Bioinformatics 5, 118 (2004)
5. Elo, L.L., Jrvenp, H., Oresic, M., Lahesmaa, R., Aittokallio, T.: Systematic construction of gene coexpression networks with applications to human T helper cell differentiation process. Bioinformatics 23, 2096–2103 (2007)
6. Ficklin, S.P., Feltus, F.A.: Gene Coexpression Network Alignment and Conservation of Gene Modules between Two Grass Species: Maize and Rice[C][W][OA]. Plant. Physiol. 156, 1244–1256 (2011)
7. Li, W., Yang, J.Y.: Comparing Networks from a Data Analysis Perspective. Complex Sciences, 1907–1916 (2009)

8. Mahanta, P., Ahmed, H.A., Bhattacharyya, D.K., Kalita, J.K.: An effective method for network module extraction from microarray data. BMC Bioinformatics 13, S4 (2012)
9. Lopez, C., et al.: Gene expression profile in response to Xanthomonas axonopodis pv. manihotis infection in cassava using a cDNA microarray. Plant Mol. Biol. 57, 393–410 (2005)
10. Irizarry, R.A., Hobbs, B., Collin, F., Beazer-Barclay, Y., Antonellis, K., Scherf, U., Speed, T.P.: Exploration, normalization, and summaries of high density oligonucleotide array probe level data. Biostatistics 4, 249–264 (2003)
11. Horvath, S., Dong, J.: Geometric Interpretation of Gene Coexpression Network Analysis. PLoS Computational Biology 4, 27 (2008)
12. Van Wijk, B.C.M., Stam, C.J., Daffertshofer, A.: Comparing Brain Networks of Different Size and Connectivity Density Using Graph Theory. PLoS ONE 5, 13 (2010)

in silico Binding Free Energy Characterization of Cowpea Chlorotic Mottle Virus Coat Protein Homodimer Variants

Armando Díaz-Valle, Gabriela Chávez-Calvillo, and Mauricio Carrillo-Tripp

Biomolecular Diversity Laboratory
Laboratorio Nacional de Genómica para la Biodiversidad (Langebio),
Cinvestav Sede Irapuato, México
trippm@langebio.cinvestav.mx

Abstract. The viral capsid's main function is to transport and protect its nucleic acid. It is formed by the self-assembly of multiple copies of one, or a few, coat proteins (CP). The molecular mechanisms of how the spontaneous self-assembly process takes place still remains obscure. Cowpea Chlorotic Mottle Virus (CCMV), an icosahedral plant pathogen, was used as model for understanding the assembly of symmetrical aggregates of biomolecules. Six potential key residues in the capsid interfaces of CCMV were identfied. *in silico* free energy of binding was estimated for two functional CP dimers; WT and the sextuple mutant. Our results show that perturbation of these specific residues will likely destabilize the capsid structure as a whole. This provides insights into how viral coat proteins recognize each other inside the cell, and suggest ways to develop mechanisms to prevent their assembly, thereby blocking the infection.

Keywords: viral self-assembly, capsid coat protein, plant icosahedral virus.

1 Introduction

Viruses are self-assembling macromolecular complexes made up of copies of the same, or a few, protein components. They assemble into symmetric closed shells encapsidating their own viral genome. Major efforts have revealed that spherical capsids display icosahedral symmetry. The assembly of simple non-enveloped viruses occurs rapidly and spontaneously with a high degree of fidelity. The latter implies that the instructions to form closed shells (capsids) are built into their components, mainly the coat protein subunits. If this is true, then it should be possible to gain insights into virus assembly on the basis of the structure and organization of the protein subunits in those types of capsids, which do not require any auxiliary or scaffolding proteins for the assembly. Many spherical capsid structures have been analyzed in terms of protein-protein interactions, quasi-equivalence, and self-assembly pathways [1][2], however, the molecular mechanisms governing these processes remain unclear.

L.F. Castillo et al. (eds.), *Advances in Computational Biology*,
Advances in Intelligent Systems and Computing 232,
DOI: 10.1007/978-3-319-01568-2_4, © Springer International Publishing Switzerland 2014

Taking advantage of the fact that many nonenveloped virus structures have been determined at near atomic resolution, different *in silico* studies at atomic level can be made. The simplest viruses that display icosahedral symmetry have 60 copies of a single coat protein subunit (single gene product) arranged in equivalent (identical) environments related by fivefold, three-fold, and twofold axes of symmetry. In addition, viral capsids provide an excellent resource for studying protein-protein interactions in homo-oligomers that form symmetric closed shells, since association and self-assembly of the capsid protein subunits are common to all viruses. Analysis of protein-protein interactions in viral capsids may reveal the molecular principles of recognition and self-assembly and the nature of interactions that lead to capsid formation, given that interactions between the coat protein subunits determine the size and the robustness of the capsid.

1.1 Capsid Assembly

It is believed that coat protein (CP) subunits should carry instructions to form a particular type of capsid. In principle, the structure and the organization of the subunits obtained from the complete capsid structures can be used to decipher these instructions. Computational approaches have identified the repeating units (building blocks) of the assembly and any potential intermediates that may be formed during the course of capsid formation [2]. A substructure of n subunits was considered a likely intermediate or a preferred substructure during the course of assembly, when the association/binding energy of formation of the substructure was favored significantly over any other combinations of n subunit associations. By monitoring the association of a full aggregate of subunits (e.g., 60, 180, etc.) forming a closed shell, it was possible to identify the preferred substructures along the way, suggesting the most likely pathway of capsid formation; the stronger the protein-protein interactions are, the greater the chance of identifying the assembly intermediates.

1.2 Bromoviridae Virus Family

In previous years, the structures of several viruses from the Bromovirus and Cucumovirus genera of the Bromoviridae family have been determined at near atomic resolution. These viruses form capsids that consist of 180 subunits. The calculated protein-protein interactions between the protein subunits of bromoviruses (e. g., Cowpea Chlorotic Mottle Virus –CCMV– and Brome Mosaic Virus –BMV–) suggest that the twofold-related coat protein dimers (CC or the equivalent AB) form stable oligomers in agreement with previous observations [3] –Fig. 1–. The derived assembly pathways for bromoviruses suggest that the second preferred structure is a trimer of dimers (6-mer), followed by a pentamer of dimers (10-mer) and a decamer of dimers (20-mer) surrounding the fivefold axes of symmetry. In other words, the pentamers and hexamers are not preformed, they are a result of the initial assembly from the A1-B5 and C1-C6 dimers. This indicates that CCMV assembly happens through dimers, and then multimers of dimers form. [4]

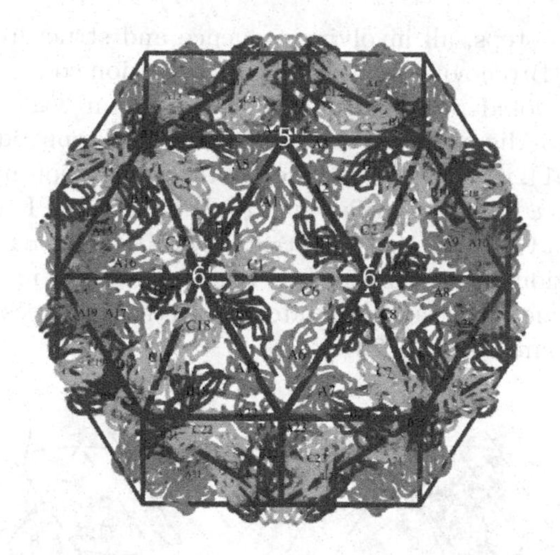

Fig. 1. Three dimensional structure of the CCMV spherical capsid. Independent subunits forming the capsid are grey shaded according to their placement in relation to the symmetry axes: *A* around the five-fold axis (5') and *B*-*C* around the six-fold axis (6') respectively.

2 Methods

We have selected the CCMV as a model of study, given that its three dimensional structure is known, making it a good candidate for molecular modeling *in silico*, and that is relatively easy to manipulate *in vitro* using molecular biology techniques. The latter is important since we are also interested in validating our computational predictions by producing the recombinant wild-type as well as a library of specific interface mutants, through heterologous systems (bacteria).

In this section a description is made on how specific residues at the protein-protein interfaces are identified, how mutant proteins where created and the steps followed to produce an estimate of the binding free energy of homodimers, all of the above using an *in silico* approach.

2.1 Hot-Spots Prediction

By definition, a hot-spot is a residue which is located in the protein-protein interface between capsid subunits, and is conserved both in linear sequence and space (tertiary and quaternary structure) among all members of a specific virus family or genus. A computational algorithm was previously developed [6] to identify this type of residues using all viral capsid structural information available at VIPERdb [7]. When applied to the Bromoviridae family, six residues where predicted as hot-spots: Proline P99, Phenylalanine F120, Glutamic Acid E176, Arginine R179, Proline P188 and Valine V189. This identification was achieved following a series of steps, all involving sequence and structure information of five members of the Bromoviridae family (VDB accession codes 1cwp, 1f15, 1js9, 1laj and 1za7): i) global residue sequence conservation was found through a Multiple Sequence Alignment (`http://viperdb.scripps.edu/alignments/Bromoviridae.html`), ii) interface residues and their position in space in spherical coordinates (ϕ, ψ) were identified using VIPERdb's API (`http://viperdb.scripps.edu/API2/api_phipsi.php?format=html&VDB=1cwp&format=csv&cons` changing the VDB accession code respectively) –Fig. 2a–, and iii) manually matching residues with the same spherical coordinates within three degrees in angular space present in all five members –Fig. 2b–.

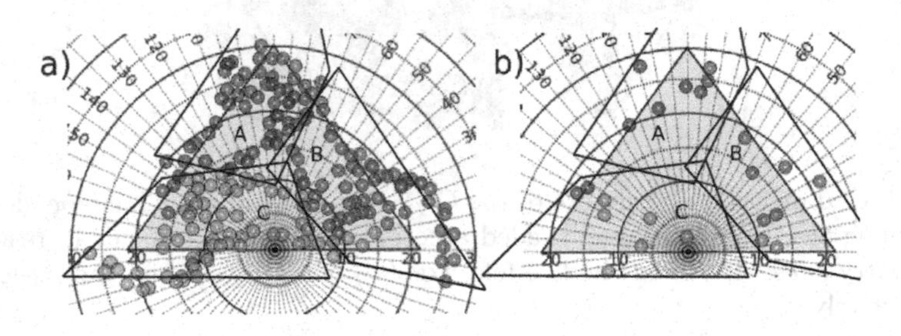

Fig. 2. ϕ-ψ maps of CCMV. a) Interface residues of the three subunits (*A*, *B* and *C*, broadly indicated by trapezoids) forming the capsid asymmetric unit. b) Hot-spot prediction.

2.2 WT and Mutant Homodimers

The starting three dimensional structure of the functional CCMV WT homodimer was aquired using the Oligomer Generator tool at VIPERdb (`http://viperdb.scripps.edu/oligomer_multi.php`). Given that dimers A1-B5, B1-A2 and C1-C6 are equivalent, only the first was chosen –Fig. 3–. It is clear that residues R179, V189 and P188 play a critical role in the homodimer interface, while residues P99 and F120 participate in the homo-pentamer and hexamer interfaces when the capsid forms. These six residues were mutated, *in silico*, to an aminoacid with contrasting physical-chemical properties as follows: P99A, F120A, E176Q, R179Q, P188A and V189N. Charged residues (E,R) where mutated to a neutral aminoacid (Glutamine, Q), while keeping a similar size, whereas others were reduce in volume (F for Alanine, A), or changed from hydrophobic to hydrophilic (Valine, V, for Asparagine, N). Following this logic, it was believed that all favoring interactions in the WT homodimer would be diminished in the mutant. The sextuple mutant, M6, was a result of consecutive single mutations starting from the WT structure using the Mutator Plugin available in VMD (`http://www.ks.uiuc.edu/Research/vmd/plugins/mutator/`).

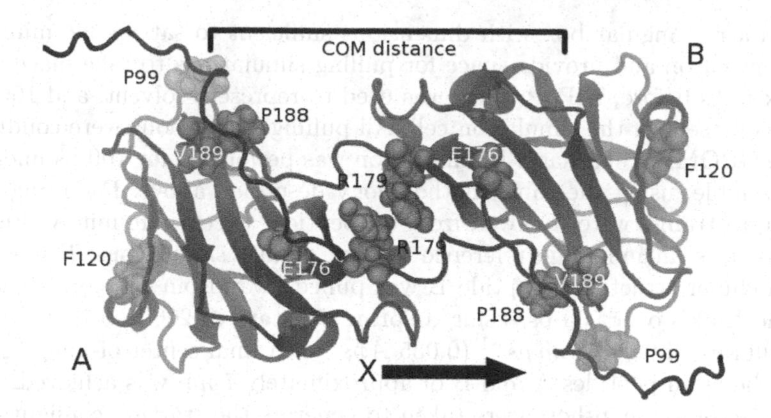

Fig. 3. CCMV WT homodimer starting configuration, showing first subunit on the left (*A*) and second subunit on the right (*B*). Also shown, corresponding hot-spots in each subunit, with residues 188 and 189 interlocking in the opposite subunit. The Center-of-Mass distance and the direction of the X axis are indicated.

2.3 System Construction

In order to generate equilibrated starting structures for the following steps, all Histidines were assigned their canonical state at physiological pH, and the CHARMM27 all-atom force field (with CMAP) - version 2.0 was used. Each homodimer (WT and M6) was placed in a cubic box with explicit TIP3P water to which 100 mM NaCl was added, including neutralizing counterions. Following steepest descents minimization, each of the homodimer systems was equilibrated in two steps, with position restraints applied to peptide heavy atoms throughout.

The first phase involved simulating for 50 ps under a constant volume (NVT) ensemble. Protein and nonprotein atoms were coupled to separate temperature coupling baths, and temperature was maintained at 310 K using the Berendsen weak coupling method. Following NVT equilibration, 50 ps of constant-pressure (NPT) equilibration were performed, also using weak coupling to maintain pressure isotropically at 1.0 bar. The Nosé-Hoover thermostat was used to maintain temperature, and the Parrinello-Rahman barostat was used to isotropically regulate pressure. This combination of thermostat and barostat ensures that a true NPT ensemble is sampled. Short-range nonbonded interactions were cut off at 1.4 nm, with long-range electrostatics calculated using the particle mesh Ewald (PME) algorithm. Dispersion correction was applied to energy and pressure terms to account for truncation of van der Waals terms. Periodic boundary conditions were applied in all directions. Simulations were conducted using the GROMACS package, version 4.5.

2.4 Free Energy Estimation

Structures from the end of each of these trajectories (WT and M6) were used as starting configurations for pulling simulations. The homodimer structures were placed in a rectangular box with dimensions sufficient to satisfy the minimum image convention and provide space for pulling simulations to take place along the X-axis. As before, TIP3P water was used to represent solvent, and 100 mM NaCl was present in the simulation cell. All pulling simulations were conducted with the GROMACS package. Equilibration was performed for 100 ps under an NPT ensemble, using the same methodology described above. Following equilibration, restraints were removed from all peptides except subunit A (Fig. 3); it was used as an immobile reference for the pulling simulations. For each of the homodimer structures, peptide B was pulled away from the core structure along the X-axis over 500 ps, using a spring constant of 2000 $kJ\ mol^{-1}\ nm^{-2}$ and a pull rate of 0.0035 $nm\,ps^{-1}$ (0.035 $\AA\,ps^{-1}$). A final center-of-mass (COM) distance between peptides A and B of approximately 7 nm was achieved. From these trajectories, snapshots were taken to generate the starting configurations for the umbrella sampling windows.

A symmetric distribution of sampling windows was used, such that the window spacing was 0.2 nm. Such spacing allowed for increasing detail at trouble COM distances, resulting in 42 windows. In each window, 5 ns of MD was performed for a total simulation time of 210 ns utilized for umbrella sampling for each homodimer. Analysis of results was performed with the weighted histogram analysis method (WHAM).

3 Results

Umbrella sampling simulations were used to determine the ΔG_{bind} of a particular event along a reaction coordinate. In this case, the reaction coordinate corresponds to the X-axis. By using 42 sampling windows along this axis, one-dimensional potential of mean force (PMF) curves were obtained for each homodimer (Fig. 4), leading to the ΔG of binding for subunit B in each of the experiments in this study –Table 1–.

It is clear from these results that disrupting the hot-spots, especially by increasing its exposure to solvent in the core of the dimer interface, would likely destabilize the structure as a whole. This finding is corroborated by the PMF data for the M6 mutant. The $\Delta\Delta$G of binding for the M6 mutant is +5.1 $kcal\ mol^{-1}$, indicating that the perturbation of the native hot-spots leads to destabilization.

Table 1. Binding free energy values, ΔG_{bind}, for wild-type and mutant M6 of CCMV homodimers

System	ΔG ($kcal\ mol^{-1}$)	$\Delta\Delta$G ($kcal\ mol^{-1}$)
WT	-38.3	-
M6	-33.2	+5.1

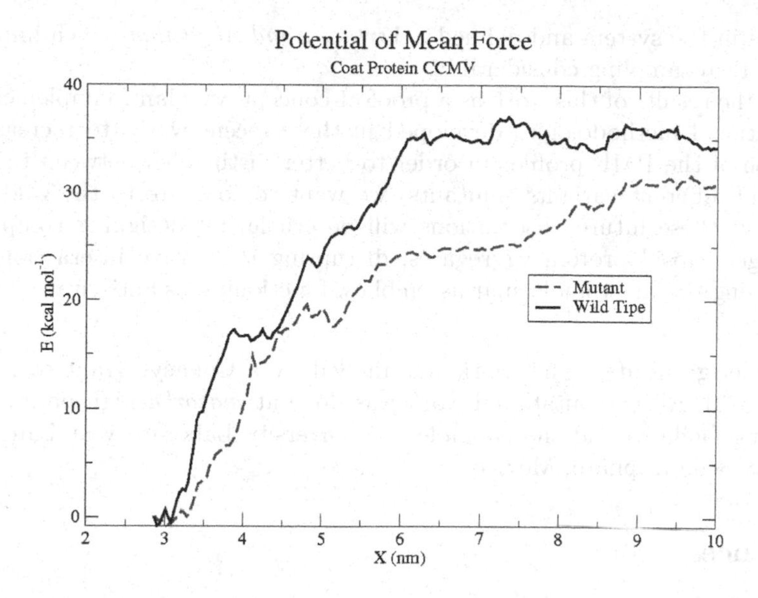

Fig. 4. Potential of mean force (PMF) of the CCMV homodimers, WT in solid black line and M6 in dashed line, along the reaction coordinate defined as the X-axis.

4 Conclusions

The results presented here suggest that the hot-spot hypothesis could hold true in the sense that there is a small subset of residues in the protein-protein interface that are critical for the capsid formation. Although the individual contribution of most residues forming the interface to the total interaction energy can be small acting as a cooperative effect, a few residues seem to stand out in the crowd, implying that there could be specific signals proteins use in order to correctly form functional macromolecular structures with a particular morphology.

This work is a first attempt to fully characterize the molecular mechanism of capsid assembly of an spherical virus at atomic level using a thermodynamic approach. Despite the fact that the statistical sampling done here can be improved, there is strong evidence that a more detailed study will provide a better understanding of the role that the so-called hot-spots play in the protein-protein recognition and self assembly process. The same approach used here has been employed successfully in the past to study other protein aggregation phenomena. Lemkul et al. analyzed the effect of mutations in the interface of Alzheimer's amyloid protofibrils to asses their stability, finding that the level of hydration around certain interface residues is crucial. The specific packing between a few of them serves to regulate the level of hydration in the core of the protofibril [5]. Another study by Periole et al. used a similar but more refined methodology to study the structural determinants of the supramolecular organization of rhodopsin in bilayers [8]. They used a coarse-grain model to represent the molecules in the system and defined a *Virtual Bond Algorithm* which improved the statistical sampling considerably.

Using the results of this work as a proof-of-concept, we plan to implement the computational methodologies developed in these recent works to increase the resolution of the PMF profiles in order to better distinguish between the high number of different variants (mutants) we want to compare to the wild type. We believe these future observations will be crucial for designing compounds that target capsid protein aggregates, disrupting the native interactions and destabilizing the macromolecular assemblies, functioning as anti-virals.

Acknowledgements. This work was funded by a Conacyt grant to M.C-T. (No. 132376). All computational work was done at *mazorka* high performance computing facilities and the Biomolecular Diversity Laboratory at Langebio - Cinvestav Sede Irapuato, México.

References

1. Damodaran, K.V., Reddy, V.S., Johnson, J.E., Brooks III, C.L.: A general method to quantify quasi-equivalence in icosahedral viruses. J. Mol. Biol. 324, 723–737 (2002)
2. Reddy, V.S., Giesing, H.A., Morton, R.T., Kumar, A., Post, C.B., Brooks III, C.L., Johnson, J.E.: Energetics of quasiequivalence: Computational analysis of protein-protein interactions in icosahedral viruses. Biophys. J. 74, 546–558 (1998)
3. Zhao, X., Fox, J.M., Olson, N.H., Baker, T.S., Young, M.J.: In vitro assembly of cowpea chlorotic mottle virus from coat protein expressed in Escherichia coli and in vitro-transcribed viral cDNA. Virology 207, 486–494 (1995)
4. Reddy, V., Johnson, J.: Structure-derived insights into virus assembly. Advances in Virus Research 64, 45–68 (2005)
5. Lemkul, J., Bevan, D.: Assessing the stability of alzheimer's amyloid protofibrils using molecular dynamics. The Journal of Physical Chemistry B 114(4), 1652–1660 (2010)
6. Carrillo-Tripp, M., Brooks, C., Reddy, V.: A novel method to map and compare protein-protein interactions in spherical viral capsids. Proteins 73(3), 644–655 (2008)
7. Carrillo-Tripp, M., Shepherd, C., Borelli, I., Venkataraman, S., Lander, G., Natarajan, P., Johnson, J., Brooks, C., Reddy, V.: VIPERdb2: an enhanced and web API enabled relational database for structural virology. Nucleic Acids Research 37(Database issue), D436–D442 (2009)
8. Periole, X., Knepp, A., Sakmar, T., Marrink, S., Huber, T.: Structural determinants of the supramolecular organization of G protein-coupled receptors in bilayers. Journal of the American Chemical Society 134(26), 10959–10965 (2012)

Analysis of Binding Residues between PDGF-BB and Epidermal Growth Factor Receptor: A Computational Docking Study

Ricardo Cabezas, Daniel Torrente, Marco Fidel Avila, Jannet González*, and George Emilio Barreto

School of Sciences, Department of Nutrition and Biochemistry, Pontificia Universidad Javeriana
janneth.gonzalez@javeriana.edu.co, ricardocabe@gmail.com

Abstract. A directed docking was performed using Cluspro between human PDGF-BB and EGFR using specific templates obtained from PDB. Various conserved residues were found to be involved in the docking interaction of the complex by means of hydrophobic interactions and hydrogen bonds. An electrostatic potential evaluation of the PDGF-BB-EGFR complex was also performed to validate if the complex is electrostatically complementary in the binding area. Results suggested a possible binding mechanism which could explain the *in vivo* evidence of formation of heterodimeric receptors EGFR-PDGFR.

Keywords: PDGF-BB, EGFR, docking, Transactivation, interactions.

1 Introduction

The human EGF receptor (EGFR) is a transmembral glycoprotein of 1186 amino acids [1,2], that belongs to the ErbB family 2 which consist of four members: The epidermal growth factor receptor (EGFR, also referred to as ErbB1 or Her1),3 ErbB2 (p185Neu or Her2),4 ErbB3 (Her3)5 and ErbB4 (Her4),6 all of which have been identified to be either over-expressed, amplified or altered in a wide range of human epithelial tumours [3]. EGFR has a 622 amino acid extracellular region, with 4 domains (I–IV), also known as the L1, S1, L2, and S2 domains respectively [4]. EGF binding by EGFR occurs at Domains I and III, which share 37% amino acid identity, while domains II and IV are homologous Cys-rich domains, denoted CR1 and CR2, respectively [5].

The transmembrane domain IV is involved in both receptor dimerization, as EGF binding [6]. This domain IV, includes residues 480-614 and contains a transmembrane domain of 23 amino acids starting at Ile-622. EGFR binding to the receptor, activates receptor dimerisation, internalisation and autophosphorylation and downstream signal transduction pathways including the RAS/MAP kinase, the prosurvival

* Corresponding author.

L.F. Castillo et al. (eds.), *Advances in Computational Biology*,
Advances in Intelligent Systems and Computing 232,
DOI: 10.1007/978-3-319-01568-2_5, © Springer International Publishing Switzerland 2014

phosphatidylinositol 3-kinase (PI-3K)/Akt, the PLCγ and the JAK-STAT pathways [7]. Aditionally, EGFR can be activated by other receptor tyrosine kinases including insulin-like growth factor-1 receptor, adhesion molecules (e.g. E- cadherin and integrins) and G-protein-coupled receptors (GPCR). Furthermore, it has been noticed that the EGFR is able to interact with other receptors and perform transactivavion signaling by virtue of the fact that it does not require binding of its ligand for tyrosine kinase activity and dimerization [8,9].

Previous research have shown that in certain types of tumors such as glioblastoma, Mul tiforme have shown overexpression of both EGFR and PDGFR (platelet derived growth factor) that can contribute to the malignant phenothype of glioblastoma [10,11]. Importantly, both EGFR and PDGF receptors have been reported to be associated and exert its signaling via caveolin-containing membranes (caveolae) in the presence of ligand [12]. Interestingly these authors also founded that pretreatment of cells with EGF prevented PDGF-dependent phosphorylation of PDGF receptors and ERK1/2 kinase activation and that cells pretreated with PDGF prevent EGF dependent phosphorylation of EGF receptors and ERK1/2 kinase activation, suggesting that both receptors are physically located in the same membrane domains and an heterologous desensitization of the other receptor by sequestration of cholesterol caveolin-containing membranes [12]. Moreover, a recent study suggested that PDGFRA, PDGFRB, and EGFR were expressed and activated, in high levels of EGFR expression inducing an autocrine loop activation of these receptors along with their coactivation [13].

Furthermore, other studies have shown that the epidermal growth factor creceptor (EGFR) may become transactivated by specific ligands of other membrane creceptors, such as angiotensin II, carbachol, thrombin, endothelin and others [9,14]. Additionally various researchers have shown that the EGFR becames transactivated in response to PDGF [9,15-17]. This last study, performed in vascular smooth muscle cells from rat also shows the formation of PDGFBR-EGFR heterodimers, induced by PDGF-BB, which are important for cell migration processes [9].

Taken together, these results suggest a fundamental cross talk between these tyrosine kinase coupled receptors, in signal transduction [9]. However, at the present moment there is no information regarding the possible interaction between the PDGF-BB and the EGFR at the molecular level. The structural analysis of complexes between EGFR in complex with PDGFB is unknown, due to the lack of suitable crystal structure and difficulties of protein multimerization in solution [1,9,18]. In the present study, molecular modeling, protein–protein docking simulations and docking validation with bioinformatics tools were assessed to predict the structure of PDGF-B and its interaction with EGFR. Furthermore, the interacting residues of the EGFR-PDGFB complex were analyzed, suggesting the existence of a possible binding between the PDGF-B and the EGFR, which could be of importance in a biological contex.

2 Materials and Methods

2.1 Data Set

The crystallized structures of PDGF-BB and EGFR, 1PDG and 3NJP respectively [1,6,18] were obtained from the PDB at the Broohaven National Laboratory [19], the details of which are given in Table 1. The EGFR structure was solved with < 3.3A° resolutions and the PDGF-BB was solved with < 3.0A°. These are the best templates for modeling the PDGF-BB and EGFR as on date.

2.2 Protein–Protein Docking

We used ClusPro's [20] algorithm for obtaining the bound structures of EGFR receptor with PDGF-BB. ClusPro has the option of selecting either DOT or ZDOCK to perform rigid body docking and both are based on the fast Fourier transform (FFT) correlation techniques [20]. We selected DOT for our work, since this selection allows for the use of an static potential in the scoring function and on the surface complementarity between the two structures. DOT runs on a 128A° × 128A° × 128A grid, using a grid spacing ° of 1 Å. It performs 13 000 rotations, initially obtaining over 2.7×10 10 structures and finally retaining only 20 000 structures with the best surface complementarity scores. All the 20 000 docked structures are then filtered using distance-dependent electrostatics and an empirical potential energy. The 2000 conformations retained after filtering is clustered based on the pairwise RMSD (root mean square deviation) and selects the best conformational structure. Finally, the representative conformation selected is refined using CHARMM minimization [21,22]. An initial set of 30 conformations were obtained, based on the coefficient weights following the application of the formula of Kozakov et. al. 2010:

$$E=0.40_{Erep}-0.40_{Eatt}+600_{Eelec}+1.00_{EDARS}$$

Biochemical and literature data was used to filter the docking results, using the residues on the surface of the interacting proteins as candidates, as well the experimental information on PDGFRβ-PDGFB and EGFR-EGF complexes previously reported [1,6,18].

2.3 Docking Validation

The best structural model for the complexes EGFRα-PDGFB obtained from the docking analysis were validated using the following methods: Ramachandran plots from Mol Probity [23], and PROCHECK [24] were employed to examine the stereochemical quality of the built model. Results indicates that more than 76.7% of the residues have phi and psi angles in the most favored regions, 21.1% in additionally allowed regions, 1.3% in generally allowed regions and 0.8% in disallowed regions. An overall G-factor of –0.09 indicates a good quality of the built model. Additionally, the residues in the disallowed regions were located outside of the binding pocket, and so, no further action was taken to further improve the backbone folding of these residues.

2.4 Docking Interface Analysis

Protein-protein interactions were analyzed with LigPlot+ v1.4 [25], using the DIMPLOT program for protein-protein interactions. This program shows hydrophobic and hydrogens bonds interactions between proteins. Electrostatic potential surface and interactions of the complexes were calculated and represented by using Pymol v1.5 [26]. Electrostatic potential surface and interactions of the complexes were calculated and represented by using Pymol v1.5 [26].

3 Results

3.1 Sequence Analysis of Human PDGF-BB and EGFR

Initially a BLAST [27] was performed in the RefSeq database to the NCBI human PDGFB. Found 3 access numbers for human PDGFB, reported for June 2011: CAG46606, CAG46606.1 and NP_002599.1. These correspond to the PDGF protein to establish a sequence of 241 amino acids which is accurate to that reported by the Uniprot database with the accession number P01127.

Meanwhile NP_002599.1 sequence corresponds to the preprotein form of PDGF-B. This was the most recent sequence (July, 2012) reported for the PDGF-B and for this reason, further analysis was conducted with this sequence. During maturation, the PDGF is a secreted protein, suffers removing its signal peptide of 20 amino acids and 2 additional regions located at the N terminus (61 amino acids) and the C terminus (51 amino acids). The mature form of this growth factor, corresponds to the central region of the preprotein, and has a size of 109 amino acids. EGFR sequence was obtained after performing a search against the whole PDB to select the templates which could be used to generate receptor model. Information regarding the templates are summarized in table 1.

Table 1. Detailed information of the human EGFR and PDGFBB used in this work as a templates

	No of aminoacids	Molecular weight	Source organism	Swiss-Pro accession number	PDB ID
EGFR	1210	120,692	Homo sapiens	Q504U8	3NJP
PDGF-BB	109	30	Homo sapiens	P01127	1PDG

3.2 Sequence Edition of Human PDGF-BB and EGFR Templates

The crystallized structures of PDGF-BB and EGFR, 1PDG and 3NJP respectively, were initially edited in Pymol [26] for the elimination of redundant and unnecessary chains, glycosylated residues and hydrogens. To validate the edited models, we used the QMEAN6 from Swiss-model structure assessment tool [28,29]. QMEAN6 is a composite scoring function used for the estimation of the global quality of the entire model edited in Pymol and for the local per-residue analysis of different regions within a model. The results showed a value of 0.546 for the PDGF-BB (fig. 1) and 0.696 for the modified EGFR (fig. 2). Those are relatively good scores in a scale ranging from 0 to 1.

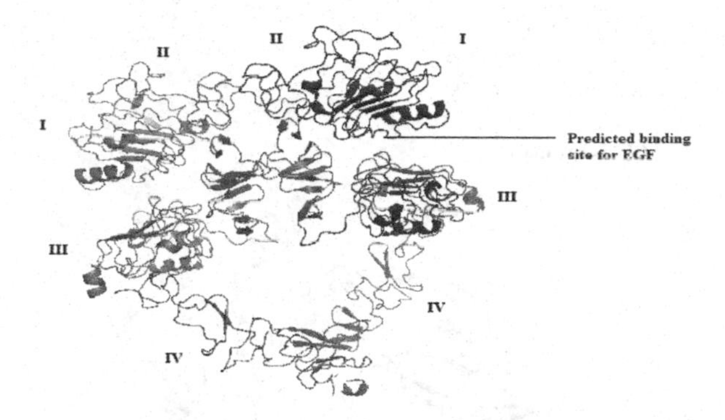

Fig. 1. Modified structure of the EGFR (From Lu et al 2010). The four domains are shown in roman numeration, along with the predicted site for EGF binding.

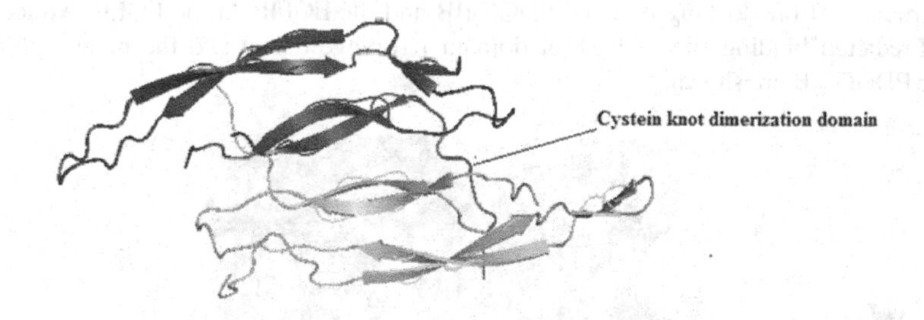

Fig. 2. Modified structure of the EGFR (From Oefner et al 1992). Different PDGF-B monomers are in grey and darker grey. The cysteine knot dimerization domain is shown.

3.3 PDGF-BB Docked with EGFR

Based on the output of ClusPro's [20] algorithm, we report here that the EGFR-PDGF complex showed a good docking behavior (fig. 3). The selected docking model was chosen based on factors such as the area of surface contact, extent of interactions and stability of the model. Additionally, an electrostatic potential evaluation of the PDGF-BB-EGFR complex was performed to validate if the complex is electrostatically complementary in the binding area. Our results show that the EGFR, in the interacting regions of domains I and III (fig. 3) has strong negative charges (+36. 616 kT/e and -36.616 kT/e;) and might be electrostatic complementary to PDGF- BB (+60.813 kT/e and -60.813 kT/e) due to the positive charges in the interacting region with the EGFR (fig. 4).

3.4 Interacting Residues

The Protein-protein interactions of the solved complex are mostly hydrophobic interactions but present some hydrogen bond interactions. The residues implicated in the complex PDGFB-EGFR are: ASN 128, SER 127, LYS 86, ASP 155, ILE 13, LEU

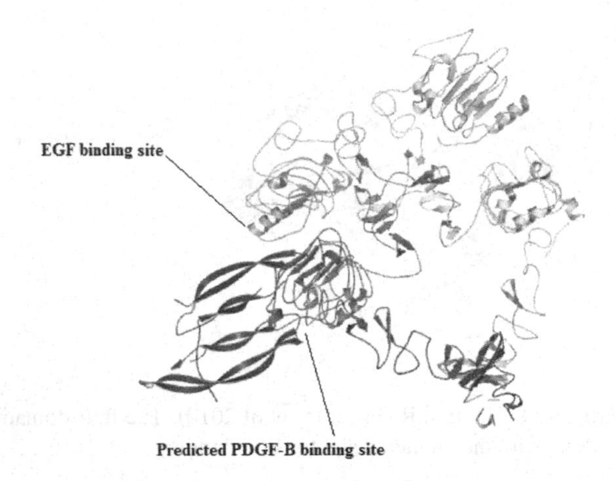

Fig. 3. Structure of the docking between PDGF-BB and the EGFR. At the EGFR extracellular domain. Predicted binding site of EGF at domain I (orange color) and the putative binding region for PDGF-BB are shown.

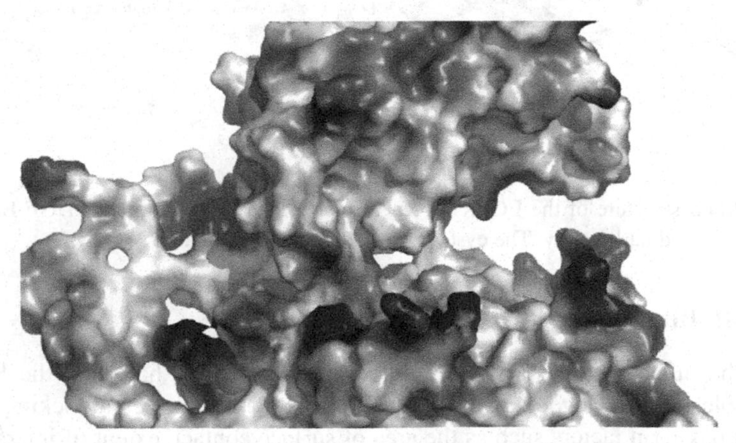

Fig. 4. Electrostatic potential generated for PDGF-BB. Electrostatic scale [+58.551 kT/e and -58.551 kT/e] .The darked colored structure represents the positive charge, where PDGFs interacts with EGFR.

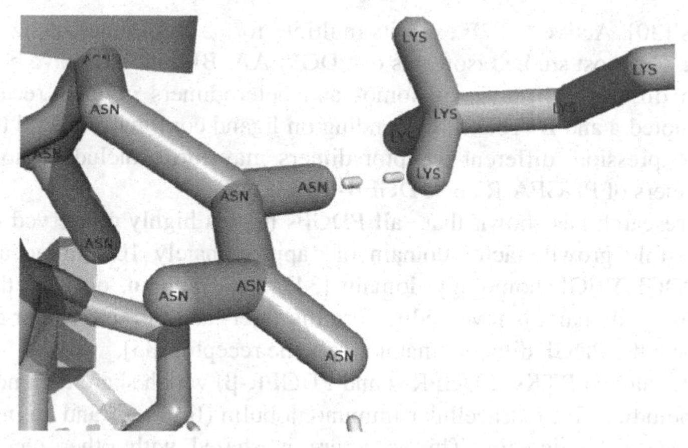

Fig. 5. PDGF-B binding to the EGFR. Interaction between ASN 151 (EGFR) and LYS 81 (PDGF-B chain B). Terminal Nitrogen of the LYS 81 interacts with an oxygen of the ASN 151 by polar contact (dotted lines) of 2.48 Å.

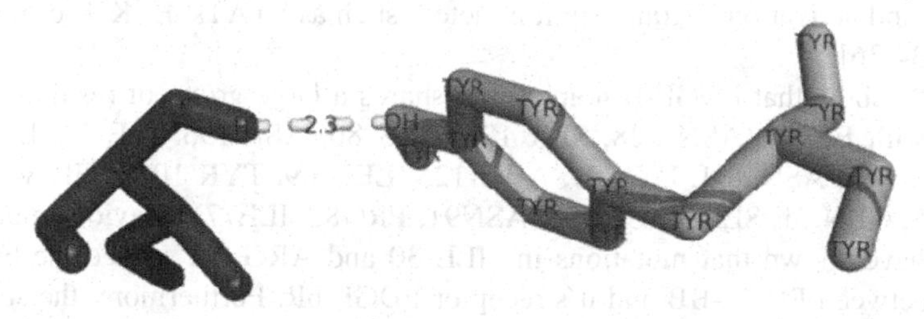

Fig. 6. PDGF-B binding to the EGFR. Interaction between TYR 42 (EGFR) and ILE 13 (PDGF-B chain A). Terminal oxigen of the TYR 42 interacts with an hydrogen of the ILE 13 by polar contact (yellow dotted lines) of 2.3 Å.

14, THR 15, TRP 40, TYR 45, ASN 151, LYS 81, ARG 125, LEU 69, TYR 101, PHE 84. ARG 73, LYS 105, GLN 71, SER92, GLU90, ASN91, PRO82, ILE77. Results show that SER 127, ASP 155, LYS 86, LYS 81, ASN 151, GLN 71 and LYS 105 are present in the hydrogen bond formation within this complex, suggesting the importance of both hydrophobic interactions and hydrogen bond for the binding of PDGF-BB to EGFR. The specific interaction between ASN 151 and LYS 81 is shown in fig. 5 and the specific interaction between TYR 42 and ILE 13 is shown in fig. 6.

4 Discussion

The main purpose of this study was to determine the complex structure of PDGF-BB homodimer with the EGFR, based on previously reported structures. The overall scope of these results is to consider computationally a suitable protocol for predicting the best protein conformational structure using resolved templates from the PDB.

PDGF is a disulfide-linked dimeric protein, which is composed of two related polypeptide chains A and B, expressed from different genes and assembled as homo or heterodimers [30]. Active PDGF exhibits multiple forms and ranges in size from 28-35 kDa [31]. The most studied isoforms of PDGF; AA, BB and AB, have been shown to bind with different affinities to homo- and heterodimers of their receptor gene products, denoted a and B [18,32]. Depending on ligand configuration and the pattern of receptor expression, different receptor dimers may form, including homodimers and heterodimers of PDGFA-R and PDGFB-R [33].

Previous research has shown that all PDGFs have a highly conserved conserved Cystine knot-fold growth factor domain of approximately 100 amino acids (aa), called the PDGF/VEGF homology domain [34]. This domain, denoted the PDGF/VEGF homology domain, is involved in forming inter- and intramolecular disulphide bridges to form the PDGF dimers that activate the receptor [35].

PDGFs act via two RTKs (PDGFR-α and PDGFR-β) which share common domain structures, including five extracellular immunoglobulin (Ig) loops and an intracellular tyrosine kinase (TK) domain. This structure is shared with other receptors like VEGFR, c-Fms, c-Kit, and Flt3 [34]. Ligand binding promotes receptor dimerization, which initiates signaling through various pathways became activated including: Mitogen activated protein kinase (MAPK), PI3K phosphatidylinositol 3- kinase, PLC-y, Src TK, and activation of transcription factors such as STATs, ELK-1, c-jun/c-fos and NFkB [34,36].

Our results show that PDGF-B homodimer shares a large group of residues in the interaction with EGFR (ASN 128, SER 127, LYS 86, ASP 155, ILE 13, LEU 14, THR 15, TRP 40, ASN 151, LYS 81, ARG 125, LEU 69, TYR 101, PHE 84. ARG 73, LYS 105, GLN 71, SER92, GLU90, ASN91, PRO82, ILE77), Previous mutational studies have shown that mutations in ILE 30 and ARG 27, reduce the binding interaction between PDGF-BB and it´s receptor PDGF-bR. Furthermore, the study of Ostman et al., [37], indicates that residues from segment 73-77, specially Arg73 and Ile77, are involved in receptor binding. Oefner et al. [18] had also reported that positively charged amino acid residues have an important role in the binding of PDGF-BB to its receptors, including residues from the loop III (Val78-Arg79-Lys8O-Lys81) which are conserved in the human PDGF-BB homodimers. Finally, the crystallographic structure of Oefner et al [18], recognizes an hydrophobic patch, adjacent to loop I, that includes Arg27, Leu38, Pro82 and Phe84 which is also conserved in the PDGF-A sequence.

Interestingly, in the present study, we have reported ILE13, ARG73, ILE77, LYS 81, PRO82 and PHE84, as part of the binding interactions between PDGF-BB and EGFR. These results suggests a possible conserved motif in the interaction between PDGFB and EGFR which could be important in the binding of PDGF-BB with other receptors such as PDGFRα, that can associate with PDGFRβ and form the heterodimeric PDGFRαβ.

It has been previously reported that both hydrophobic interactions and hydrogen bonds play an important role in the recognition of the PDGFR to its ligand [35], Our results show that both of these interactions are important for the binding interaction of PDGF-BB to EGFR. Importantly, both of these interactions are also important in the interaction between EGF and its receptor [1]. It´s important to notice that many tyrosine kinase receptors including PDGFR or EGFR are often able to form heterodimeric or hybrid receptors with differential affinity for their ligands [38,39]. Regarding EGFR, in addition to its cognate ligands, this receptor has been reported to become activated by stimuli that do not directly interact with the EGFR ectodomain including other RTK agonists, chemokines, cytokines GPCR ligands, and cell adhesion elements that involves important transactivation mechanisms [40]. The existence of PDGFBR-EGFR heterodimers have been previously reported in a vascular model from rat [9]. In it, EGFR transactivation does not require PDGFBR kinase activity, and the disruption of this heterodimeric receptor significantly inhibited PDGF-mediated ERK activation, which could have an important effect in biological processes such as cell migration cell during wound healing [9,14].

Results presented in this study, shown the possibility of an interaction between EGFR and PDGF-BB in the interacting region of the EGFR (domains I and III), by means of conserved aminoacids from the PDGF homology domain (cysteine Knot). Additionally, these results may be used as an output for a better understanding of the signaling mechanism exerted by the PDGF-B in the context of human pathologies that shown aberrant expression of the PDGF and EGF ligands and receptor such as brain diseases or cancer. Further biological information will be needed in order to corroborate this model and it´s regulation over cell signaling.

5 Conclusions

A directed docking was performed using Cluspro between human PDGF-BB and EGFR using the templates 1PDG and 3NJP from PDB. Various conserved residues were found to be involved in the docking intereaction of the complex. Additionally, an electrostatic potential evaluation of the PDGF-BB-EGFR complex was performed to validate if the complex is electrostatically complementary in the binding area. Results suggested a possible binding mechanism which could explain the in vivo evidence of formation of heterodimeric receptors EGFR-PDGFRB.

References

1. Ogiso, H.R., Ishitani, O., Nureki, S., Fukai, M., Yamanaka, J.H., Kim, K., Saito, A., Sakamoto, M., Inoue, M., Shirouzu, N., Yokoyama, S.: Crystal structure of the complex of human epidermal growth factor and receptor extracellular domains. Cell 110, 775–787 (2002)
2. Ullrich, A., Coussens, L., Hayflick, J.S., Dull, T.J., Gray, A., Tam, A.W., Lee, J., Yarden, Y., Libermann, T.A., Schlessinger, J., Downward, J., Mayes, E.L.V., Whittle, N., Waterfield, M.D., Seeburg, P.H.: Human epidermal growth factor receptor cDNA sequence and aberrant expression of the amplified gene in A431 epidermoid carcinoma cells. Nature 309, 418–425 (1984)
3. Gan, H.K., Kaye, A.H., Luwor, R.B.: The EGFRvIII variant in glioblastoma multiforme. J. Clin. Neurosci. 16(6), 748–754 (2009)
4. Bajaj, M., Waterfield, M.D., Schlessinger, J., Taylor, W.R., Blunprogram, C.N.S.: On the tertiary structure of the extracellular domains of the epidermal growth factor and insulin receptors. Biochim. Bio. Phys. Acta. 916, 220–226 (1987)
5. Ward, C.W., Hoyne, P.A., Flegg, R.H.: Insulin and epidermal growth factor receptors contain the cysteine repeat motif found in the tumor necrosis factor receptor. Proteins 22, 141–153 (1995)

6. Lu, C., Mi, L.Z., Grey, M.J., Zhu, J., Graef, E., Yokoyama, S., Springer, T.A.: Structural evidence for loose linkage between ligand binding and kinase activation in the epidermal growth factor receptor. Mol. Cell Biol. 22, 5432–5443 (2010)
7. Haas-Kogan, D.A., Prados, M.D., Tihan, T., Eberhard, D.A., Jelluma, N.,: Epidermal growth factor receptor, protein kinase B/Akt, and glioma response to erlotinib. J. Natl. Cancer Inst. 97, 880–887 (2005)
8. Burgaud, J.L., Baserga, R.: Intracellular transactivation of the insulin-like growth factor I receptor by an epidermal growth factor receptor. Exp. Cell Res. 223, 412–419 (1996)
9. Saito, Y., Haendeler, J., Hojo, Y., Yamamoto, K., Berk, B.C.: Receptor heterodimerization: essential mechanism for platelet-derived growth factor-induced epidermal growth factor receptor transactivation. Mol. Cell Biol. 21(19), 6387–6394 (2001)
10. Nazarenko, I., Hede, S.M., He, X., Hedrén, A., Thompson, J., Lindström, M.S., Nistér, M.: PDGF and PDGF receptors in glioma. Ups. J. Med. Sci. 117(2), 99–112 (2012)
11. Brennan, C., Momota, H., Hambardzumyan, D., Ozawa, T., Tandon, A., Pedraza, A., Holland, E.: Glioblastoma subclasses can be defined by activity among signal transduction pathways and associated genomic alterations. PLoS ONE 4, e7752 (2009)
12. Matveev, S.V., Smart, E.J.: Heterologous desensitization of EGF receptors and PDGF receptors by sequestration in caveolae. Am. J. Physiol. Cell Physiol. 282(4), C935–C946 (2002)
13. Perrone, F., Da Riva, L., Orsenigo, M., Losa, M., Jocollè, G., Millefanti, C., Pastore, E., Gronchi, A., Pierotti, M.A., Pilotti, S.: PDGFRA, PDGFRB, EGFR, and downstream signaling activation in malignant peripheral nerve sheath tumor. Neuro Oncol. 11(6), 725–736 (2009)
14. Li, J., Kim, Y.N., Bertics, P.J.: Platelet-derived growth factor stimulated migration of murine fibroblasts is associated with epidermal growth factor receptor expression and tyrosine phosphorylation. J. Biol. Chem. 275, 2951–2958 (2000)
15. Countaway, J.L., Girones, N., Davis, R.J.: Reconstitution of epidermal growth factor receptor transmodulation by platelet-derived growth factor in Chinese hamster ovary cells. J. Biol. Chem. 264, 13642–13647 (1989)
16. Walker, F., Burgess, A.W.: Reconstitution of the high affinity epidermal growth factor receptor on cell-free membranes after transmodulation by platelet-derived growth factor. J. Biol. Chem. 266, 2746–2752 (1991)
17. Walker, F., de Blaquiere, J., Burgess, A.W.: Translocation of pp60csrc from the plasma membrane to the cytosol after stimulation by platelet derived growth factor. J. Biol. Chem. 268, 19552–19558 (1993)
18. Oefner, C., D'Arcy, A., Winkler, F.K., Eggimann, B., Hosang, M.: Crystal structure of human platelet-derived growth factor BB. EMBO J. (11), 3921–3926 (1992)
19. Berman, H.M., Bhat, T.N., Bourne, P.E., Feng, Z., Gilliland, G., Weissig, H., Westbrook, J.: The Protein Data Bank and the challenge of structural genomics. Nature Structural Biology (11), 957–959 (2000)
20. Comeau, R.S., Gatchell, W.D., Vajda, S., Camacho, J.C.: ClusPro: an automated docking and discrimination method for the prediction of protein complexes. Bioinformatics 20(1), 45–50 (2004)
21. Brooks, B.R., Bruccoleri, R.E., Olafson, B.D., States, D.J., Swaminathan, S., Karplus, M.: CHARMM: a program for macromolecular energy, minimization, and dynamics calculations. J. Comput. Chem. 4, 187–217 (1983)
22. Kozakov, D., Hall, D.R., Beglov, D., Brenke, R., Comeau, S.R., Shen, Y., Li, K., Zheng, J., Vakili, P., Paschalidis, I.C., Vajda, S.: Achieving reliability and high accuracy in automated protein docking: Cluspro, PIPER, SDU, and stability analysis in CAPRI rounds 13–19. Proteins: Structure, Function, and Bioinformatics 78, 3124–3130 (2010)

23. Davis, et al.: MolProbity: structure validation and all-atom contact analysis for nucleic acids and their complexes. Nucleic Acids Research 32, W615–W619 (2004)
24. Laskowski, R.A., MacArthur, M.W., Thornton, J.M.: PROCHECK: validation of protein structure coordnates. In: Rossmann, M.G., Arnold, E. (eds.) International Tables of Crystallography, Volume F. Crystallography of Biological Macromolecules, pp. 722–725. Kluwer Academic Publishers, Dordrecht (2001)
25. Laskowski, R.A., Swindells, M.B.: LigPlot+: multiple ligand-protein interaction diagrams for drug discovery. J. Chem. Inf. Model 51(10), 2778–2786 (2011)
26. Warren, L., DeLano.: The PyMOL Molecular Graphics System. DeLano Scientific LLC, San Carlos, CA, USA, http://www.pymol.org
27. Altschul, S.F., Madden, T.L., Schäffer, A.A., Zhang, J., Zhang, Z., Miller, W., Lipman, D.J.: Gapped BLAST and PSI-BLAST: a new generation of protein database search programs. Nucleic Acids Res. 25(17), 3389–3402 (1997)
28. Arnold, K., Bordoli, L., Kopp, J., Schwede, T.: SWISS-MODEL Workspace: A web-based environment for protein structure homology modelling. Bioinformatics 22, 195–201 (2006)
29. Benkert, P., Biasini, M., Schwede, T.: Toward the estimation of the absolute quality of individual protein structure models. Bioinformatics 27, 343–350 (2011)
30. Johnsson, A., Heldin, C.H., Westmark, B., Wasteson, A.: Platelet-derived growth factor: Identification of constituent polypeptide chains Biochem. Biophys. Res. Commun. 104, 66–71 (1982)
31. Deuel, T.F., Huang, J.S., Proffitt, R.T., Baenziger, J.U., Chang, D., Kennedy, B.B.: Human platelet-derived growth factor. Purification and resolution into two active protein fractions. J. Biol. Chem. 256, 8896–8899 (1981)
32. Matsui, T., Heidaran, M., Miki, T., Popescu, N., LaRochelle, W., Kraus, M., Pierce, J., Aaronson, S.: Isolation of a novel receptor cDNA establishes the existence of two PDGF receptor genes. Science 243, 800–804 (1989)
33. Heldin, C.H., Westermark, B.: Mechanism of action and in vivo role of platelet-derived growth factor. Physiol. Rev. 79, 1283–1316 (1999)
34. Andrae, J., Gallini, R., Betsholtz, C.: Role of platelet-derived growth factors in physiology and medicine. Genes Dev. 22(10), 1276–1312 (2008)
35. Shim, A.H., Liu, H., Focia, P.J., Chen, X., Lin, P.C., He, X.: Structures of a platelet-derived growth factor/propeptide complex and a platelet-derived growth factor/receptor complex. Proc. Natl. Acad. Sci. U S A 107(25), 11307–11312 (2010)
36. Tallquist, M., Kazlauskas, A.: PDGF signaling in cells and mice. Cytokine Growth Factor Rev. 15(4), 205–213 (2004)
37. Ostman, A., Andersson, M., Hellman, U., Heldin, C.H.: Identification of three amino acids in the platelet-derived growth factor (PDGF) B-chain that are important for binding to the PDGF beta-receptor. J. Biol. Chem. 266, 10073–10077 (1991)
38. Pacifici, R.E., Thomason, A.R.: Hybrid tyrosine kinase/cytokine receptors transmit mitogenic signals in response to ligand. J. Biol. Chem. 269(3), 1571–1574 (1994)
39. Slaaby, R., Schäffer, L., Lautrup-Larsen, I., Andersen, A.S., Shaw, A.C., Mathiasen, I.S., Brandt, J.: Hybrid receptors formed by insulin receptor (IR) and insulin-like growth factor I receptor (IGF-IR) have low insulin and high IGF-1 affinity irrespective of the IR splice variant. J. Biol. Chem. 281(36), 25869–25874 (2006)
40. Gschwind, A., Zwick, E., Prenzel, N., Leserer, M., Ullrich, A.: Cell communication networks: epidermal growth factor receptor transactivation as the paradigm for interreceptor signal transmission. Oncogene 20(13), 1594–1600 (2001)

Software as a Service for Supporting Biodiversity Conservation Decision Making

Maria Cecilia Londoño-Murcia[1], Camilo Moreno[1], Carolina Bello[1], David Méndez[2], Mario Villamizar[2], and Harold Castro[2]

[1] Instituto de Investigación de Recursos Biológicos Alexander von Humboldt, Bogotá, Colombia
mlondono@humboldt.org.co
[2] Grupo COMIT, Departamento de Ingeniería de Sistemas y Computación, Universidad de los Andes, Bogotá, Colombia
hcastro@uniandes.edu.co

Abstract. This paper presents e-clouds as a tool to support biodiversity decision making, offering a Software as a Service (SaaS) paradigm to execute computing and technic intensive applications such as species distribution models. But mere access to these tools is not enough if usability and economy are not aligned with users interests. This article presents a friendly interface hiding all the complexities of using a public cloud infrastructure, containing an application supported by expert researchers, and an architecture behind the scenes that minimizes the cost of using such a computational and technical infrastructure, what results in a very attractive option for researchers and other stakeholders to get the most out of public clouds for their tasks.

Keywords: Biodiversity Conservation, Spatial analysis, e-clouds, SaaS.

1 Introduction

The development of geographic information systems, the availability of data, both remote sensing and species records, and the improvements of modeling techniques has resulted in the use of spatial analyses such as species distribution modeling to support conservation decision making. Species distribution modeling is a theoretical and methodological approach that uses different mathematical techniques such as general linear models, artificial networks or machine learning, to develop an environmental existing niche for a certain species and project it on a geographical space, obtaining as output a map of the potential distribution areas for a particular species. Explicit geographical information about the presence or potential presence of a species have been used in many applications concerning biodiversity conservation, such as invasive species risk [1], identification and evaluation of conservation areas [2] and climate change [3], besides others.

Species distribution models use two types of information: environmental information that includes topography and climate, and biological information based mainly on

L.F. Castillo et al. (eds.), *Advances in Computational Biology*,
Advances in Intelligent Systems and Computing 232,
DOI: 10.1007/978-3-319-01568-2_6, © Springer International Publishing Switzerland 2014

species records. Both environmental and biological information has increased in amount and availability imposing digital storage necessities and infrastructure framework that can mobilize and integrate diverse types of data [4].

In Colombia many of the governmental institutions that need to make conservation decisions, or small regional academic universities, lack the human capital or computing infrastructure to develop species distribution models. This situation imposes a challenge for a country rich in biodiversity that needs to take urgent conservation actions.

Computing and data intensive applications have been traditionally executed in HPC (High Performance Computing) infrastructure, as cluster and grid computing, allowing researchers to take advantage of tens of thousands of dedicated servers to execute application tasks. Although this infrastructure can provide many benefits to large research groups, when small and medium research groups try to use them, they often face the following constraints: (1) the total cost of infrastructure ownership is high; (2) executing applications may involve complex processes related to IT management; (3) there are different models to parallelize the execution of applications and in many cases parallelization models require complex configurations; (4) researchers require broad computing capabilities during peak periods and results can take weeks or months to be generated.

At the business level, the Software as a Service (SaaS) model, allows businesses to adopt ready-to-use applications, abstracting problems associated with the installation, configuration, monitoring, management and updating of applications. We present an e-Cloud solution that allows the delivery of scientific applications under the Software as a Service (SaaS) model, which opens opportunities for research groups and decision makers with high computation requirements to develop biodiversity spatial analysis, such as species distribution modeling, in an easy, fast and cheap way, without needing to buy and support big computing infrastructure.

2 Materials and Methods

This proposal is based on the computing necessities of the Humboldt Institute in order to improve its capacity to process and make public the research done in species distribution modeling. The methodological design is proposed by the Systems and Computing Engineering Department of the Universidad de Los Andes. The aim of this proposal is to offer the country a service with high scientific quality and easy accessibility that allows information and analysis capacities to support decision making in biodiversity conservation.

The Humboldt Institute has proceeded with species distribution modeling as follows: species records are download from GBIF, a taxonomic and geographic verification process takes place; records that passed the filters are incorporated into the distribution modeling using WorldClim Database [5] an MaxEnt [6]. Procedures were written in R code in order to dispose of a script ready to configure on e-Clouds.

e-Clouds was designed to permit researchers to access their private workspace through a web portal, each researcher can manage submissions of new jobs, the upload and download of data, the monitoring of jobs in execution and cost reports associated to the used computing resources. If a user wants to execute a set of jobs of a scientific application, he/she selects the application, enters the jobs parameters (numbers, strings and files), selects the virtual machines required to execute the jobs, and finally accepts the estimated cost for the execution of the jobs. e-Clouds begin with the execution of the jobs in a transparent way on a public cloud, computing infrastructure using as much computing resources as needed. During job executions, researchers can monitor the execution time and costs of each job and once the jobs have finished, researchers can download the results.

Behind the scenes, e-Clouds uses the computing resources provided by a public cloud computing provider. e-Clouds architecture has many scalable components to allow that it can be used by many researchers at the same time. e-Clouds is composed of different components, including a resource manager designed to operate in cloud infrastructure, a relational database, an application repository, a scalable data repository (for researchers), a repository of virtual machines, and a repository to store the agent executed on the virtual servers of the public cloud infrastructure to allow that applications can be installed on-demand. e-Clouds was designed to guarantee the security and privacy of the data stored by researchers. All of these components are hidden to researchers but they do the job allowing researchers to use large computing capacities on-demand without the need of maintain complex configurations and applications and paying only by the resources they consumed.

Besides the services provided to researchers, e-Clouds has an administration web portal used by the team responsible for operating and managing e-Clouds. Through this portal the management team can monitor the resources consumed by all of the researchers, the costs associated to the resources consumed on the public cloud infrastructure, the audit of all operations performed by researchers and the management of quotes of resources that can be used by researchers. This web portal also allows creating, testing and publishing new scientific applications in the marketplace.

3 Results

e-Clouds was developed using the Ruby on Rails (RoR) framework and the PostgreSQL database engine. The e-Clouds Web portal is being executed on the Heroku cloud Platform and Amazon Web Services (AWS) cloud infrastructure. The resource manager was developed using Ruby and it is executed on a virtual server in AWS. Jobs sent by researchers are executed on AWS using scalable services such as EC2, S3 y SQS. The agent installed on each virtual server that allows the on-demand installation of scientific applications was developed in Ruby. Currently researchers can access e-Clouds doing a previous authorization step. A requirement for the execution of jobs is that researchers use an amount of credit approved by the e-Clouds team to do the first tests with e-Clouds. Then the researchers need to pay the consumed resources. In the near future the goal is that any researcher can access e-Clouds and pays by using his/her credit card or the credit approved by his/her research group.

The R script from the Humboldt Institute has been set and displayed in e-Clouds and is ready to make species distribution models. Once you run data in the application and set the parameters, the output will be a distribution map for each species in the database. e-Clouds allows the execution of Bag of Task (BoT) applications and during the next months new scientific applications related to biodiversity conservation will be published on e-Clouds to allow that researchers can access a large number of ready-to-use applications. We will continue working in different areas such as: the development of a platform that allows researchers to publish new applications to e-Clouds using a self-publishing portal; to support the execution of parallel applications such as MPI; to improve the Graphical User Interfaces provided to researchers, focusing on providing the best and easy to us experience; and to execute large amounts of jobs to test the performance of e-Clouds with larger computing workloads.

4 Discussion and Conclusions

This paper presents a proposal to create a SaaS marketplace for scientific application (scientific SaaS). The SaaS marketplace, called e-Clouds, is built on a public IaaS infrastructure, allowing researchers of medium and small groups to access ready-to-use scientific applications which can be used on-demand whenever they need them, without having to incur high TCO costs and complex IT tasks, and having technical and scientific support about the applications used. e-Clouds allows that jobs, applications and data can be easily managed by researchers and the e-Clouds team. The first implementation shows that e-Clouds may be easily extended to include new applications and more researchers, which will be reflected in new research results.

Some analyses, such as The Magellan report [7], have shown that cloud computing is a viable model for executing MTC and MPI scientific applications, where small or medium groups without large or optimized cluster or grid infrastructure can adopt cloud computing to execute scientific workload in a cost effective manner. The main challenges for running scientific applications and workflows in the cloud involve: heterogeneous resource management; cost-effective executions; installation, configuration and management of scientific applications in virtual machine images; data management; performance and security of cloud infrastructure; and tools to facilitate executing the applications, because clouds require significant configuration and administration tasks [7].

Different efforts have been developed to facilitate the use and execution of scientific applications in the cloud [8], but management and maintenance of an image per application are complex in these options, others [9] allows researchers to download and install preconfigured virtual machines with scientific applications in their desktop, but researchers can only can take advantage of the resources of a single desktop computer.

Different projects have analyzed and executed migration tests of legacy cluster, grid applications or workflows to cloud environments for different scientific fields [10], [11], and [12], and Projects such as [13] have developed different approaches to integrate traditional grid infrastructures with on-demand cloud services. Other projects have incorporated cloud features (auto-scaling, on-demand access, pay-per-use, etc.) to grid infrastructure [14].

The above results show that throughout the next years, cloud computing will be adopted by a large number of scientific groups as current cloud providers will improve their infrastructure to support scientific workloads. However, in all existing options, researchers executing scientific workloads need to know important IT tasks. Thus if non-IT researchers would like to use cloud options they face similar problems than those found in a grid/cluster infrastructure.

At business level, SaaS marketplaces such as ZOHO [15], SuccesssFactors [16], among many others, allow businesses of all sizes to access complex and ready-to-use cloud enterprise applications (with low or non-initial configuration efforts). We propose to adopt the same model of business SaaS marketplaces in the scientific field, where central cloud providers or research groups configure and maintain complex and commonly-used applications and offer them to a high number of researchers, on-demand, very fast and at a low cost. This issue is the main motivation of this article whose main goal is to provide new opportunities to small and medium research groups.

Some commercial scientific SaaS options such as Cyclone [17] and Cloud Numbers [18] are offering scientific SaaS applications, built on private or public IaaS infrastructure, which can be easily used by researchers of different fields. In contrast to these commercial SaaS offers, e-Clouds is conceived as an open-source SaaS marketplace that is maintained by specialized research groups (initially Humboldt Institute) that dedicate IT and non-IT researchers of different fields to maintain the e-Clouds SaaS option. Small and medium research groups use e-Clouds to execute applications on public cloud IaaS infrastructure. e-Clouds users only will have to pay the resources consumed of the public IaaS, without incurring extra costs by the use of e-Clouds services.

Finally e-clouds can be incorporated into proposed frameworks for species distribution data management and analysis currently under implementation [4], allowing issues about upload and storage data, workspace for user, automated data integration and multiples applications for data analysis be dealt with.

Acknowledgments. We are thankful to Laura Nägele for comments on this manuscript and J. C. Bello, O. Ramos and J. Velásquez for support and ideas. This project is supported by Ministerio de Ambiente y Desarrollo Sostenible, biotic component for "Mapa de Ecosistemas Continentales, Marinos y Costeros a escala 1:100000" and by Proyecto "Plan de Fortalecimiento Institucional-PFI 2011-2013" supported by Colciencias.

References

1. Peterson, A.T., Robins, C.R.: Using ecological-niche modeling to predict barred owl invasions with implications for spotted owl conservation. Conservation Biology 17, 1161–1165 (2003)
2. Urbina-Cardona, J.N., Flores-Villela, O.: Ecological-niche modeling and prioritization of conservation-area networks for Mexican herpetofauna. Conservation Biology 24, 1031–1041 (2010)

3. Ihlow, F., Dambach, J., Engler, J.O., Flecks, M., Hartmann, T., Nekum, S., Rajaei, H., Rödder, D.: On the brink of extinction? How climate change affect global chelonian species richness and distribution. Global Change Biology 18, 1520–1530 (2012)
4. Jetz, W., McPherson, J.M., Guralnick, R.P.: Integrating biodiversity distribution knowledge: toward a global map of life. Trends in Ecology and Evolution 27, 151–159 (2012)
5. Hijmans, R.J., Cameron, S.E., Parra, J.L., Jones, P.G., Jarvis, A.: Very high resolution interpolated climate surfaces for global land areas. International Journal of Climatology 25, 1965–1978 (2005)
6. Phillips, S.J., Anderson, R.P., Schapire, R.E.: Maximum entropy modeling of species geographic distributions. Ecological Modelling 190, 231–259 (2006)
7. Office of Advanced Scientific Computing Research (ASCR): The Magellan Report on Cloud Computing for Science. U.S. Department of Energy (2011)
8. Krishnan, S., Clementi, L., Jingyuan, R., Papadopoulos, P., Li, W.: Design and Evaluation of Opal2: A Toolkit for Scientific Software as a Service. In: World Conference on Services, pp. 709–716 (2009)
9. Chen, W., Cao, J., Li, Z.: Customized Virtual Machines for Software Provisioning in Scientific Clouds. In: Second International Conference on Networking and Distributed Computing (ICNDC), pp. 240–243 (2011)
10. Saripalli, P., Oldenburg, C., Walters, B., Radheshyam, N.: Implementation and Usability Evaluation of a Cloud Platform for Scientific Computing as a Service (SCaaS). In: Fourth IEEE International Conference on Utility and Cloud Computing (UCC), pp. 345–354 (2011)
11. Doddavula, S.K., Rani, M., Sarkar, S., Vachhani, H.R., Jain, A., Kaushik, M., Ghosh, A.: Implementation of a Scalable Next Generation Sequencing Business Cloud Platform. In: IEEE 4th International Conference on Cloud Computing, pp. 598–605 (2011)
12. Xiaoyong, B.: High performance computing for finite element in cloud. In: International Conference on Future Computer Sciences and Application (ICFCSA), pp. 51–53 (2011)
13. Calatrava, A., Molto, G., Hernandez, V.: Combining Grid and Cloud Resources for Hybrid Scientific Computing Executions. In: Third IEEE International Conference on Coud Computing Technology and Science, pp. 494–501 (2011)
14. Fisher, S.: The Architecture of the Apex Platform, salesforce.com's Platform for Building On-Demand Applications. In: 29th International Conference on Software Engineering - Companion, p. 3 (2007)
15. Zoho Corporation Pvt. Ltd. ZOHO Work Online, http://www.zoho.com/
16. SuccessFactors, Inc. SuccessFactors, http://www.successfactors.com/
17. Silicon Graphics International. Cyclone, http://www.sgi.com/products/hpc_cloud/cyclone/
18. Cloudnumbers.com, http://cloudnumbers.com

Structural and Functional Prediction of the Hypothetical Protein Pa2481 in *Pseudomonas Aeruginosa Pao1*

David Alberto Díaz*, George Emilio Barreto, and Janneth González Santos

Nutrition and Biochemistry Department. Faculty of Science. Pontificia Universidad, Javeriana, Bogotá D.C., Colombia
{diaz.david,janneth.gonzalez,gsampaio}@Javeriana.edu.co

Abstract. Pseudomonas aeruginosa is a bacterium resistant to a large number of antibiotics and disinfectants. Since the antigens are responsible for producing this resistance, and most of these are proteins, the objective of this work was to predict by computational means the 3D structure and function of PA2481 protein, determining whether is involved in the antibiotic resistance of this bacterium. In order to do this, the primary structure was analyzed computationally by using servers PROSITE, PFAM, BLAST, PROTPARAM, GLOBPLOT, and PROTSCALE. The secondary structure was obtained by the consensus of algorithms SOPM, PREDATOR, DPM, DSC, and GOR4. The 3D structure was predicted with the I-TASSER server and its stereochemical conformation was evaluated with the STRUCTURE ASSESSMENT tool. The final model was visualized with PyMol program. As a result, the 3D structure of PA2481 is proposed according to its stereochemical conformation; two domains are identified as cytochrome c. The function of the protein may be related to electron transport and proton pumping to generate ATP in *Pseudomonas aeruginosa*. These results allow a better understanding of the role this protein plays in the physiology of this bacterium.

Keywords: hypothetical protein, cystic fibrosis, external otitis, structural modeling, PA2481, *Pseudomonas aeruginosa*.

1 Introduction

Pseudomonas aeruginosa is a Gram-negative bacterium characterized by its environmental versatility, growing in soil, marshes and coastal marine habitats, including plant and animal tissues [1]. Because of its resistance to antibiotics and disinfectants, P. aeruginosa is a bacterium that is characterized as a major opportunistic pathogen in humans, causing serious complications caused by infections in patients particularly susceptible like people with immune system deficiencies, victims of skin burns, catheterized patients who suffer urinary tract infections and patients with respirators, causing nosocomial pneumonia. It is also the predominant cause of morbidity and mortality in patients with cystic fibrosis colonizing the lungs [2].

* Corresponding author.

L.F. Castillo et al. (eds.), *Advances in Computational Biology*,
Advances in Intelligent Systems and Computing 232,
DOI: 10.1007/978-3-319-01568-2_7,

The structural genomics field determines the three dimensional structure of all proteins of a given organism by experimental methods such as X-ray crystallography, nuclear magnetic resonance spectroscopy, or by computational approaches such as homology modeling. This raises new challenges in structural bioinformatics, focused on determining protein function from its 3D structure. At present there are plenty of genome sequencing projects, which are producing linear sequences of amino acids that are stored in data bases, however, for an understanding of the biological function, knowledge of the structure and function is required [3].

The ultimate goal of structural genomics is to contribute to the organization principles of the structure of proteins in biology and medicine through functional annotation [4] and the application of protein structure such as virtual drug screening [5], whereas bioinformatic tools play a crucial role in the evaluation and classification of new structural data obtained, and also benefit directly from the data stream generated by structural genomics projects, resulting in improved algorithms, software and databases [6]. On the other hand, functional genomics makes use of the vast richness of data produced by genome sequencing projects to describe functions and interactions of genes and proteins. Its goal is to discover the biological function of individual genes and how groups of genes and their products work together in health and disease [7]. During the last decades many complete genomes of several bacteria, archea, and eukaryotes have been sequenced. Stover et al [2], described the complete genome sequence of P. aeruginosa strain PAO1, which was noted for its diverse metabolic capacity and large size with 6.3 million base pairs, being the largest bacterial genome sequenced, however, about 30 to 40% of the genes have not been assigned a function. Furthermore, in studies of gene annotation, a large fraction of open reading frames are labeled as conserved hypothetical proteins and many of these hypothetical proteins are found in more than one bacterial species [8].To assign functions to novel proteins, homology based on gene annotations has been the standard during the last years, as it infers molecular characteristics through transfer of information of experimentally characterized proteins [9].

Knowing the complete genome sequence including hypothetical proteins, together with encoding processes, provide a lot of information to the discovery and exploitation of new targets for antibiotics, and hope for the development of more effective strategies for the treatment of threatening opportunistic infections caused by P. aeruginosa in humans [2]. Therefore, in this study the prediction of the three dimensional structure is assessed, and a functional approximation of hypothetical protein PA2481 of Pseudomonas aeruginosa PAO1 is performed through various bioinformatic tools and software, to determine whether it is involved in the antibiotic resistance of this microorganism, in order to better understand the physiology of this bacterium.

2 Materials and Methods

2.1 Computational Analysis of the Primary Structure of PA2481

Sequence analysis of PA2481 of 291 amino acid residues with unknown function and structure were performed using various bioinformatic tools and databases available on the network. The sequence of the hypothetical protein, PA2481 (Acc. No. AAG05869.1), was obtained through the GenBank at the National Center for Biotechnology Information (NCBI) [10]. A similarity search with other reported sequences was performed using BLAST [11] available in NCBI.

Since proteins are generally composed of one or more functional regions commonly known as domains, and the identification of these allow to infer their possible function, the prediction of families, motifs, domains and functional sites was made by reference to matches found in the database PROSITE [12] from the ExPASy bioinformatics resource portal [13] and the database of protein families PFAM [14] of England Trust Sanger Institute. These databases represent protein families by multiple sequence alignments and hidden Markov models. Additionally, an alignment with proteins of known structure in PDB (http://www.rcsb.org/pdb/home/home.do) was performed to determine potential homologues with known structure and function.

For the physicochemical characterization of the protein, the isoelectric point, molecular weight, extinction coefficient [15], instability index [16], aliphatic index [17] and overall average of hydropathy [18] were determined by the PROTPARAM tool [19] from the ExPASy bioinformatics resource portal (http://expasy.org/tools), which performs calculations based on the amino acid composition and the N-terminal residue. Regarding the hydrophobicity analysis, the PROTSCALE tool and the Kyte and Doolittle algorithm from the ExPasy Proteomic tools (http://expasy.org/tools) were used based on the physical and chemical properties of the amino acids of the protein. The prediction of globular and disordered regions or nonstructural regions of the protein was determined with GLOBPLOT [20].

2.2 Secondary Structure Prediction

The server NSP@ [21] was used to predict the secondary structure of hypothetical protein PA2481, where the consensus of five algorithms was used, within which two approaches were selected relying on the use of probability parameters determined by the relative frequencies of occurrence of each amino acid in each type of secondary structure (SOPM and PREDATOR) and three methods based on Bayesian inference probability (GOR4, DPM and DSC).

2.3 Three-Dimensional Structure Prediction

The three dimensional structure of hypothetical protein PA2481 was predicted with the ITASSER algorithm [22]. 3D models were built from multiple alignments of protein sequences with known structure and function. Model evaluation was performed by means of the tool "structure assessment" of SWISSMODEL [23], determining the correct geometric conformation of the protein by stereochemical analysis presented in the Ramachandran plot. The final model was visualized with PyMol (http://www.pymol.org).

3 Results and Discussion

3.1 Computational Analysis of the Primary Structure of PA2481

When performing local alignment of sequences with BLAST, six proteins were identified with 99% and 98% similar amino acid sequence to the sequence of PA2481 (Table 1). Since the identity of each protein is based on the specific order in each of its amino acids [24], the identity percentage of the PA2481 protein with each of the six identified proteins, infers that these are probably homologous, almost identical in their 3D structure and therefore in their function. Given that proteins are hypothetical, they do not have a defined structure and function despite having annotations in the NCBI regarding conserved domains corresponding to cytochrome c, which is a good indication to predict its possible biological function.

Regarding the prediction of families, motifs, domains, and functional sites, the results obtained out of the PROSITE and PFAM databases from the amino acid sequence of the PA2481 protein (Table 2) agree that there are two domains corresponding to cytochrome c superfamily, where the highest score domain is between residues 165 and 250. In addition, two sites of covalent binding to the heme group were found in the residues (178 y 181) y (70 y 73), and one axial ligand binding (iron for this protein) in the residues 182 and 74 respectively in each domain, common characteristic of c-type cytochromes, which have covalent binding to the heme group in one or more motifs Cys-XX-Cys-H [25]. Furthermore, the inclusion of these heme groups is a requirement for the oxidase assembly [26]. The PFAM server presented significant overlaps with the cytochrome c oxidase, cbb3 type, subunit III, which is corroborated by performing BLAST sequence alignment using only proteins of known structure deposited in the PDB, wherein the sequence that obtained the best alignment with a score of 37.4, E-value of 0.008 and 50% identity, was precisely the chain c of the structure of the cytochrome c oxidase cbb3, sequence identified with 3MK7_C in PDB. These results indicate that the function of the hypothetical protein PA2481 could be related to the electron transfer and proton pumping in Pseudomonas aeruginosa to generate ATP, as the cytochrome c oxidase cbb3 fulfills these functions in bacteria and mitochondria [27].

In proteins belonging to the family of cytochrome c, the heme group is covalently attached by thioether bonds to two conserved cysteine residues located in the cytochrome c center. Cytochrome c play an important role in the electron transfer [28]; in addition, in the centers of the Cytochrome c are also active sites of many enzymes that have a significant function in eukaryotic cell apoptosis [29].In the known structures of c-type Cytochromes, there are about 6 classes that vary in their folds [30].

The species of the genus Pseudomonas were traditionally considered non-fermenting organisms with aerobic respiration. The elements involved in breathing in these species comprise a pool of 17 respiratory dehydrogenases responsible for the passage of electrons from the coenzyme to the quinones. P. aeruginosa has five terminal oxidases that catalyze the reduction of oxygen in water: three of them are cytochrome c oxidase (Cbb3-1, Cbb3-2, and Aa3), whereas the remaining two

Cytochrome bo3 (Cyo) and insensitive oxidase to cyanide, are quinol oxidases. Diverse terminal oxidases have different affinity for oxygen and capacity for pumping protons [31], as might be the case of the PA2481 protein.

The physicochemical properties obtained from PROTPARAM tool (Table 3) allowed to determine the molecular weight of PA2481 that coincides with the molecular weight of the chain c of the cytochrome c, oxidase cbb3, subunit III, reported by Zufferey [26] which is 31 kDa. The Isoelectric point is where positive and negative charges are equal annulling the existence of motion in an electric field, representing the pH of the PA2481 protein, which could present a minimum solubility in experimental tests providing isolation in an electric field [32]. Regarding the life time of PA2481, this refers to the time (in vitro) that takes for a protein to disappear once it is synthesized by the cell. According to the lifetime of the PA2481 protein, it can be determined that the N-terminal residue corresponds to methionine, serine or alanine, threonine, valine or lysine, because these amino acids are specific for proteins with lifetime greater than 20 hours, since it has been shown that the last amino acid in the chain determines the survival of the protein [33]. N-terminal residue can also be determined experimentally to be able to know exactly which one of the above-mentioned amino acids is the N terminal [34].

Aliphatic index determines the relative volume occupied by aliphatic side chains, where the proteins with a high aliphatic index, such as the PA2481 protein, tend to be more thermostable. As for the overall average of hydropathy for PA2481, this indicates the ability of the protein to establish interaction with water, suggesting that it may be highly hydrated in aqueous media, since the lower this value the greater possibility of such interaction. This measure is based on the amount of energy (kg / mol) that is used for transferring a segment of sequence of defined length from a hydrophobic environment to a hydrophilic environment [35].

Results from PROTSCALE (Figure 1) identified hydrophobic and hydrophilic regions of the PA2481 protein, showing the score based on the residue. The highest peaks indicate hydrophobic regions and the lowest, near -2, indicate hydrophilic regions. Using a window size 9, it was determined that PA2481 is a highly hydrophilic protein, meaning that most of its structure is in contact with an aqueous environment, which coincides with the results of hydropathy overall average. PA2481 also has three hydrophobic regions that could be in contact with the inside of the cell membrane.

Regions, where there are low levels of hydrophobic amino acids with long stretches of hydrophilic residues but with high net charge and with neutral pH, are theoretically associated with lack of compaction of the proteins under normal physiological conditions resulting in structurally disordered native structures [36]. It is known that disordered segments of proteins contain important functional sites predicted as linear motifs [36]. The disordered regions in the PA2481 protein were analyzed to predict their behavior in solution and identify regions that may be functionally important. The prediction of globular domains and disordered regions of the protein was performed with GLOBPLOT which allows the prediction of structural domains and is not limited by the search for homologous sequences known structurally [37]. Disordered regions revealed by GLOBPLOT (Figure 2) show 7 regions with intrinsic disorders between amino acid position 1-7, 24-40, 119-123, 145-165, 201-206, 225-236 and 251-257, which are possibly involved in important functions of Pseudomonas aeruginosa. In addition, 2 of the 7 disordered regions (residues 201-206 and 225-236) are in the domain of cytochrome oxidase cbb3 revealed in PFAM results (Table 2), which confirms the importance of this domain.

3.2 Secondary Structure Prediction

The secondary structure prediction was performed with the server NPSA, which gives the results based on a consensus of several algorithms (SOPM, PREDATOR GORIV, DSC and DPM), since the quality of the prediction of the secondary structure may be better if it is done by consensus-based classifications to obtain a more reliable prediction than using only one method [38]. PREDATOR and SOPM algorithms are based on the conditional probability that a given amino acid becomes part of a secondary structure considering that adjacent amino acids acquire the same structure, while the algorithms GORIV, DSC, and DPM use training sets whose elements are solved structures to identify common motif sequences associated with particular arrangements of secondary structures [39].

According to the above, PA2481 secondary structure (Figure 3) revealed the existence of eight α helices between amino acid residues: 5-14, 42-57, 104-113, 129-141, 171-179, 215 - 222, 239-247 and 263-269 represented with 29.90%;11 loops between amino acids 11: 17-41, 57-62, 65-94, 118-128, 142-165, 182-189, 197-213, 225 -238, 248-261, 270-276 and 278-284 with 60.82%; and extended strands represented with only 4.12%. The prevalence of loop conformation makes these molecules more ductile and it can be explained by the functional need to easily establish interactions with proteins [40].

3.3 Three Dimensional Structure Prediction

The prediction of the three dimensional structure of the PA2481 protein was performed using the I-TASSER server, which is a unified platform for automated prediction of the structure and function of proteins. The three dimensional structure of the PA2481 protein (Figure 4), was determined by the homology modeling method, which uses template proteins derived from sequence alignments of known structure in PDB. Validation criteria of the three dimensional structure of a stable protein, depend on stereochemical parameters corresponding to torsion angles which determine its folding. Since the allowable range of angles within a structure is very restrictive, variations in the angle of torsion are very few and have made possible the identification of angles which allow an approximation to the probability that the structural conformation of the protein is correct [41].

As stated before, the PA2481 three-dimensional model validation was performed through evaluation using the "structure assessment" of SWISSMODEL which essentially fulfilled stereochemical parameters of a stable structure with 204 (86.1%) residues in the most favorable regions, 30 (12.7%) residues in additional permitted regions, 2 (0.8%) residues in generously allowed regions and only one (0.4%) residue is located in disallowed regions (Figure 5). Sites where there is greater probability that the angles that form the protein structure are possible correspond to the most favorable regions and to those additionally permitted while generously allowed and disallowed regions correspond to sites that are less likely to approach the correct angle. Analyzing the results of the evaluation, it is possible to deduce that the model provides a good approximation to the actual structure of the PA2481 protein, the model also largely coincides with the secondary structure predicted above.

4 Conclusions

The predicted three-dimensional model is a good approach to the possible real structure of protein PA2481, as it coincides mostly with the predicted secondary structure and also its correct conformation was confirmed through geometric evaluation (Graphic Ramachandran). Through different bioinformatics tools used in the present study, it was determined that the hypothetical protein PA2481 of *Pseudomonas aeruginosa PAO1* belongs to the superfamily of cytochrome c, showing the most significant characteristics of this type of proteins such as the two sites of covalent binding to the heme group. This indicates that PA2481 possibly participates in biological processes involved in cellular metabolism, of oxidation-reduction, in the electron transfer, and in the generation of metabolites and energy precursors, contributing to the mechanisms of antibiotic resistance of the bacteria, by pumping substances to the exterior, which requires ATP, or by redox processes leading to their degradation.

Acknowledgements. We thank to the Department of Nutrition and Biochemistry, Pontificia Universidad Javeriana for academic and infrastructure support in carrying out this work.

References

1. Hardalo, C., Edberg, S.C.: Pseudomonas aeruginosa: assessment of risk from drinking water. Crit. Rev. Microbiol (1997), doi:10.3109/10408419709115130
2. Stover, C.K., Pham, X.Q., Erwin, A.L., Mizoguchi, S.D., Warrener, P., Hickey, M.J., Brinkman, F.S., Hufnagle, W.O., Kowalik, D.J., Lagrou, M., Garber, R.L., Goltry, L., Tolentino, E., Westbrock-Wadman, S., Yuan, Y., Brody, L.L., Coulter, S.N., Folger, K.R., Kas, A., Larbig, K., Lim, R., Smith, K., Spencer, D., Wong, G.K., Wu, Z., Paulsen, I.T., Reizer, J., Saier, M.H., Hancock, R.E., Lory, S., Olson, M.V.: Complete genome sequence of Pseudomonas aeruginosa PAO1: an opportunistic pathogen. Nature 406, 959–964 (2000)
3. Baker, D.A., Sali, A.: Protein structure prediction and structural genomics. Science 294, 93–96 (2001)
4. Skolnick, J., Fetrow, J.S., Kolinski, A.: Structural genomics and its importance for gene function analysis. Nature Biotechnologic 18, 283–287 (2000)
5. Makino, S., Ewing, T.J., Kuntz, I.D.: Dream++: Flexible docking program for virtual combinatorial libraries. Journal Comput. Aided Mol. Des. 13, 513–532 (1999)

6. Terwilliger, T.C., Waldo, G., Peat, T.S., Newman, J.M., Chu, K., Berendzen, J.: Class-directed structure determination: foundation for a protein structure initiative. Protein Science (1998), doi:10.1002/pro.5560070901
7. Hieter, P., Boguski, M.: Functional Genomics: It's All How You Read It. Science (1997), doi:10.1126/science.278.5338.601
8. Dunham, I.: Genomics-the new rock and roll? Trends in Genetics (2000), doi:10.1016/S0168-9525(00)02109-0
9. Roberts, R.J.: Identifying Protein Function—A Call for Community Action. PLoS Biology (2004), doi:10.1371/journal.pbio.0020042
10. Pruitt, K., Tatusova, T., Maglott, D.: NCBI Reference Sequence (RefSeq): a curated non redundant sequence database of genomes, transcripts and proteins. Nucleic Acids Research (2007), doi:10.1093/nar/gkl842
11. Altschul, S.F., Madden, T.L., Schaffer, A.A., Zhang, J., Zhang, Z., Miller, W., Lipman, D.J.: Gapped BLAST and PSI-BLAST: a new generation of protein database search programs. Nucleic Acids Res. (1997), doi:10.1093/nar/25.17.3389
12. Sigrist, C.J.A., Cerutti, L., de Castro, E., Langendijk-Genevaux, P.S., Bulliard, V., Bairoch, A., Hulo, N.: PROSITE, a protein domain database for functional characterization and annotation. Nucleic Acids Res. (2010), doi:10.1093/nar/gkp885
13. Artimo, P., Jonnalagedda, M., Arnold, K., Baratin, D., Csardi, G., Castro, E., Duvaud, S., Flegel, V., Fortier, A., Gasteiger, E., Grosdidier, A., Hernandez, C., Ioannidis, V., Kuznetsov, D., Liecht, R., Moretti, S., Mostaguir, K., Redaschi, N., Rossier, G., Xenarios, L., Stockinger, H.: ExPASy: SIB bioinformatics resource portal. Nucleic Acids Research (2012), doi:10.1093/nar/gks400
14. Bateman, A., Coin, L., Durbin, R., Finn, R., Hollich, V., Griffiths-Jones, S., Khanna, A., Marshall, M., Moxon, S., Sonnhammer, E., Studholme, D., Yeats, C., Eddy, S.: The PFAM protein families database. Nucleic Acid Research (2012), doi:10.1093/nar/gkh121
15. Gill, S., Von Hippel, P.: Calculation of protein extinction coefficients from amino acid sequence data. Analytical Biochemistry 182, 319–326 (1989)
16. Guruprasad, K., Reddy, B.V., Pandit, M.V.V.: Correlation between stability of a protein and its dipeptide composition: a novel approach for predicting in vivo stability of a protein from its primary sequence. Protein Eng. Dec. 4, 155–161 (1990)
17. Ikai, A.J.: Thermostability and aliphatic index of globular proteins. Journal Biochem. 88, 1895–1898 (1980)
18. Kyte, J., Doolittle, R.F.: A simple method for displaying the hydropathy character of a protein. J. Mol. Biol. 157, 105–132 (1982)
19. Gasteiger, E., Hoogland, C., Gattiker, A., Duvaud, S., Wilkins, M.R., Appel, R.D., Bairoch, A.: In: Walker, J.M. (ed.) Protein Identification and Analysis Tools, Totowa, NJ (2005)
20. Imer, O., Cavas, L.: The Bioinformatics Tools for the Estimation of Disordered. Biomedical Engineering Meeting (2009), doi:10.1109/BIYOMUT.2009.5130377
21. Combet, C., Blanchet, C., Geourjon, C., Deléage, G.: NPS@: Network Protein Sequence Analysis TIBS 25, 147–150 (2000)
22. Ambrish, R., Kucukural, A., Zhang, Y.: I-TASSER: a unified platform for automated protein structure and function prediction. Nature Protocols (2010), doi:10.1038/nprot.2010.5
23. Arnold, K., Bordoli, L., Kopp, J., Schwede, T.: The SWISS-MODEL Workspace: A web-based environment for protein structure homology modelling. Bioinformatics (2006), doi:10.1093/bioinformatics/bti770
24. Claverie, J., Notredame, C.: Bioinformatics for Dummies. Indianapolis, Indiana (2007)

25. Barker, P.D., Ferguson, S.J.: Still a puzzle: why is covalently attached in c-Type cytochromes? Structure 7, 281–290 (1999)
26. Zufferey, R., Preising, O., Hennecke, H., Thony-meyer, L.: Assembly and function of the cytochrome cbb3 oxidase subunits in *Bradyrhizobium japonicum*. . J. Biol. Chem. 271, 9114–9119 (1996)
27. Buschmann, S., Warkentin, E., Xie, H., Langer, J.D., Ermier, U., Michel, H.: The structure of cbb3 cytochrome oxidase provides insights into proton pumping. Science (2010), doi:10.1126/science.1187303
28. Lee, B.: Physical properties of cytochromes c: 1. role of cytochrome c' as a nitric oxide carrier 2. Folding of cytochrome c. New York (2009)
29. Martinou, J.C., Desagher, S., Antonsson, B.: Cytochrome c release from mitochondria: all or nothing. Nat. Cell Biol. 2, 41–43 (2000)
30. Allen, J.W., Daltrop, O., Stevens, J.M., Ferguson, S.J.: C-type cytochromes: diverse structures and biogenesis Systems pose evolutionary problems. Philios. Trans. R. Soc. 358, 255–266 (2003)
31. María, T.P.: Influencia del regulador global Anr en la fisiología de Pseudomonas extremaustralis, una bacteria productora de polihidroxibutirato. Tesis Doctoral. Facultad de Ciencias Exactas y Naturales Universidad de Buenos Aires (2011)
32. Werner, M.S.: Biología: Fundamentos para Medicina y Ciencias de la Vida. Barcelona, España (2007)
33. Bachmair, A., Finley, D., Varshavsky, A.: In vivo half-life of a protein is a function of its amino-terminal residue. Science 234, 179–186 (1986)
34. Jinling, C., Dandan, Z., Pei, S., Wei, S., Gengfu, X., Yinong, D.: Bioinformatics analysis on ORF1 protein of Torque teno virus (SANBAN isolate). Asian Pacific Journal of Tropical Medicine (2011), doi:10.1016/S1995-7645(11)60207-1
35. Lieberman, M., Marks, A.D., Smith, C.M.: Marks. Basic medical biochemistry: a clinical approach, China (2009)
36. Uversky, V.N., Dunker, A.K.: Understanding protein non-folding. Biochimica et Biophysical Acta (2010), doi:10.1016/j.bbapap.2010.01.017
37. Uversky, V.N., Radivojac, P., Lakoucheva, L.M., Obradovic, Z., Dunker, A.K.: Prediction of intrinsic disorder and its use in functional proteomics. Methods in Molecular Biology 408, 69–92 (2007)
38. Pollastri, G., Martin, A.J.M., Mooney, C., Vullo, A.: Accurate prediction of protein secondary structure and solvent accessibility by consensus combiners of sequence and structure information. BMC Bioinformatics (2007), doi:10.1186/1471-2105-8-201
39. Barreto, L.V., Barreto, G.E., Morales, L., Acevedo, O.E., Santos, J.G.: Proteína LIC10494 de *Leptospira interrogans serovar copenhageni*: modelo estructural y regiones funcionales asociadas. Universitas Scientiarum 17, 16–27 (2012)
40. Homero, S.S., Lareo, L.R., Oribio-quinto, C., Martinez-Mendoza, J., Chavez-Zobel, A.: Predicci6n computacional de estructura terciaria de las proteínas humanas Hsp27, αB-cristalina y HspB8. Universitas Scientiarum 16, 15–28 (2011)
41. Tosatto, S., Battistutta, R.: TAP score: torsion angle propensity normalization applied to local protein structure evaluation. BMC Bioinformatics (2007), doi:10.1186/1471-2105-8-155

Exploration of the Effect of Input Data on the Modeling of Cellular Objective in Flux Balance Analysis (FBA)

Carlos Eduardo García Sánchez[1] and Rodrigo Gonzalo Torres Sáez[2]

[1] Universidad Industrial de Santander, Escuela de Ingeniería Química, Bucaramanga, Colombia
[2] Universidad Industrial de Santander, Escuela de Química, Bucaramanga, Colombia
carlos.garcia6@correo.uis.edu.co, rtorres@uis.edu.co

Abstract. Flux Balance Analysis (FBA) is a technique that allows estimation of metabolic fluxes in established conditions, focusing the flux determination as an optimization problem. For this reason, it is important to use an appropiate objective function in order to adjust estimations of FBA with the real behaviour of the cell. The aim of this work was to examine the effect of a set of input data fluxes (among five diferent sets) on the accuracy of predictions obtained with FBA with application in seven different objective functions. In this study, ***Saccharomyces cerevisiae*** was selected as model microorganism, and its metabolism was represented at genomic scale using the model iMM904. Accuracy of obtained predictions was evaluated and compared with eight set of experimental data. Results showed that the objective function representing in a better way the cellular behaviour depends on the set of fluxes used as input data.

Keywords: Flux Balance Analysis, input data, objective functions, estimation accuracy.

1 Introduction

Flux Balance Analysis (FBA) is a technique of mathematical modeling, which allows determining a distribution of fluxes in a steady state through maximization of a supposed objective function [1]. Obtaining of a suited estimation by using of FBA requires a good metabolic model, a set of adjusted restrictions to real conditions and an objective function that generates realistic results. In other words, a mathematical function should model a cellular objective [1-5]. It also requires a set of input data that allows a feasible region for the model data [6]. Nevertheless, both quality of the obtained estimation and more appropriate objective function can depend on the number of fluxes used as input data in the FBA and chosen fluxes.

Different set of input data can result interesting for getting distinct objective in the cellular modeling. If it is wanted to analyze a cell as a biological system, the set of input data will consist probably of all uptake metabolic fluxes. On the other hand, if it is needed to predict production of secondary metabolites in a continuous culture, maybe it is necessary to add the specific growth rate to the input data, because this

L.F. Castillo et al. (eds.), *Advances in Computational Biology*,
Advances in Intelligent Systems and Computing 232,
DOI: 10.1007/978-3-319-01568-2_8, © Springer International Publishing Switzerland 2014

will be controlled by both mode of design and operation of the chemostat, and so on. From this way, it is necessary to broaden the searching for appropriate cellular objectives to conditions with different set of input data.

The aim of the present study was to explore the possible relationship between the set of selected data as input data in the application of FBA and the best objective function in this technique (that is, the one that allows lower errors in the estimation of exchange fluxes and specific growth rate). Therefore, based on sets of known experimental data, we compared the performance of different objective functions in the FBA varying input data and using the rest of measured fluxes for the evaluation of the quality of the estimations.

2 Data and Methods

We decided to use the yeast *Saccharomyces cerevisiae* as cell model, because is one of the most important microorganism for biotechnological applications. For the cell modeling, we selected the iMM904 model [7], which is a model at genomic scale and that includes 1228 metabolites in 1577 reactions with eight cell compartments.

As experimental data for carrying out evaluation of the performance of objective functions, we selected two dataset from continuous cultures. The first dataset was obtained from a number of aerobic experiments with μ = 0,1 h^{-1} but with different oxygen uptake rates [8], and the second one was taken from anaerobic culture with four distinct specific growth rates [9]. These two dataset do not represent all different environmental conditions and growth that can be found. However, they attempt a range of conditions broad enough to obtain some conclusions from them, considering the objectives of the present study.

Data of aerobic growth contain measurements of both glucose and oxygen uptakes, production of ethanol, glycerol and carbon dioxide, and growth rate. The experimental configurations differ in the O_2 inlet percentages to the cell culture, being 0,5%, 1,0%, 2,8% y 20,9% of saturation of O_2 [8]. On the other hand, anaerobic experiments include measurements of substrate and oxygen uptake, and production of ethanol, glycerol, acetate, succinate, private and carbon dioxide, and specific growth rate (μ). The experiments were carried out with different μ (0,1, 0,2, 0,3 and 0,4 h^{-1}) [9].

Experimental data were classified in two categories: (i) the values used as equality restrictions in the FBA optimization problem were named 'input data', and (ii) the rest of fluxes were used in order to compare predictions obtained by FBA with experimental data, and named 'output data'. Different categories of input data were defined in order to determine if distinct initial conditions cause differences in the objective functions necessary to define cellular modeling. The categories were the following: uptake, uptake and μ, uptake and production, production and μ, and production. Each category makes reference to those fluxes that were taken as input data.

As objective functions were taken a set of representations of cellular objective that have been proposed in distinct previous studies on comparison of performance and generation of objective functions in FBA [3-4], [10-11]. The selected functions were

(I) maximization of biomass production, (II) minimization of glucose uptake, (III) minimization of NADH and NADPH in cytosol, (IV) maximization of biomass production plus minimization of carbon dioxide production, (V) maximization of biomass production plus minimization of NADH production in cytosol, (VI) maximization of biomass production plus minimization of NADH production in cytosol plus minimization of NADH and NADPH uptake in mitochondria, and (VII) minimization of ATP consumption in cytosol plus maximization of ATP transport from mitochondria to cytosol. By shortness, in Fig. 1 y Fig. 2 is kept roman numeration shown for objective functions.

The accuracy of estimations was evaluated with the error percentage on the estimation of the specific growth rate, and named Biomass error (see equation (1)), while the error percentage on the estimation of exchange fluxes that were experimentally measured, but not used as input data, were named Fluxes error (see equation (2)).

$$\text{Biomass error} = \left(\frac{\mu_{exp}-\mu_{est}}{\mu_{exp}}\right) * 100 \quad (1)$$

$$\text{Fluxes error} = \left(\frac{\|\vec{v}_{exp}-\vec{v}_{est}\|}{\|\vec{v}_{exp}\|}\right) * 100 \quad (2)$$

In the equations (1) and (2), the subscript "est" represents the estimations, and the subscript "exp" represents experimental data. In some categories of input data it was no possible to compute both errors, because of use of μ as input data, or of all exchange fluxes. Similarly, some objective functions do not appear in some Figures; this happens when some objective cannot be applied to some category of input data (for instance, maximization of biomass production cannot be used if μ is among the input data). It is important to note that sign Biomass error indicates if FBA overestimated (negative sign) or underestimate (positive sign) the specific growth rate of the cell.

3 Results and Discussion

In Fig. 1 are presented results obtained by simulation of aerobic experiments. For aerobic continuous cultures at low growth rates, it was found that among compared functions, maximization of biomass (function I) presents the lower errors using as input data the set 'uptake', while 'minimization of glucose uptake' (function II) showed better results using as input data 'production' and 'production and μ'. The function VII was clearly the second better in the category 'production'. Strangely, maximization of biomass (function I) gave worst estimations if only metabolite excretion fluxes are used as input data (that is, category 'production'). In the remaining category, 'uptake and μ', function VI showed the best performance.

The results for the anaerobic experiments are showed in Fig. 2. Again, maximization of biomass production (function I) is one of the best objective functions in the categories 'uptake' and 'uptake and production'; however, in the first category the function V gave slightly better estimations, and in the second with the functions IV

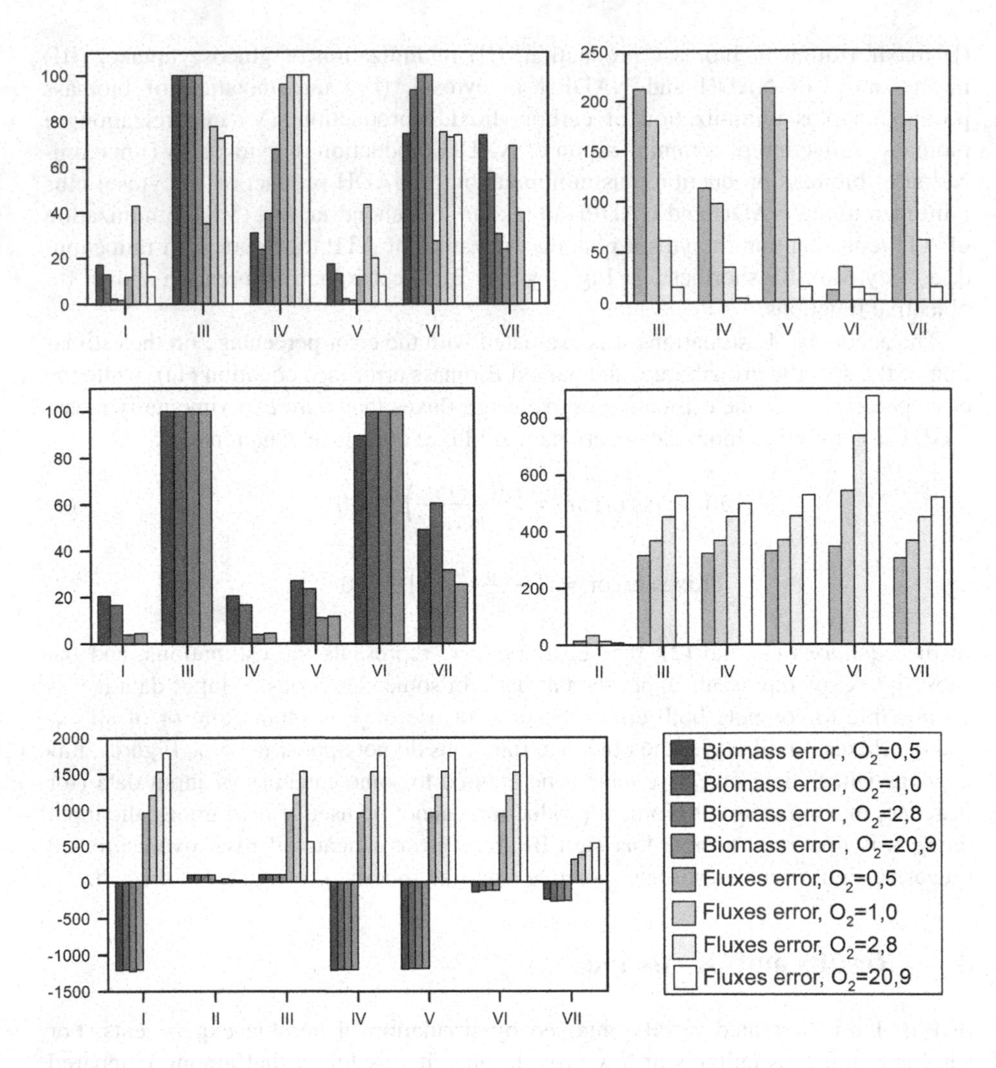

Fig. 1. Error percentages of the estimations obtained with FBA in aerobic growth conditions. Roman numbers represent seven different objective functions tested (See Data and Methods). The different panels show the obtained errors using the following input data: (**A**) uptake fluxes, (**B**) specific growth rate and uptake fluxes, (**C**) uptake and production fluxes, (**D**) specific growth rate and production, and (**E**) production fluxes.

and VII were obtained similar errors. In this case, with the function VII was obtained the best estimation for the category 'uptake and μ'. In the categories 'production' and 'production and μ', again the best function was the minimization of the glucose uptake (function II). In both situations, the function VII reached the second place on the quality of estimations.

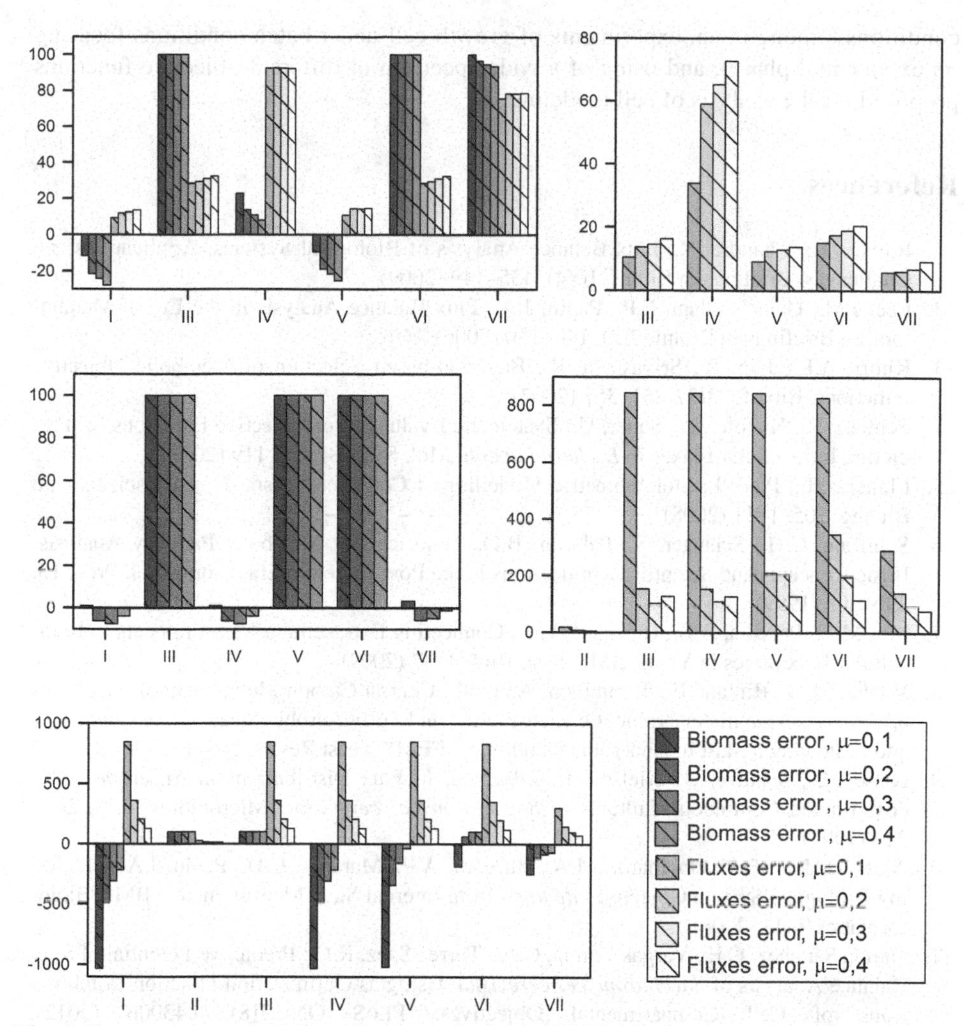

Fig. 2. Error percentages of the estimations obtained with FBA in anaerobic growth conditions. Roman numbers represent seven different objective functions tested (See Data and Methods). The different panels show the obtained errors using the following input data: (**A**) uptake fluxes, (**B**) specific growth rate and uptake fluxes, (**C**) uptake and production fluxes, (**D**) specific growth rate and production, and (**E**) production fluxes.

4 Conclusions

The results of this study indicate that the best objective function for representing the cellular behavior in FBA depends on the set of fluxes used as input data. Therefore, depending on the objective of the application of FBA (that is, according to the experimental condition to be represented, or the objective to be reached with the modeling) the objective function can be changed, being important to carry out a more detailed study of this problem, including more experimental data and different cell conditions (among them, experiments of growth cell under batch conditions focusing on exponential phase), and using of a wide spectrum of different objective functions proposed for the analysis of cell modeling.

References

1. Raman, K., Chandra, N.: Flux Balance Analysis of Biological Systems: Applications and Challenges. Briefings in Bioinf. 10(4), 435–449 (2009)
2. Lee, J.M., Gianchandani, E.P., Papin, J.A.: Flux Balance Analysis in the Era of Metabolomics. Briefings in Bioinf. 7(2), 140–150 (2006)
3. Knorr, A.L., Jain, R., Srivastava, R.: Bayesian-based Selection of Metabolic Objective Functions. Bioinf. 23(3), 351–357 (2007)
4. Schuetz, R., Kuepfer, L., Sauer, U.: Systematic Evaluation of Objective Functions for Predicting Intracellular Fluxes in *Escherichia coli*. Mol. Syst. Biol. 3, 119 (2007)
5. Llaneras, F., Picó, J.: Stoichiometric Modelling of Cell Metabolism. J. of Bioscience and Bioeng. 105, 1–11 (2008)
6. Schilling, C.H., Schuster, S., Palsson, B.Ø., Heinrich, R.: Metabolic Pathway Analysis: Basic Concepts and Scientific Applications in the Post-Genomic Era. Biotechnol. Prog. 15, 296–303 (1999)
7. Mo, M.L., Palsson, B.Ø., Herrgård, M.J.: Connecting Extracellular Measurements to Intracellular Flux States in Yeast. BMC Syst. Biol. 3, 37 (2009)
8. Wiebe, M.G., Rintala, E., Tamminen, A., et al.: Central Carbon Metabolism of *Saccharomyces cerevisiae* in Anaerobic, Oxygen-limited and Fully Aerobic Steady-state Conditions and Following a Shift to Anaerobic Conditions. FEMS Yeast Res. 8, 140–154 (2008)
9. Nissen, T., Schulze, U., Nielsen, J., Villadsen, J.: Flux Distributions in Anaerobic, Glucose-limited Continuous Cultures of *Saccharomyces cerevisiae*. Microbiology 143, 203–218 (1997)
10. Gianchandani, E.P., Oberhardt, M.A., Burgard, A.P., Maranas, C.D., Papin, J.A.: Predicting Biological System Objectives *de novo* from Internal State Measurements. BMC Bioinformatics 9, 43 (2008)
11. García Sánchez, C.E., Vargas García, C.A., Torres Sáez, R.G.: Predictive Potential of Flux Balance Analysis of *Saccharomyces cerevisiae* Using as Optimization Function Combinations of Cell Compartmental Objectives. PLoS ONE 7(8), e43006 (2012), doi:10.1371/journal.pone.0043006

Prediction of Potential Kinase Inhibitors in *Leishmania* spp. through a Machine Learning and Molecular Docking Approach

Rodrigo Ochoa[1], Mark Davies[2], Andrés Flórez[3], Jairo Espinosa[4], and Carlos Muskus[1]

[1] Programa de Estudio y Control de Enfermedades Tropicales- PECET, Universidad de Antioquia, Colombia
{rodrigoochoa,carlos.muskus}@udea.edu.co

[2] European Bioinformatics Institute, Wellcome Trust Genome Campus, United Kingdom
mdavies@ebi.ac.uk

[3] DKFZ, German Cancer Research Center, Heidelberg, Germany
a.florez@dkfz-heidelberg.de

[4] Grupo de Automática - GAUNAL, Universidad Nacional Sede Medellín, Colombia
jairo.espinosa@ieee.org

Abstract. Currently, tropical diseases are a major research objective in biomedical sciences due to the overall impact on vulnerable populations ignored by pharmaceutical companies. For that reason, the search for new therapeutic treatments is essential in the fight against tropical parasites such as *Leishmania* spp. The proposed approach will involve collecting the set of kinases from both the parasite and other organisms (except human), attempting to identify compounds and approved drugs, which are selective to the parasite, based on *in silico* methodologies. ChEMBL, Therapeutic Target Database (TTD) and DrugBank were used as sources for a list of compounds and kinase drug targets, which were represented using fingerprints based on patterns detected in the protein sequence, and a set of descriptors based on physic-chemical properties of the catalytic domains. The enzymes were used as a training set for a Support Vector Machine in conjunction with Feature Selection techniques, looking to predict druggable kinases, found in five sequenced *Leishmania* species. Following the target selection, a list of compounds was inferred and filtered according to some cheminformatics protocols. Finally, to support the predictions, some *Leishmania* kinases and their associated compounds were 3D modeled, and docked to each other according to a consensus docking schema based on the open packages AutoDock 4, AutoDockVina and DOCK.

Keywords: Bioinformatics, Docking, Machine Learning, *Leishmania*, Kinases.

1 Background

The tropical disease called Leishmaniasis is caused by at least 20 species of the protozoan parasite *Leishmania*, and is classified generally in three clinical forms: visceral, cutaneous and mucocutaneous. This disease is a major health problem in

L.F. Castillo et al. (eds.), *Advances in Computational Biology*,
Advances in Intelligent Systems and Computing 232,
DOI: 10.1007/978-3-319-01568-2_9, © Springer International Publishing Switzerland 2014

approximately 98 countries, affecting 12 million people worldwide [1]. Currently the treatment of this Neglected Tropical Disease (NTD) is focused in chemotherapy strategies. Nevertheless, the number of drug alternatives is very limited, with the disadvantage of several adverse effects ranging from mild to severe [2]. In Colombia there are reports mentioning treatment failures based on antimonial therapies and other therapeutical alternatives like Miltefosine, Paromomycin or Sitamaquine[3-4]. For all the reasons above, one important task nowadays is to find new effective treatments against tropical diseases like leishmaniasis. One way is through computational methodologies, using as input the vast amount of molecular information derived from the sequencing of a diverse range pathogen genomes.

One accepted and widely used approach is based on the application of predictive models, using as input different features provided by biological or chemical datasets. To execute such kind of models on the pharmaceutical field, is important to take into account which molecular entities are the most likeable to be potential therapeutic targets or treatments. With regard to the targets, one of the most popular and studied protein families in drug discovery are the kinases. For that reason, there is a large amount of data available on this family of enzymes, which could be used to create and execute the aforementioned computational models [5]. In *Leishmania*, there are reports about a set of these proteins that belongs only to this parasite, and could be exploited as molecular targets [6]. In addition, homology studies have been conducted between *Leishmania* kinases and human kinases, looking to detect substantial differences among them [7].

At the end, all these facts have increased the interest of certain academic and industrial groups to identify approved and experimental kinase inhibitors with potential anti-*Leishmania* activity profile. For that reason, we have proposed a Machine Learning approach, which aims to detect druggable kinases in different *Leishmania* species, according to kinase targets reported in different species, except human. After that, a list of inhibitors with potential action against the parasite was associated, avoiding connections with human kinases that could influence in adverse effects. In addition, some Molecular Docking protocols were executed based on some *Leishmania* kinases protein models. The results allowed us to prioritize the most viable compounds for posterior experimental validations.

2 Methods

2.1 Data Extraction

Interactions between drugs or compounds against kinases (except human) were extracted from three different sources: DrugBank (version 3.0) [8], TTD (version 4.3.02) [9] and ChEMBL (version 14) [10]. The first two databases were used only to extract approved drug data with validated pharmacological action. The compounds were extracted from the ChEMBL database, which contains experimental binding affinity measures such as IC50, Ki, EC50 and inhibition percentages. In addition, only interactions with a confidence score greater than 0.7 (from a 0 to 1 scale) were selected.

After the extraction process, the positive and negative training sets were established. The criteria for inclusion in the positive kinases set were, as follows: standard concentration and constant values (IC50, EC50 and Ki) less than 1uM, and inhibition percentages greater or equal than 70%. For the negative set, the criteria were concentration and constant values greater than 10uM and inhibition percentages lower than 30%. Additionally, in order to complement the negative set, all the kinases without compounds associated in the databases, and with an identity value lower than 60% against the positive set were included. The sequence alignments were performed using the BLAST algorithm [11]. Finally a group of reviewed and predicted *Leishmania* kinases were obtained from the UniProt database, looking to test the predictive model.

2.2 Model Set Up and Performance

With the training group established, a 70-30 learning protocol was performed. 70% of the data was implemented to exclusively train the classifier, and the remaining 30% to test the predictive capabilities of the model. A Support Vector Machine together with the Recursive Feature Elimination techniques was configured, using as features both, descriptors and fingerprints extracted from the amino acid sequences of the selected kinases. The Machine Learning (M.L) models were tested within the WEKA package [12] under the Linux Operative System. At the end, some metrics were taken into account (accuracy, precision, recall), looking to approve the model performance for subsequent predictions.

2.3 *Leishmania* Kinase Targets and Associated Compounds Selection

All the *Leishmania* spp. kinases extracted from UniProt were submitted to the chosen model, using as features the data stored in their amino acid sequences. Consequently, a group of potential kinase targets was selected based on the predicted classes. The parasite kinases chosen were aligned against the kinases used in the positive training group through BLAST protocols. Those with the higher identity percentages were related to each other to infer subsequently potentially active compounds, using a transitivity criterion between the original target and the parasite kinase. Nevertheless, a thorough search of the Ensembl database [13] and the OrthoMCL database [14] was carried out, looking to select only the other organism kinases and *Leishmania* kinases that do not report orthologue with the human host respectively.

With regard to the compounds, an initial list was filtered based on a set of chemin formatics strategies. Initially, all the compounds with favorable interactions against human kinases were dismissed. Then a substructure search through the RDKit chemoinformatics libraries (version 2012-09) was performed, using as input toxic fragments reported in the PAINS dataset [15]. Finally, the commercial availability of the compounds was verified based on information stored in the ZINC database [16].

2.4 Molecular Docking of Some Predicted *Leishmania* Kinase Targets

After the selection process, some of the parasite kinase structures were modeled using the Modeller package, or looking directly into the ModBase database [17]. Furthermore, different evaluations per model were executed, based on statistics such as the

RMSD C-alpha between the models and their templates, the identity percentage and the z-DOPE curves generated residue per residue. In addition, for each modeled kinase, a comparison against the template structure was done with the aim to identify key residues involved in the potential active site or ligand binding site.

All the molecules were subsequently prepared for Docking through the AutoDock Tools software. After that, three different docking protocols: AutoDock 4 [18], AutoDockVina [19] and DOCK [20] were implemented, looking to test different score functions and search algorithms, with the aim to obtain a consensus result per protein. According to the results, a list of priority compounds was selected for posterior experimental validations.

3 Results and Discussion

3.1 Machine Learning Set Up and Execution

According to the protocol, a significant number of interactions between kinases from other organisms (except human) and small molecules were extracted. After that, based on bioactivity values and reliability scores, a total of 98 and 49 kinases were used as positive and negative training sets for the predictive model respectively.

After the training set up, two machine learning configurations were applied: Support Vector Machines with and without Feature Selection Techniques. Specifically, the Recursive Feature Elimination Technique through Cross Validations (RFECV) was implemented as the Feature selection technique, looking to improve the classification capabilities of the model. Based on this strategy, the Table 1 describes the results obtained for the training set (70% of the original data) and the testing set (the remaining data set).

Table 1. Machine learning statistical results

Machine Learning	Train Acc.	Test Acc.	Train Pre.	Test Pre.	Train Rec.	Test Rec.
SVM	1.00	0.75	1.00	0.74	1.00	0.97
SVM + RFECV	1.00	0.94	1.00	0.95	1.00	1.00

Acc: Accuracy. Pre: Precision. Rec: Recall. Train: Training set. Test: Testing set.

According to Table 1, the performance of the Support Vector Machine with and without feature selection reported optimal results in the training set for all the metrics used. Nevertheless, the performance on the testing set changed considerably when no relevant features were eliminated. Using RFECV, the accuracy reached levels close to 95%, fact that was crucial for posterior predictions using the *Leishmania* kinases.

3.2 Predictions on the *Leishmania* Spp. Kinases

After the predictions, 504 potential kinase targets were classified as positives. That means these kinases have more probability to be druggable, and potentially could be inhibited by one or more compounds which are associated to the kinases used on the positive training set. To avoid as most as possible side effects, the orthologue filter (respect to the human enzymes) let a reduced list of 446 exclusive parasite targets. Furthermore, this number was reduced using filters based on the list of compounds associated to the selected targets. For that purpose, selecting compounds without toxic substructures, and leaving only compounds that can be purchased from approved vendors was the applied strategy. Consequently, we get a list of 46 compounds related with 10 exclusive *Leishmania* spp. potential targets. The list of selected kinases is detailed in Table 2.

Table 2. List of *Leishmania* kinases selected as potential targets

UniProt ID	GeneDB ID	Specie
A4H3W3	LbrM.04.0480	*L. braziliensis*
A4HS38	LinJ.04.0420	*L. infantum*
A4IAU7	LinJ.35.0480	*L. infantum*
E9ANH1	LmxM.11.0250	*L. mexicana*
E9ARY8	LmxM.19.0360	*L. mexicana*
E9B936	LdbPK.070410	*L. donovani*
E9BE32	LdbPK.190360	*L. donovani*
E9BLD1	LdbPK.300370	*L. donovani*
Q4QEB9	LmjF.17.0670	*L. major*
Q4QH55	LmjF.11.0250	*L. major*

Most of the kinases detected are putative protein kinases and some of them are classified as serine/threonine kinases. Nevertheless, one of the detected kinases (UniProt:E9BLD1) is a putative MAP kinase, which according to external sources (KEGG, literature) could be implicated as an element of the leishmaniasis infectious process [21].

3.3 Molecular Docking Protocols Results

Looking to prioritize the detected compounds, some of the predicted *Leishmania* kinase targets were modeled with the aim of using their 3D structures for Molecular Docking simulations. At the end only 3 protein kinases were successfully modeled based on different evaluations measures such as significant identity percentages and energy profiles curves (DOPE) suitable for *in silico* validations [22]. In addition, for selecting the pockets inside the protein with more probability of being binding or active sites, different key residues that would be involved in the conformation of those pockets were identified per each protein (Fig. 1).

Moreover, other preparations were implemented on the structures to facilitate potential computational interactions between the proteins and their compounds, using the three Docking protocols chosen in this work. After the simulations, the overall results were compared. For that purpose, all the Docking scores were normalized (0 to 1 scale), with the aim to recognize those compounds with higher affinities. As an example, the Docking results comparison of one parasite modeled kinase (UniProt:E9BLD1) against their 9 associated compounds (named as N1 to N9) are shown in Fig. 2.

Fig. 1. Detection of potential active sites within the *Leishmania* modeled kinases. First the protein structure used as template was analyzed, looking to detect which residues are responsible for interacting with the ligand (A). Then a pairwise alignment was done between each *Leishmania* protein and its template, looking to detect if key residues observed before are conserved in the parasite sequence (B). Finally a box or sphere search space was created, using as centroid the coordinates of any of the previous residues detected (C).

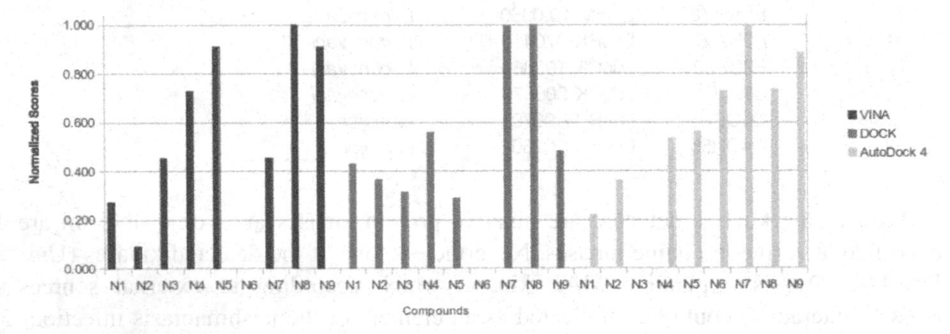

Fig. 2. Performance comparison of three Docking software using the *Leishmania* protein kinase (UniProt:E9BLD1)

According to the consensus score, the compounds with highest affinities tend to maintain this behavior within all the protocols, especially N8. Nevertheless, some others changed significantly their final results from one algorithm to another. This depends largely on the Docking packages internal differences such as the protocols used for scoring the interactions and optimizing the better ligand poses. At the end, a list of compounds per protein was selected based on the higher consensus scores.

4 Conclusion

In computational drug discovery is necessary the continuous addition of new functionalities, capable to support in a more reliable way predictions about new treatments alternatives. Specifically, in the field of Neglected Tropical Diseases is recommended the inclusion of additional tools, with the aim to overpass current treatment inconveniences such as adverse side effects or parasite resistance. The combination of machine learning methodologies with classic computational protocols based on orthologue detection and structural simulations, allowed us to include deeper levels of complexity with regard to select druggable proteins based on inherent physic-chemical properties, and consequently associate potential leishmanicidal compounds.

Nevertheless, is mandatory the use of posterior experimental validations, looking to support the current results and open the possibilities to apply these kind of protocols in other diseases.

Acknowledgements. This work was supported by the Administrative Department of Science, Innovation and Technology in Colombia, together with the University of Antioquia [1115-519-29015].

References

1. World Health Organization: Leishmaniasis: worldwide epidemiological and drug access update (2012)
2. Maltezou, H.: Drug resistance in visceral leishmaniasis. J. Biomed. Biotechnol. 2010 (2010)
3. Goyeneche-Patino, D., Valderrama, L., Walker, J., Saravia, N.: Antimony resistance and trypanothione in experimentally selected and clinical strains of *Leishmania* panamensis. Antimicrob. Agents. Ch. 52(12), 4503–4506 (2008)
4. Choudhury, K., Zander, D., Kube, M., Reinhardt, R., Clos, J.: Identification of a *Leishmania infantum* gene mediating resistance to miltefosine and SbIII. Int. J. Parasitol. 38(12), 1411–1423 (2008)
5. Brooijmans, N., Mobilio, D., Walker, G., Nilakantan, R., Denny, R., Feyfant, E., Diller, D., et al.: A structural informatics approach to mine kinase knowledge bases. Drug. Discov. Today 15(5-6), 203–209 (2010)
6. Naula, C., Parsons, M., Mottram, J.: Protein kinases as drug targets in trypanosomes and *Leishmania*. Biochim. Biophys. Acta. 1754(1-2), 151–159 (2005)
7. Ochoa, R., Florez, A., Muskus, C.: Detección *in silico* de segundos usos de medicamentos con potencial acción leishmanicida. Ingeniería Biomédica 10, 10–16 (2011)
8. Knox, C., Law, V., Jewison, T., Liu, P., Ly, S., Frolkis, A., Pon, A., et al.: DrugBank 3.0: a comprehensive resource for “omics” research on drugs. Nucleic. Acids. Res. 39, D1035–D1041 (2011)
9. Zhu, F., Shi, Z., Qin, C., Tao, L., Liu, X., Xu, F., Zhang, L., et al.: Therapeutic target database update 2012: a resource for facilitating target-oriented drug discovery. Nucleic. Acids. Res. 40, D1128–D1136 (2012)
10. Gaulton, A., Bellis, L., Bento, P., Chambers, J., Davies, M., Hersey, A., Light, Y., et al.: ChEMBL: a large-scale bioactivity database for drug discovery. Nucleic. Acids. Res. 40, D1100–D1107 (2012)
11. Altschup, S., Gish, W., Pennsylvania, T., Park, U.: Basic Local Alignment Search Tool. J. Mol. Biol. 215, 403–410 (1990)
12. Hall, M., National, H., Frank, E., Holmes, G., Pfahringer, B., Reutemann, P., Witten, I.: The WEKA Data Mining Software: An Update. SIGKDD Explorations 11(1), 10–18 (2009)
13. Flicek, P., Aken, B.L., Beal, K., Ballester, B., Caccamo, M., Chen, Y., Clarke, L., et al.: Ensembl 2008. Nucleic. Acids. Res. 36, D707–D714 (2008)
14. Li, L., Stoeckert, C., Roos, D.: OrthoMCL: identification of ortholog groups for eukaryotic genomes. Genome. Res. 13(9), 2178–2189 (2003)

15. Baell, J., Holloway, G.: New substructure filters for removal of pan assay interference compounds (PAINS) from screening libraries and for their exclusion in bioassays. J. Med. Chem. 53(7), 2719–2740 (2010)
16. Irwin, J., Sterling, T., Mysinger, M., Bolstad, E., Coleman, R.: ZINC: A Free Tool to Discover Chemistry for Biology. J. Chem. Inf. Model. 52(7), 1757–1768 (2012)
17. Pieper, U., Eswar, N., Braberg, H., Madhusudhan, M., Davis, F., Stuart, A., Mirkovic, N., et al.: MODBASE, a database of annotated comparative protein structure models, and associated resources. Nucleic. Acid. Res. 32, D217–D222 (2004)
18. Morris, G., Huey, R., Lindstrom, W., Sanner, M., Belew, R., Goodsell, D., Olson, A.: AutoDock4 and AutoDockTools4: Automated docking with selective receptor flexibility. J. Comput. Chem. 30(16), 2785–2791 (2009)
19. Trott, O., Olson, A.: AutoDockVina: improving the speed and accuracy of docking with a new scoring function, efficient optimization, and multithreading. J. Comput. Chem. 31(2), 455–461 (2010)
20. Lorber, D., Shoichet, B.: Flexible ligand docking using conformational ensembles Despite important successes. Protein. Sci. 7, 938–950 (1998)
21. Kaye, P., Scott, P.: Leishmaniasis: complexity at the host-pathogen interface. Nat. Rev. Microbiol. 9(8), 604–615 (2011)
22. Eswar, N., Webb, B., Marti-Renom, M., Madhusudhan, M., Eramian, D., Shen, M., Pieper, U., et al.: Comparative protein structure modeling using MODELLER. Current protocols in Bioinformatics 2(15), 1–30 (2007)

Functional Protein Prediction Using HMM Based Feature Representation and Relevance Analysis

Diego Fabian Collazos-Huertas, Andres Felipe Giraldo-Forero, David Cárdenas-Peña, Andres Marino Álvarez-Meza, and Germán Castellanos-Domínguez

Signal Processing and Recognition Group, Universidad Nacional de Colombia, Campus la Nubia, Km 7 vía al Magdalena, Manizales - Caldas, Colombia
{dfcollazosh,afgiraldofo,dcardenasp,amalvarezme, cgcastellanosd}@unal.edu.co

Abstract. The prediction of subcellular location aims to understand the biological processes being carried out within the cell. Here, a feature representation methodology is proposed to identify subcellular locations in gram-positive bacteria. Regarding this, each considered class is employed to train a hidden Markov model, and the probability of a sequence of amino acids, being generated by each of the trained models is employed as a feature in further classification stage. Our proposal is tested on a well known database, containing amino acids sequences of bacteria. For concrete testing, a percentage of less than 80% identity is studied, using a multi-label Support Vector Machines with soft margin classifier. Attained results show that our approach improves issues raised in *PfamFeat*. Moreover, it seems to be an appropriate tool for predicting subcellular location proteins.

Keywords: HMM, Multiclass SVM, Protein, Subcellular Localization.

1 Introduction

One of the biggest challenges in the field of bioinformatics has been the prediction of subcellular localizations of proteins, because of the increased interest for studying sequenced genomic data. For instance, computational analysis of protein sequences in bacterial level allows to develop new drugs. Additionally, the emergence of new sequencing techniques have led to the need for seeking prediction systems that provide the ability to analyze data in a massive way.

Since the location of known proteins can be determined by their role in the cell [1], protein prediction systems based on function analysis have been proposed, such as: CELLO, PSORT-B, PSORT, SubLoc, and PfamFeat [2–5]. CELLO uses four types of sequence coding schemes: the amino acid, the dipeptide, the partitioned amino acid, and the sequence composition based on the physico-chemical properties of amino acids. CELLO takes place locally while using hidden Markov Models (HMM) to analyze the dynamics of each sequence.

Advances in Intelligent Systems and Computing 232,
DOI: 10.1007/978-3-319-01568-2_10,

PSORT examines the owners protein for the amino acid composition. However, PSORT requires many modules, each of them based on different prediction tools, yielding a very complex system. SubLoc is based on an array of profiles, which are predicted previously to improve the estimation of subcellular localization of proteins. Finally, PfamFeat, which is based on Pfam adaptations, is applied for predicting subcellular localizations of proteins. Nonetheless, since PfamFeat is based on Pfam, it could lead to an unstable system performance when the standard Pfam models are not enough to represent specific classes of interest.

This paper presents a HMM-based feature representation methodology for predicting subcellular localization of Gram-Positive bacterial proteins. In the proposed approach, input protein sequences are represented by its inferred generation probability values given a set of one-sequence trained HMMs. The aim of such approach is to enhance, as well as possible, the dynamic relationships into the provided data, while dealing with variable length sequences. The set of considered bacterial Gram-Positive proteins corresponds to Celwall, Cytoplasmic, Cytoplasmic membrane, and Extracellular. Afterwards, the classification problem is boarded from a proposal based on Multi-class Support Vector Machine, which unlike most of the SVM multi-class approaches that are based on multiple independent binary classification tasks, it considers a notion of margin that yields a direct method for training multi-class predictors [6].

The remainder of this paper is organized as follows. In section II is specified each of the materials and methods used in the development of the work. Section III describes the experimental set-up. Section IV shows the obtained results and the discussion. Finally, in section V the conclusions of the work are presented.

2 Materials and Methods

First, the proposed HMM based feature representation (characterization) scheme is described, which is employed to analyze protein sequences. Then, given such feature representation space, a multi-class SVM classifier is presented to deal properly with the HMM based highlighted data dynamics.

2.1 Hidden Markov Models Based Feature Representation

HMMs are double stochastic models, composed of a hidden layer θ leading the evolution, along a given axis, of stochastic features on an observable layer, noted as φ. In the discrete case, a HMM is made up of a set of S states $\theta = \{\theta_1, \cdots, \theta_S\}$, associated with a set of K possible observations regarded as the representation symbols. Denoted as $\lambda = \{\pi, \mathbf{A}, \mathbf{B}\}$, the model parameters include: (i) an initial state probability π with elements $\{\pi(\theta_i) \in \mathbb{R}[0,1]\}$, describing the distribution over the initial state set; (ii) a transition matrix $\mathbf{A} \in \mathbb{R}^{S\times S}$ with elements $a_{ij} \in \mathbb{R}^+$, $i, j \in [1, S]$ for the transition probability from node i to node j; and (iii) an observation matrix $\mathbf{B} \in \mathbb{R}^{S\times K}$ with elements $\{b_{ik} \in [0,1]\}$ referring the probability of each observed symbol $k \in [1, K]$, given that the system remains at the state i [7].

Within the considered framework, the commonly employed statistical method used for fitting a stochastic model is the Maximum Likelihood Estimation (MLE), which is grounded upon the following objective function of evaluation:

$$f(\lambda) = \mathbb{E}\{\log p(\hat{X}_n|\lambda) : \forall n\} \tag{1}$$

where $p(\hat{X}_n|\lambda)$ is the likelihood of the observed sequence $\hat{X}_n$, being generated by the model λ. Forward-backward algorithm is employed for computing the $\log p(X|\lambda)$ term. Since each protein sequence fits a different HMM, the total set of HMM parameters comprises N models, $\{\lambda_n : n \in [1, N]\}$, with λ_n denoting the respective parameter set for the sequence n.

2.2 SVM Based Multi-class Classifier

After evaluating all the N sequences using the above mentioned HMM representation models, a feature representation matrix $\mathbf{Z} \in \mathcal{X} \subseteq \mathbb{R}^{N \times N}$ is obtained, where each column vector $\mathbf{z}_n$ is an input sample. From such matrix, and given the label vector $\boldsymbol{y} \in \mathcal{Y} \subseteq \mathbb{R}^{N \times 1}$, with $y_n \in [1, 2, \ldots N_C]$, being N_C the number of classes, a multi-class SVM classifier is used to predict each sample label. Thus, traditional SVM optimization problem can be written as

$$\min \ \frac{1}{2} \parallel w \parallel^2 + C \sum \xi_n, \tag{2}$$

subject to $y_n(\mathbf{w}^\top \mathbf{z}_n + b) \geq 1 - \xi_n$ and $\xi_n \geq 1$, being ξ_n a slack variable, $\mathbf{w}$ is a projection vector, b is a bias term, and C is a regularization parameter. Based on the so-called "kernel trick", equation (2) can be treated as a quadratic problem [8]. Now, inspired by [6], the SVM multi-class classifier can be written as a function $H : \mathcal{X} \rightarrow \mathcal{Y}$ that maps an instance $\mathbf{z}_n$ to an element y_n, where H is estimated by

$$H_M(\mathbf{z}) = \arg \max_{r=1}^{N_C} \{\mathbf{M}_r \mathbf{z}\}, \tag{3}$$

where $\mathbf{M}_r \in \mathbb{R}^{N_C \times n}$. The value of the inner-product of the *r-th* row of $\mathbf{M}$ with the instance $\mathbf{z}$ is named the confidence and the similarity score for the r class. Therefore, according to (3), the predicted label is the index of the row attaining the highest similarity score with $\mathbf{z}$. To solve (3), a misclassification error is defined with a piecewise linear bound, and afterwards, based on the kernel trick, a quadratic problem is formulated (see [6] for details).

3 Experimental Set-Up and Results

The general scheme of the proposed prediction system is presented through a flow diagram based on three main components: database, characterization (feature representation), and classification. Fig. 1 shows the flowchart of the work, which presents each of the main system components. Datasets are represented by ovals and the computational processes by rectangles.

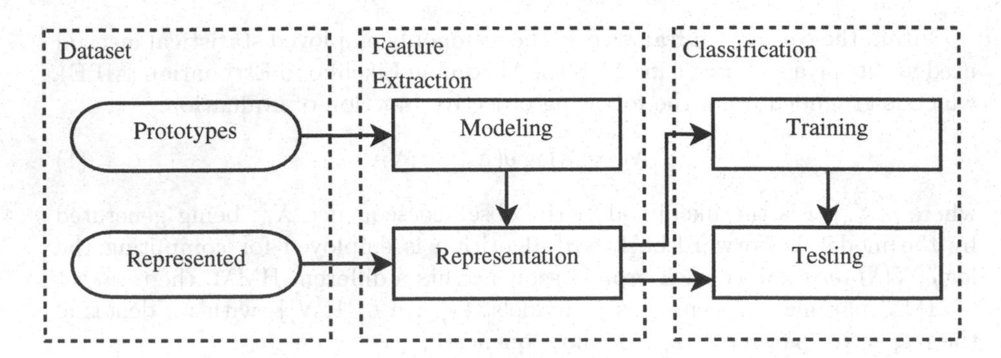

Fig. 1. Proposed prediction system general scheme

3.1 Database

In order to test the described approach, the well known "**PSORTb v.3.0 Gram-positive sequences**" dataset is used. Thus, our data is build by the local feature descriptor of Gram-Positives, choosing them as proposed in [9]. Additionally, to avoid over fitting problems, an identity filter is conducted with an 80 % identity cut-off between training and testing sets. Note that latter procedure was done with CD-HIT software [10]. Therefore, the dataset is composed of a total of 1729 sequences, distributed as follows: 80 of Celwall, 708 of Cytoplasmic, 674 of Cytoplasmic membrane, and 267 of Extracellular. Each class is made up of both prototype and represented sequence, with a 70 % and 30 % of the sequences, respectively.

3.2 Characterization

In order to represent the protein sequence, the proposed HMM based feature representation framework is used. Particularly, 70 % of the sequences are selected as mentioned above for generating the models. In this regard, a random initiation and a maximum verisimilitude estimation (MLE) are employed for training each HMM. Parameter tuning of the number of states, S, is carried out by using the Bayesian information criterion (BIC) from 5 to 12 states, with 4 repetitions for each configuration, such that the larger the number of states, the larger the penalty (regularization) term. The aim of BIC is to find the largest likelihood model while avoiding over fitting. Finally, forward-backward algorithm is used to calculate the probability of each data stream being generated by each of the trained models.

3.3 Classification

The classification of the models are performed through SVM based multi-class classifier. The 30 % of the sequences selected to represent the data set are employed using a 10 - fold cross-validation scheme, using a linear kernel. The regularization parameter C is tuned by heuristic search in a mesh grid from 1 to 1000, with a step of 1 unit, this process was used for both characterization methodologies, HMM and PfamFeat. In this case, the maximization of geometric mean between sensitivity and specificity is used as cost function.

$$S_n = \frac{n_{TP}}{n_{TP}+n_{FN}} \quad S_p = \frac{n_{TN}}{n_{FP}+n_{TN}} \quad G_m = \sqrt{S_n S_p} \tag{4}$$

4 Results and Discussion

For the sake of comparison, the same classification scheme was employed for both characterization methodologies, proposed HMM and PfamFeat. Multiclass kernel SVM system was used because it is a classification system robust, commonly used in pattern recognition applications, which is available at[1]. Obtained classification results for 10-fold cross-validation framework are presented in Table 1, in terms of average class specificity, sensitivity and geometric mean, along with their respective standard desviation.

Table 1. Results for the Gram-positive dataset

Class	**HMM**			**Pfamfeat**		
	$S_p\%$	$S_n\%$	$G_m\%$	$S_p\%$	$S_n\%$	$G_m\%$
Cellwall	93.7 ± 5.9	87.1 ± 12.5	90.2 ± 8.3	88.0 ± 30.4	19.7 ± 19.1	35.5 ± 22.5
Cytoplasmic	87.6 ± 9.9	58.0 ± 14.5	70.5 ± 8.0	72.0 ± 3.4	37.7 ± 59.0	51.9 ± 3.8
Cytoplasmic Membrane	80.1 ± 7.2	57.1 ± 29.1	65.6 ± 17.0	72.2 ± 3.8	37.1 ± 4.7	51.6 ± 3.1
Extracellular	84.1 ± 10.9	66.8 ± 16.9	74.0 ± 5.0	78.5 ± 3.8	40.7 ± 7.3	56.3 ± 4.9

From Table 1, it is clear that the proposed HMM-based methodology outperforms the prediction results of PfamFeat. The above implies that, own design of class models is more efficient than use models already identified of protein families, as the pfam database. Because, not always able to represent so reliable information about new classes. on the other hand the HMM representation described effectively dynamics of the sequence.

5 Conclusions and Future Work

In this paper, an analysis about the prediction of Subcellular localizations of Gram-Positive proteins was carried out and a new sequence representation using hidden Markov models was introduced. Obtained classification results, using a Multiclass SVM, show a better performance of proposed approach than the PfamFeat baseline results. Therefore, the HMM are more adequate for representing sequences than those already identified in Pfam. As future work, relevant regions of protein families will be employed to build the HMM representation

[1] `http://svmlight.joachims.org/svm_multiclass.html`

aiming to improve the performance of the system with respect to sensitivity. It is also interesting to validate the proposed scheme on other taxonomic categories.

Acknowledgements. This work was partially funded by the Research office (DIMA) at the Universidad Nacional de Colombia at Manizales and the Colombian National Research Centre (COLCIENCIAS) through grant No.111952128388.

References

1. Gardy, J.L., Brinkman, F.S.L.: Methods for predicting bacterial protein subcellular localization. Nature Reviews Microbiology 4(10), 741–751 (2006)
2. Gardy, J.L., Spencer, C., Wang, K., Ester, M., Tusnady, G.E., Simon, I., Hua, S., Lambert, C., Nakai, K., Brinkman, F.S., et al.: Psort-b: Improving protein subcellular localization prediction for gram-negative bacteria. Nucleic Acids Research 31(13), 3613–3617 (2003)
3. Yu, C.S., Lin, C.J., Hwang, J.K.: Predicting subcellular localization of proteins for Gram-negative bacteria by support vector machines based on n-peptide compositions. Protein Science 13(5), 1402–1406 (2004)
4. Lu, Z., Szafron, D., Greiner, R., Lu, P., Wishart, D., Poulin, B., Anvik, J., Macdonell, C., Eisner, R.: Predicting subcellular localization of proteins using machine-learned classifiers. Bioinformatics 20(4), 547–556 (2004)
5. Punta, M., Coggill, P.C., Eberhardt, R.Y., Mistry, J., Tate, J., Boursnell, C., Pang, N., Forslund, K., Ceric, G., Clements, J., Heger, A., Holm, L., Sonnhammer, E.L.L., Eddy, S.R., Bateman, A., Finn, R.D.: The Pfam protein families database. Nucleic Acids Research 40(Database issue), D290–D301 (2012)
6. Crammer, K.: On the algorithmic implementation of multiclass kernel-based vector machines. The Journal of Machine Learning Research 2, 265–292 (2002)
7. Rabiner, L.R.: A tutorial on hidden markov models and selected applications in speech recognition. Proceedings of the IEEE 77(2), 257–286 (1989)
8. Scholkopg, B., Smola, A.J.: Learning with Kernels. The MIT Press, Cambridge (2002)
9. Rey, S., Acab, M., Gardy, J.L., Laird, M.R., Lambert, C., Brinkman, F.S., et al.: Psortdb: a protein subcellular localization database for bacteria. Nucleic Acids Research 33(suppl. 1), D164–D168 (2005)
10. Li, W., Godzik, A.: Cd-hit: a fast program for clustering and comparing large sets of protein or nucleotide sequences. Bioinformatics (Oxford, England) 22(13), 1658–1659 (2006)

High Throughput Location Proteomics in Confocal Images from the Human Protein Atlas Using a Bag-of-Features Representation

Raúl Ramos-Pollán[1], John Arévalo[2], Ángel Cruz-Roa[2], and Fabio González[2]

[1] Universidad Industrial de Santander
[2] MindLab Research Group, Universidad Nacional de Colombia
{rramosp,jearevaloo,aacruzr,fagonzalezo}@unal.edu.co

This work addresses the problem of predicting protein subcellular locations within cells from confocal images, which is a key issue to reveal information about cell function. The Human Protein Atlas (HPA) is a world-scale project addressed at proteomics research. The HPA stores immunohistological and immunofluorescence images from most human tissues. This paper concentrates on the problem of analyzing HPA immunofluorescence images from immunohistochemically stained tissues and cells to automatically identify the subcellular location of particular proteins expression. This problem has been previously tackled using computer vision methods which train classification models able to discriminate subcellular locations based on particular visual features extracted form images. One of the challenges of applying this approach is the high computational cost of training the computer vision models, which includes the cost of visual feature extraction from multichannel images, classifier training and evaluation, and parameter tuning. This work addresses this challenging problem using a high-throughput computer-vision approach by (1) learning a visual dictionary of the image collection for representing visual content through a bag-of-features histogram image representation, (2) using supervised learning process to predict subcellular locations of proteins and (3) developing a software framework to seamlessly develop machine learning algorithms for computer vision and harness computing power for those processes.

1 Introduction

The goal of this work is to automatically find the location of protein expression in immunoflorescence images of immunohistochemically stained tissues and cells. This images come from the Human Protein Atlas (HPA, www.proteinatlas.org), which is a comprehensive database that provides the protein expression profiles for a large number of human proteins, presented as immunohistological and immunofluorescence images from most human tissues [13,14]. This database includes more than 5 million images of immunohistochemically stained tissues and cells, based on 6,122 antibodies, which represents 5,011 human proteins encoded by approximately 25% of the human genome. Recently, the HPA was extended including confocal immunofluorescence images from cultured cells [1,2]. These set of images comprises 5,915 gray scale images of 1728×1728 pixels with four channels, three reference channels with well known dyes (nucleus, endoplasmic reticulum (ER) and cytoskeleton) and the last channel contains

L.F. Castillo et al. (eds.), *Advances in Computational Biology*,
Advances in Intelligent Systems and Computing 232,
DOI: 10.1007/978-3-319-01568-2_11, © Springer International Publishing Switzerland 2014

the expression of the target protein, leading to 23,660 total images occupying 66 GB of storage. These immunofluorescence images show subcellular locations, visually and manually annotated by experts, for thousands of proteins. Within such collections, the combinatorial explosion of data from different cell lines and proteins requires increasingly larger computational resources to extract and discover useful knowledge.

Traditionally the image based subcellular location approaches has been to explore a huge number of visual feature descriptors to describe visual content of confocal images and train a classifier to predict which subcellular locations are expressed by the proteins, through ad-hoc techniques to use the available computing resources [4,3,6,11,10,7,8]. In contrast, this work addresses this challenging problem using a high-throughput computer-vision approach by developing a distributed computing framework to seamlessly develop learned image representation and machine learning algorithms to harness computing power and ease image based location proteomics research workflow. This work is related to a new emerging research area called bioimage informatics [15,12], which comprises image processing, data mining and database visualization, extraction, searching, comparison and management of biomedical knowledge inside massive image collections.

2 Materials and Methods

We used a set of confocal fluorescence images from the HPA comprising image fields with three standard dyes used in fluorescence images to highlight three kinds of cell organelles (nucleus, endoplasmic reticulum and cytoskeleton) and a fourth image channel with the protein expression profile [9]. We adopted a parameter exploration strategy for the entire experimental cycle to build bag-of-features image representations and perform supervised learning processes through different cross-validation modalities. A framework for large-scale machine learning was also built to efficiently use the required distributed computing resources in our experiments, based on Big Data techniques. The performance measures used were average overall accuracy from confusion matrices and efficiency measures for the computing resources.

2.1 Automatic Image Based Subcellular Location

The paper proposes a two-phases method: first, the confocal fluorescence images are represented by a state-of-the-art method used in computer vision known as Bag of Features; second, the automatic subcellular location is performed by a straightforward probabilistic classifier known as Naïve Bayes. These two phases are implemented within our own distributed computing framework.

Bag of Features (BOF) Representation of Confocal Fluorescence Images. BOF represent the visual content of a given image in terms of the occurrence of a set of fundamental, and learned, visual patterns/words (a.k.a. visual dictionary) as a histogram. Figure 1 depicts the overall BOF scheme and its corresponding stages: 1) visual words extraction and description, 2) visual dictionary learning, and 4) histogram image representation.

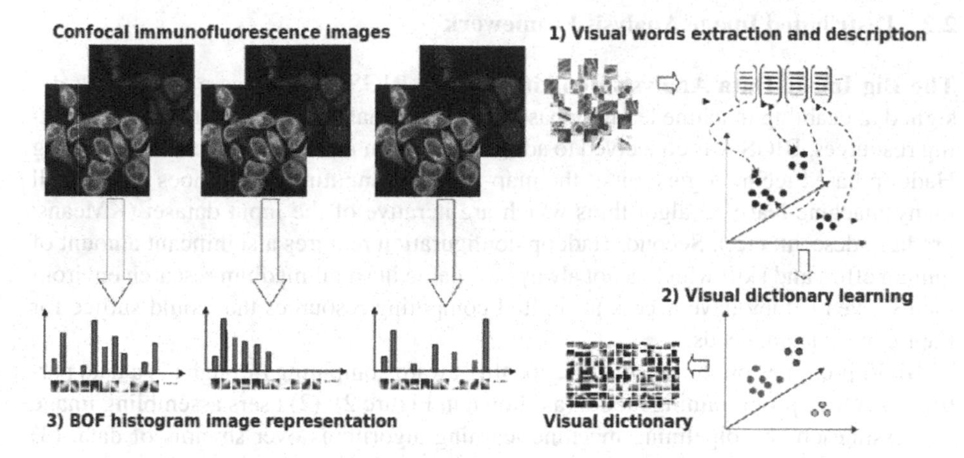

Fig. 1. Overall scheme of BOF approach for confocal fluorescence image representation. Adapted from [5].

Visual words extraction and description: The patches were extracted for each image from a regular image partition of 16×16 pixels without overlapping. Each patch is represented by the discrete cosine transform (DCT) coefficients as visual descriptor to capture the local texture patterns.

Visual dictionary learning: This stage builds a k-word visual dictionary using k-means clustering algorithm over a sample of visual word descriptors of the previous stage taken randomly from the whole image collection.

Histogram image representation: Finally, each image is represented by a k-bin histogram, capturing the frequency of occurrence of each visual codeword in the image.

Subcellular Location Using Naïve Bayes Classifier. In this work the problem of finding the subcellular location of a particular protein is approached as a binary classification problem where a classification model is built for each subcellular location in the dataset. Classification is performed using a maximum-a-posteriori (MAP) strategy as follows:

$$C \leftarrow \underset{c_j \in \{0,1\}}{\arg\max} P(C = c_j | W),$$

where $C = 1$ indicates that the protein is present in a particular subcellular location and $W = (W_1, \ldots, W_k)$ is the BOF representation of the corresponding image. The posterior probability $P(C = c_j|W)$ is estimated using a naïve Bayes approach where features, BOF histogram bins, are considered as independent:

$$P(C = c_j|W) = P(C = c_j) \prod_i P(W_i|C = c_j)$$

2.2 Distributed Image Analysis Framework

The Big Image Data Analysis Toolkit (BIGS). BIGS is a software framework designed at enabling machine learning based Big Data analytics over distributed computing resources. BIGS was conceived to address two main issues encountered when using Hadoop based technologies. First, the map-reduce computing model does not fit well many machine learning algorithms which are iterative of the input dataset (KMeans, gradient descent, etc.). Second, Hadoop configuration requires a significant amount of human effort and skill which is not always available in small/medium research environments, even if they have access to limited computing resources that could suffice for their experimental needs.

BIGS promotes opportunistic, data locality aware computing through (1) a data partition iterative programming model (as shown in Figure 2). (2) users assembling image processing jobs by pipelining machine learning algorithms over streams of data, (3) BIGS workers are software agents deployed over the actual distributed computing resources in charge of resolving the computing load, (4) a NoSQL storage model with a reference NoSQL central database, (5) removing the need of a central control node so that workers contain the logic to coordinate their work through the reference NoSQL database, (6) simple and opportunistic deployment model for workers, requiring only connectivity to the reference NoSQL database, (7) redundant data replication throughout workers, (8) a two level data caching in workers in memory and disk, (9) a set of strategies for workers for data access so that users can enforce data locality aware computing or only-in-memory computing; and (10) a set of APIs through which BIGS can be extended with new algorithms, storage and data import modules. More information can be found at `http://www.3igs.org`.

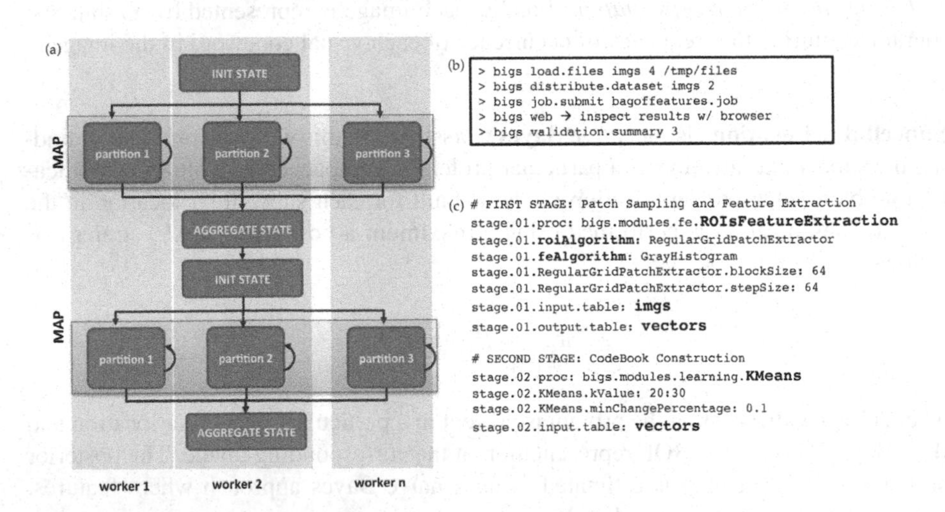

Fig. 2. (a) BIGS computing model, (b) user session, (c) example job definition

Usage. Exprimenters use BIGS by (1) launching workers over the available computing resources, (2) loading the dataset to be processed in the reference NoSQL database, (3) optionally distributing the dataset partitions into the existing workers and (4) defining their data processing pipelines into a job definition file and submitting it for workers to start taking over and executing its operations. BIGS exposes its functionality mainly through a shell command line interface, through which workers can easily be launched over computing resources (`bigs worker`) or also through Java Web Start for the less experienced users. Figure 2 also shows (b) a user session loading and distributing data redundantly (two copies of each partition would be replicated in different workers) , submitting a job and inspecting the results; and (c) an example job definition file composed of two stages, the first one extracting patches and features vectors from input images and the second one performing a KMeans on the features vectors extracted in the first phase.

Note as well how job definitions may include parameter explorations (in this case, KMeans with 20 or 30 centroids). BIGS will handle such exploration by creating two parallel pipelines of operations which can be taken over in parallel if there are enough workers. Job definition files are the primary user interface element with BIGS. Through them, users express declaratively their data processing pipelines and parameter explorations and rely on BIGS for their parallel execution through the available workers. This enables an agile experimental life cycle where users tune up their job definition files and submit them for parallel execution with virtually no sysadmin logistics through a command cycle such as in Figure 2(b).

Additionally, users can tune workers memory and disk cache sizes according to their datasets so that computing could even take place using only in-memory data in cases where datasets fit into the aggregated memory of the workers available.

3 Results

Preliminary results show that our high-throughput computer-vision based approach rapidly obtains classification performance measures comparable to those in the literature through an agile and easy to use experimental life cycle. This enhances the researcher's capabilities for scientific discovery in location proteomics based on confocal images through easily developing distributed machine learning algorithms and harnessing distributed computing power for computer vision processes.

4 Discussion and Conclusion

Automatic methods to determine subcellular location of proteins through exploring increasingly larger confocal image databases are increasingly crucial for human proteomics research. This work shows how automatic image analysis processes demanding large amounts of computational resources can be seamlessly integrated within image-based location proteomics experiments, enabling researchers to focus on their knowledge discovery processes in the field.

Acknowledgments. This work has been supported by Colciencias through the project "Anotación Automática y Recuperación por Contenido de Imágenes Radiológicas usando Semántica Latente'' (number 110152128803).

References

1. Barbe, L., Lundberg, E., Oksvold, P.: Toward a confocal subcellular atlas of the human proteome. Molecular & Cellular ... (2008)
2. Berglund, L., Björling, E., Oksvold, P., Fagerberg, L., Asplund, A., Al-Khalili Szigyarto, C., Persson, A., Ottosson, J., Wernérus, H., Nilsson, P., et al.: A genecentric Human Protein Atlas for expression profiles based on antibodies. Molecular & Cellular Proteomics 7(10), 2019 (2008)
3. Boland, M.V., Murphy, R.F.: A neural network classifier capable of recognizing the patterns of all major subcellular structures in fluorescence microscope images of HeLa cells. Bioinformatics 17(12), 1213–1223 (2001)
4. Chen, X., Velliste, M.: Automated interpretation of subcellular patterns in fluorescence microscope images for location proteomics. Cytometry 69A, 631–640
5. Cruz-Roa, A., Caicedo, J.C., González, F.A.: Visual pattern mining in histology image collections using bag of features. Journal Artificial Intelligence in Medicine 52(2), 91–106 (2011)
6. Hamilton, N.A., Teasdale, R.D.: Visualizing and clustering high throughput sub-cellular localization imaging. BMC Bioinformatics 9, 81 (2008)
7. Hamilton, N.A., Pantelic, R.S.: Fast automated cell phenotype image classification. BMC ... (2007)
8. Li, J., Murphy, R.F., Cohen, W.W., Kovacevic, J.: Automated Learning of Subcellular Location Patterns in Confocal Fluorescence Images from Human Protein Atlas (2012)
9. Li, J., Newberg, J.Y., Uhlén, M., Lundberg, E., Murphy, R.F.: Automated Analysis and Reannotation of Subcellular Locations in Confocal Images from the Human Protein Atlas. PloS One (2012)
10. Lin, Y.-S., Huang, Y.-H., Lin, C.-C., Hsu, C.-N.: Feature space transformation for semi-supervised learning for protein subcellular localization in fluorescence microscopy images. In: Proceedings of the Sixth IEEE International Conference on Symposium on Biomedical Imaging: From Nano to Macro, ISBI 2009, Piscataway, NJ, USA, pp. 414–417. IEEE Press (2009)
11. Newberg, J.Y., Li, J., Rao, A.: Automated analysis of human protein atlas immunofluorescence images. ... Imaging: From Nano ... (2009)
12. Peng, H.: Bioimage informatics: a new area of engineering biology. Bioinformatics 24(17), 1827–1836 (2008)
13. Persson, A., Hober, S., Uhlen, M.: A human protein atlas based on antibody proteomics. Current Opinion in Molecular Therapeutics 8(3), 185 (2006)
14. Pontén, F., Schwenk, J.M.: The Human Protein Atlas as a proteomic resource for biomarker discovery. Journal of Internal ... (2011)
15. Jason, R., Swedlow, I.G.: Goldberg, and Kevin W. Eliceiri. Bioimage informatics for experimental biology*. Annual Review of Biophysics 38(1), 327–346 (2009), PMID: 19416072

Measuring Complexity in an Aquatic Ecosystem

Nelson Fernández[1,2] and Carlos Gershenson[3,4]

[1] Laboratorio de Hidroinformática, Facultad de Ciencias Básicas, Univesidad de Pamplona, Colombia
nfernandez@unipamplona.edu.co
http://unipamplona.academia.edu/NelsonFernandez
[2] Centro de Micro-electrónica y Sistemas Distribuidos, Universidad de los Andes, Mérida, Venezuela
[3] Departamento de Ciencias de la Computación, Instituto de Investigaciones en Matemáticas Aplicadas y en Sistemas, Universidad Nacional Autónoma de México
cgg@unam.mx
http://turing.iimas.unam.mx/~cgg
[4] Centro de Ciencias de la Complejidad, Universidad Nacional Autónoma de México

Abstract. We apply formal measures of emergence, self-organization, homeostasis, autopoiesis and complexity to an aquatic ecosystem; in particular to the physiochemical component of an Arctic lake. These measures are based on information theory. Variables with an homogeneous distribution have higher values of emergence, while variables with a more heterogeneous distribution have a higher self-organization. Variables with a high complexity reflect a balance between change (emergence) and regularity/order (self-organization). In addition, homeostasis values coincide with the variation of the winter and summer seasons. Autopoiesis values show a higher degree of independence of biological components over their environment. Our approach shows how the ecological dynamics can be described in terms of information.

Keywords: Complex Systems, Information Theory, Complexity, Self-organization, Emergence, Homeostasis, Autopoiesis.

1 Introduction

Water bodies have always been relevant. In particular, lakes provide a broad source of water, food, and recreation. Arctic lakes are one of the most vulnerable aquatic ecosystems on the planet since they are changing rapidly, due to the effects of global warming.

The water column (limnetic zone) of an Arctic lake is well-mixed; this means that there are no layers with different temperatures. During winter, the surface of the lake is ice covered. During summer, ice melts and the water flow and evaporation increase. Consequently, the two climatic periods (winter and summer) in the Arctic region cause a typical hydrologic behavior in lakes. This behavior

L.F. Castillo et al. (eds.), *Advances in Computational Biology*,
Advances in Intelligent Systems and Computing 232,
DOI: 10.1007/978-3-319-01568-2_12,

influences the physiochemical subsystem of the lake. One or more components or subsystems can be an assessment for the Arctic lakes dynamics, for example: physiochemical, limiting nutrients and photosynthetic biomass for the planktonic and benthic zones.

In recent years, the scientific study of complexity in ecological systems, including lakes, has increased the understanding of a broad range of phenomena, such as diversity, abundance, and hierarchical structure [6]. It is important to consider that lakes exhibit properties like emergence, self-organization, and life. Lake dynamics generate novel information from the relevant interactions among components. Interactions determine the future of systems and their complex behavior. Novel information limits predictability, as it is not included in initial or boundary conditions. It can be said that this novel information is emergent since it is not in the components, but produced by their interactions. Interactions can also be used by components to self-organize, i.e. produce a global pattern from local dynamics. The balance between change (chaos) and stability (order) states has been proposed as a characteristic of complexity [5,4]. Since more chaotic systems produce more information (emergence) and more stable systems are more organized, complexity can be defined as the balance between emergence and self-organization. In addition, there are two properties that support the above processes: homeostasis refers to regularity of states in the system and autopoiesis that reflects autonomy.

Recently, abstract measures of emergence, self-organization, complexity, homeostasis and autopoiesis based on information theory have been proposed [2,1], with the purpose of clarifying their meaning with formal definitions. In this work, we apply these measures to an aquatic ecosystem. The aim of this application to an Arctic lake is to clarify the ecological meaning of these notions, and to show how the ecological dynamics can be described in terms of information. With this approach, the complexity in biological and ecological systems can be studied.

In the next section, we present a brief explanation of measures of self-organization, emergence, complexity, homeostasis, autopoiesis. Section 3 describes our experiments and results with the Arctic lake, which illustrate the usefulness of the proposed measures, closing with conclusions in Section4.

2 Measures

Emergence refers to properties of a phenomenon that are present in one description and were not in another description. In other words, there is emergence in a phenomenon information is produced. Shannon [9] proposed a quantity which measures how much information was produced by a process. Therefore, we can say that the emergence is the same as the Shannon's information $I = -K \sum_{i=i}^{n} p_i \log p_i$ where K is a positive constant and p_i is the probability of a symbol from a finite alphabet from appearing in a string. Thus $E = I$.

Self-organization has been correlated with an increase in order, i.e. a reduction of entropy [3]. If emergence implies an increase of information, which is analogous to entropy and disorder, self-organization should be anti-correlated with emergence. We propose as the measure $S = 1 - I = 1 - E$.

We can define complexity C as the balance between change (chaos) and stability (order). We can use emergence and self-organization which respectvely measure that. Hence we propose: $C = 4 \cdot E \cdot S$. The constant 4 is added to normalize the measure to $[0, 1]$.

For homeostasis H, we are interested on how all variables of a system change or not in time. A useful function for comparing strings of equal length is the Hamming distance. The normalized Hamming distance d measures the percentage of different symbols in two strings X and X'. Thus, $1 - d$ indicates how similar two strings are. To measure H, we take the average of these state similarities.

As it has been proposed, adaptive systems require a high C in order to be able to cope with changes of its environment while at the same time maintaining their integrity [5,4]. X can represent the trajectories of the variables of a system and Y can represent the trajectories of the variables of the environment of the system. If X has a high E, then it would not be able to produce its own information. With a high S, X would not be able to adapt to changes in Y. We propose $A = \frac{C(X)}{C(Y)}$, so that higher values of A indicate a higher C of a system relative to their environment.

Details of these measures can be found in [1].

3 Results

The data from an Arctic lake model used in this section was obtained using The Aquatic Ecosystem Simulator [8]. Table 1 shows the variables and daily data we obtained from the Arctic lake simulation. The model used is deterministic, so there is no variation in different simulation runs. There are a higher dispersion for variables such as temperature (T) and light (L) at the three zones of the Arctic lake (surface=S, planktonic=P and benthic=B; Inflow and outflow ($I\&O$), retention time (RT) and evaporation (Ev) also have a high dispersion, Ev being the variable with the highest dispersion.

3.1 Emergence, Self-organization, and Complexity

Figure 1 shows the values of E, S, and C of the physiochemical subsystem[1]. Variables with a high complexity $C \in [0.8, 1]$ reflect a balance between change/chaos (E) and regularity/order (S). This is the case of benthic and planktonic pH (BpH; PpH), $I\&O$ (Inflow and Outflow) and RT (Retention Time). For variables with high emergencies ($E > 0.92$), like Inflow Conductivity (ICd) and Zone Mixing (ZM), their change in time is constant; a necessary condition for exhibiting chaos. For the rest of the variables, self-organization values are low ($S < 0.32$), reflecting low regularity. It is interesting to notice that in this system there are no variables with a high S nor low E.

[1] The variables were normalized to base 10 using the method described in [1].

Table 1. Physiochemical variables considered in the Arctic lake model

Variable	Units	Acronym	Max	Min	Median	Mean	std. dev.
Surface Light	MJ/m2/day	*SL*	30	1	5.1	11.06	11.27
Planktonic Ligth	MJ/m2/day	*PL*	28.2	1	4.9	10.46	10.57
Benthic Light	MJ/m2/day	*BL*	24.9	0.9	4.7	9.34	9.33
Surface Temperature	Deg C	*ST*	8.6	0	1.5	3.04	3.34
Planktonic Temperature	Deg C	*PT*	8.1	0.5	1.4	3.1	2.94
Benthic Temperature	Deg C	*BT*	7.6	1.6	2	3.5	2.29
Inflow and Outflow	m3/sec	*I&O*	13.9	5.8	5.8	8.44	3.34
Retention Time	days	*RT*	100	41.7	99.8	78.75	25.7
Evaporation	m3/day	*Ev*	14325	0	2436.4	5065.94	5573.99
Zone Mixing	%/day	*ZM*	55	45	50	50	3.54
Inflow Conductivity	uS/cm	*ICd*	427	370.8	391.4	396.96	17.29
Planktonic Conductivity	uS/cm	*PCd*	650.1	547.6	567.1	585.25	38.55
Benthic Conductivity	uS/cm	*BCd*	668.4	560.7	580.4	600.32	40.84
Surface Oxygen	mg/litre	*SO2*	14.5	11.7	13.9	13.46	1.12
Planktonic Oxygen	mg/litre	*PO2*	13.1	10.5	12.6	12.15	1.02
Benthic Oxygen	mg/litre	*BO2*	13	9.4	12.5	11.62	1.51
Sediment Oxygen	mg/litre	*SdO2*	12.9	8.3	12.4	11.1	2.02
Inflow pH	ph Units	*IpH*	6.4	6	6.2	6.2	0.15
Planktonic pH	ph Units	*PpH*	6.7	6..40	6.6	6.57	0.09
Benthic pH	ph Units	*BpH*	6.6	6.4	6.5	6.52	0.07

Since $E, S, C \in [0, 1]$, these measures can be categorized into five categories described on the basis of an adjective, a range value, and a color. The categories are: *Very Low* $\in [0, 0, 2]$: red (dark gray in grayscale), *Low* $\in (0.2, 0.4]$: orange (mid gray), *Fair* $\in (0.4, 0, 6]$: yellow (almost white), *High* $\in (0.6, 0.8]$: green (light gray) and *Very High* $\in (0.8, 1]$: blue (almost black). This categorization is inspired on the Colombian water pollution indices. These indices were proposed by [7].

We can divide the variables in the following complexity categories:

Very High Complexity. $C \in [0.8, 1]$. The following variables balance S and E: benthic and planktonic pH (BpH, PpH), inflow and outflow ($I\&O$), and retention time (RT). It is remarkable that the increasing of the hydrological regime during summer is related in an inverse way with the dissolved oxygen (SO_2; BO_2). It means that an increased flow causes oxygen depletion. Benthic Oxygen (BO_2) and Inflow Ph (IpH) show the lowest levels of the category. Between both, there is a negative correlation: a doubling of IpH is associated with a decline of BO_2 in 40 percent.

High Complexity. $C \in [0.6, 0.8)$. This group includes 11 of the 21 variables and has a high E and a low S. These 11 variables that showed more chaotic than ordered states are highly influenced by the solar radiation that defines the winter and summer seasons, as well as the hydrological cycle. These variables were: Oxygen (PO_2, SO_2); surface, planktonic and benthic temperature (ST,

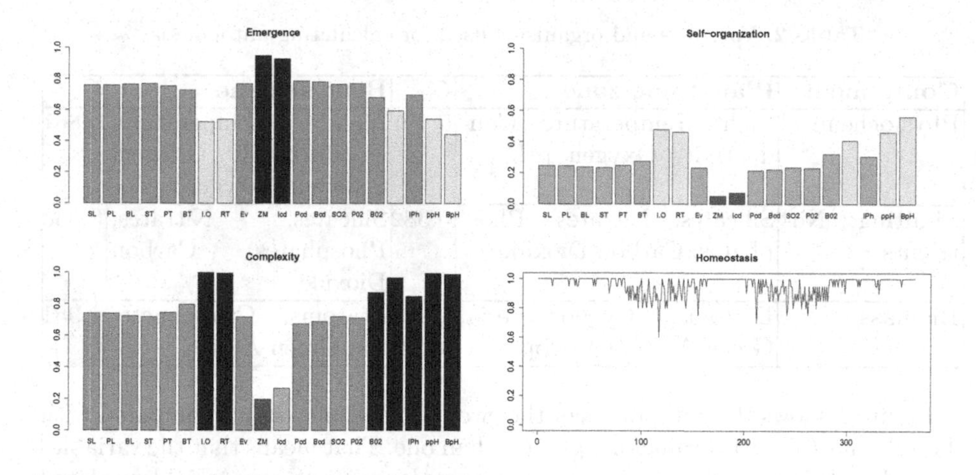

Fig. 1. E, S, and C of physiochemical variables of the Arctic lake model and daily variations of homeostasis H during a simulated year.

PT, BT); conductivity (ICd, PCd, BCd); planktonic and benthic light (PL, BL); and evaporation (Ev).

Very Low Complexity. $C \in [0, 0.2)$. In this group, E is high, and S is very low. This category includes the inflow conductivity (ICd) and water mixing variance (ZM). Both are correlated.

3.2 Homeostasis

The homeostasis was calculated by comparing the variation of all variables, representing the state of the Arctic subsystem every day. The timescale is very important, because H can vary considerably if we compare states every minute or every month. The h values have a mean (H) of 0.957 and a standard deviation of 0.065. The minimum h is 0.60 and the maximum h is 1.0. In an annual cycle, homeostasis shows four different patterns, as shown in Figure 1, which correspond with the seasonal variations between winter and summer. These four periods show scattered values of homeostasis as the result of transitions between winter and summer and winter back again.

3.3 Autopoiesis

Autopoiesis was measured for three components (subsystems) at the planktonic and benthic zones of the Arctic lake. These were physiochemical (PC), limiting nutrients (LN) and biomass (BM). They include the variables and organisms related in Table 2, where the averaged C of the variables is shown.

Table 2. Variables and organisms used for calculating autopoiesis

Component	Planktonic zone	C	Benthic zone	C
Physiochemical	Light, Temperature, Conductivity, Oxygen, pH	0.771	Light, Temperature, Conductivity, Oxygen, Sediment Oxygen, pH	0.861
Limiting Nutrients	Silicates, Nitrates, Phosphates, Carbon Dioxide	0.382	Silicates, Nitrates, Phosphates, Carbon Dioxide	0.338
Biomass	Diatoms, Cyanobacteria, Green Algae, Chlorophyta	0.937	Diatoms, Cyanobacteria, Green Algae	0.951

Figure 2 shows the autopoiesis of the two biomass subsystems compared with the LN and PC. All A values are greater than one. That means that the variables related to living systems have a greater complexity than the variables related to their environment. While we can say that some PC and LN variables have different effects on the planktonic and benthic biomass, we can also estimate that planktonic and benthic biomass are more autonomous compared to their physiochemical and nutrient environments. The very high values of C of biomass imply that these living systems can adapt to the changes of their environments because of the balance between E and S that they have.

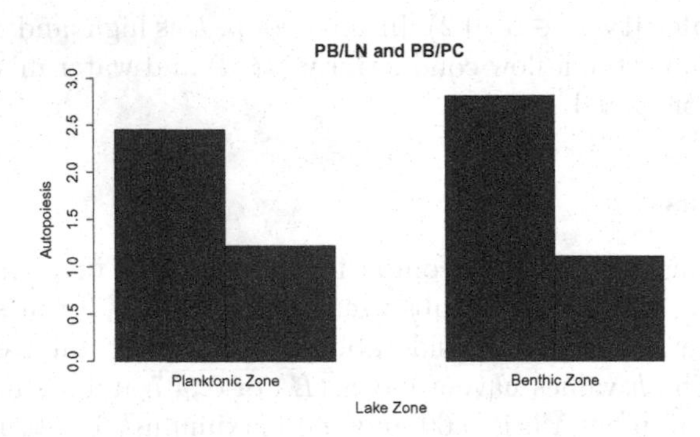

Fig. 2. A of biomass depending on limiting nutrients and physiochemical components

4 Conclusions

Measuring the complexity of ecological systems has a high potential. Current approaches focus on specific properties of ecosystems. With a general measure, different ecosystems can be compared at different scales, increasing our understanding of ecosystems and complexity itself.

We applied measures of emergence, self-organization, complexity, homeostasis, and autopoiesis based on information theory to an aquatic ecosystem. The generality and usefulness of the proposed measures will be evaluated gradually, as these are applied to different ecological systems. The potential benefits of general measures as the ones proposed here are manifold. Even if with time more appropriate measures are found, aiming at the goal of finding general measures which can characterize complexity, emergence, self-organization, homeostasis, autopoiesis, and related concepts for any observable ecosystem is a necessary step to make.

References

1. Fernández, N., Maldonado, C., Gershenson, C.: Information measures of complexity, emergence, self-organization, homeostasis, and autopoiesis. In: Prokopenko, M. (ed.) Guided Self-Organization: Inception. Springer (in Press, 2013), http://arxiv.org/abs/1304.1842
2. Gershenson, C., Fernández, N.: Complexity and information: Measuring emergence, self-organization, and homeostasis at multiple scales. Complexity 18(2), 29–44 (2012), http://dx.doi.org/10.1002/cplx.21424
3. Gershenson, C., Heylighen, F.: When can we call a system self-organizing? In: Banzhaf, W., Ziegler, J., Christaller, T., Dittrich, P., Kim, J.T. (eds.) ECAL 2003. LNCS (LNAI), vol. 2801, pp. 606–614. Springer, Heidelberg (2003), http://arxiv.org/abs/nlin.AO/0303020
4. Kauffman, S.A.: The Origins of Order. Oxford University Press, Oxford (1993)
5. Langton, C.: Computation at the edge of chaos: Phase transitions and emergent computation. Physica D 42, 12–37 (1990)
6. Lizcano, D.J., Cavelier, J.: Using GPS collars to study mountain tapirs (*Tapirus pinchaque*) in the central andes of colombia. The Newsletter of the IUCN/SSC Tapir Specialist Group 13(2), 18–23 (2004)
7. Ramírez, A., Restrepo, R., Fernández, N.: Evaluación de impactos ambientales causados por vertimientos sobre aguas continentales. Ambiente y Desarrollo 2, 56–80 (2003), http://www.javeriana.edu.co/fear/ins_amb/rad12-13.htm
8. Randerson, P., Bowker, D.: Aquatic Ecosystem Simulator (AES) — a learning resource for biological science students (2008), http://www.bioscience.heacademy.ac.uk/resources/projects/randerson.aspx
9. Shannon, C.E.: A mathematical theory of communication. Bell System Technical Journal 27, 379–423, 623–656 (1948), http://tinyurl.com/6qrcc

Possible Antibiofilm Effect of Peptides Derived from IcaR Repressor of *Staphylococcus epidermidis* Responsible for Hospital-Acquired Sepsis

Liliana Muñoz[1], Luz Mary Salazar[2], Stefany Botero[1], Jeannette Navarrete[1], and Gladys Pinilla[1]

[1] Universidad Colegio Mayor de Cundinamarca, Bogotá, Colombia
(lcmunoz,jnavarreteo,gpinillab)@unicolmayor.edu.co, lucyucmc@gmail.com
[2] Universidad Nacional de Colombia, Bogotá, Colombia
lmsalazarpu@unal.edu.co

Abstract. Antimicrobial peptides are alternatives to inhibit biofilm formation, which is mediated by ica ADBC and regulated by IcaR, as a mechanisms of virulence of *S. epidermidis* involved in hospital-acquired infections. Peptides with biological activity analog to the IcaR protein were designed using the prediction program Antibp; their chemical synthesis was carried out by Nα-Fmoc and the peptides were purified and characterized by RP-HPLC and MALDI-TOF. Red cells were used to determine their hemolytic activity. Peptides named IR1, IR2 and IR3, derived from IcaR, were characterized with a high degree of purity; their hemolytic activity was found to be less than 6 % and they were postulated as candidates with analog activity to the native IcaR repressor against the biofilm formation of *S. epidermidis*. Their low hemolytic potential, allow them to be use in future in vitro trials for therapeutic use.

Keywords: cationic peptides, antibiofilm activity, IcaR, *S. epidermidis*.

1 Introduction

Antimicrobial peptides are nowadays considered an important part in clinical strategy, since they inhibit bacterial biofilm formation. Bacterial biofilms are resistant to most of the commonly available antibacterial agents [1, 2]. *Staphylococcus epidermidis*, known as a leading cause of sepsis in newborns, produces a polysaccharide intercellular adhesion (PIA) which in the presence of glucose favors biofilm formation [3]. Biofilm synthesis is mediated by ica ADBC operon and a negative regulator gen, icaR, with a helix-turn-helix (HTH) structural motif [4]. According to the simulation carried out in the SRING server (http://string.embl.de/newstring_cgi/show_input_page.pl?UserId=IskH9N1aPbTO&sessionId=x6NILBgnr2_m) of the IcaR with other proteins, the mechanism of action ofIcaR and the antimicrobial peptides could unstabilize the biofilm or act on the bacterial cell walls or membranes by polarizing them [5]. Due to the need that exists to meet the demands of effective

L.F. Castillo et al. (eds.), *Advances in Computational Biology*,
Advances in Intelligent Systems and Computing 232,
DOI: 10.1007/978-3-319-01568-2_13, © Springer International Publishing Switzerland 2014

antibiofilm therapy, cationic compounds such as peptides, could be promising candidates for development of antimicrobial agents. Their principal mechanism of action is thought to be either by t he disruption of the cytoplasmic membrane, which kills bacteria swiftly and thoroughly,or by inhibiting the biofilm formation, hence this antimicrobial peptides capability to intervene in the mechanisms of virulence involved in hospital-acquired infections associated with *S. epidermidis*.

2 Materials and Methods

2.1 Design of Peptides

Using the *Antibp Server*(http://www.imtech.res.in/raghava/antibp/) [1] and the original reported sequence for IcaR, a peptide named IR1 was designed. The original peptide sequence was modified and peptide parameters were recalculated utilizing The Antimicrobial Peptide Database (APD), which contains up to 2221 antimicrobial peptides(http://aps.unmc.edu/AP/main.php) [2], for the purpose of increasing the antimicrobial score of two new additional peptides to be synthesized, named IR2 and IR3. In order to improve the accuracy of secondary structure predictions, the following bioinformatic tools were used: GOR(*http://gor.bb.iastate.edu/cdm/*), Fragment Database Mining (FDM), GOR V, PSIPRED (*http://bioinf.cs.ucl.ac.uk/psipred/submit*) and *HelicalWheel* [7]. The latter was employed to confirm the amphiphile configuration of the peptides [8].

2.2 Synthesis, Purification and Characterization of Peptides

The peptides IR1, IR2 and IR3 were synthesized by Fmoc (9-fluorenylmethoxycarbonyl) solid-phase chemistry using a Rink amide resin; method previously described by Merrifield and modified by HoughtenSarin [5]. The crude peptides were purified by RP-HPLC (L7400 LaChrom Merck Hitachi) using a Waters RP-18 (5 μm) column. The purity of the synthetic peptides was confirmed by mass spectrometry (AutoflexBrukerDaltonicsMaldi-Tof) and they were analyzed in a circular dichroismspectropolarimeter in order to predict their predominant secondary structure [9].

2.3 Hemolytic Activity

Hemolytic activity of the peptides was determined using red blood cells (RBCs),O-Rh positive. The determination was carried out as follows: 8 serial dilutions of a 50μg/μL solution of each peptide were prepared in two sets. Negative controls: insulin (non-hemolytic peptide) and RBCsin a physiological saline solution; positive control: RBCs in distillated water. The results from the trials in a MS Excel Worksheet were analyzed with the GraphPad Prism 5.0 Trial Version® program(LA Jolla, CA, USA).Average hemolytic concentration was determined (CH_{50}), which is the concentration of the peptide that produced 50 % of hemolysis of RBCsin suspension [10].

3 Results and Discussion

3.1 Selection and Synthesis of Peptides

Using the prediction program *Antibp Server*, antimicrobial peptides based on the IcaR sequence were designed, in which peptide IR1 (TLDDISKSVNIKKAS) was found to be the one with the highest antimicrobial score possible, and from which peptides IR2 (TLIDKLKSVGRKKAK) and IR3 (TLIDKLKSKGRKKAS) were obtained by structural modification. Furthermore, it was intended for the peptides to have a more helical and anphiphilic structure, in order to favor the interaction between these peptides and the charged components of the bacterial membrane to obtain a higher net charge, visualized by *Helical Wheel.* To this respect, it is intended for the residues to have a very similar disposition to that shown in Fig. 1.

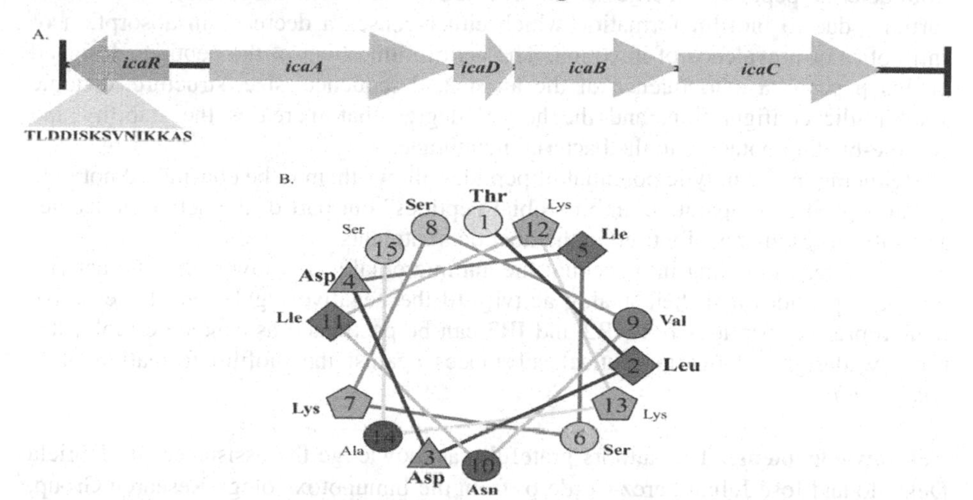

Fig. 1. Peptide-binding site on *icaR-ADBC* of *Staphylococcus epidermidis* and Helical Structure Prediction of IR1. A)Peptide-binding site. B)Helical Structure Prediction.

Characterization of the peptides shows that IR1 has an isoelectric point (IP) of 9.81 and a positive charge of +1 and peptides IR2 y IR3 have an IP of 11,0 and a positive charge of +5, which is the confirmation of the basic characterization of the synthesized peptides. The antimicrobial activity score for each one of the peptides were found to be 0.930, 3.057 and 3.426 respectively. The chromatographic technic employed shows that the peptides have a degree of purity of more than 90 %, and the mass spectrophotometric analysis MALDI-TOF confirmed the correct molecular weight compared to the ones calculated theoretically (data not shown).

3.2 Hemolytic Activity

The hemolysis assay demonstrated that these peptides have a low ability to lyse red blood cells. The hemolytic activity of the peptides, at the evaluated concentrations, were found to be between 1,83% and 6,60%, with CH_{50} calculated with a superior value of the maximum concentration used in the test. The best result of the trials was found for IR3, showing a CH_{50} similar to that of the insulin, peptide known to be non-hemolytic. A hemolytic activity of less than 6 % is evidence of reliability for the use of these peptides for in vitro trials. These findings would allow them to be used as active molecules or substances in the antimicrobial therapy and specifically as antibiofilm agents to be used as a possible medical treatment in human beings. High hemolytic activity of some peptides prevents their use from therapeutic treatment, evidencing the need for improvements in peptide design processes aiming to minimize this effect.

4 Conclusions

Antibacterial peptides emerge as an alternative for multi-drug resistant bacteria, partially due to biofilm formation which either causes a decrease in absorption of antibiotics or prevalence of infection. The antibiofilm effect of the peptides designed in this paper is a consequence of the aminoacid sequence, size, structure, cationic amphiphilic configuration, and the helical degree that increases the stability and peptide-binding potential to the bacterial membrane.

Reducing the hemolytic potential of peptides allows them to be considered not only as "nonspecific comparative antimicrobial peptides" but part of the active molecules and substances universally used in therapeutic treatments.

Therefore, and taking into account the antimicrobial score, low hemolytic activity for these peptides and their analog activity to the negative regulation of the native IcaR repressor; peptides IR1, IR2 and IR3 can be postulated as candidate molecules for new designs of pharmaceutical substances against the biofilm formation of *S. epidermidis.*

Acknowledgements. The authors gratefully acknowledge the assistance of Gabriela Delgado and José Julián Perez-Cordero from the Inmunotoxicology Research Group, Department of Pharmacy, Faculty of Sciences, Universidad Nacional de Colombia (National University of Colombia) and Paola Andrea Arévalo, from the Programa Jóvenes Investigadores de Colciencias 2012- 2013 (Young Investigator Programme of Colciencias 2012- 2013).

References

[1] Lata, S., Sharma, B.K., Raghava, G.P.: Analysis and prediction of antibacterial peptides. BMC Bioinformatics 8, 263 (2007)

[2] Wang, G., Li, X., Wang, Z.: APD2: the updated antimicrobial peptide database and its application in peptide design. Nucleic Acids Research 37, D933–D937 (2009)

[3] Fitzpatrick, F., Humphreys, H., O'Gara, J.P.: The genetics of staphylococcal biofilm formation—will a greater understanding of pathogenesis lead to better management of device-related infection? Clin. Microbiol. Infect. 11, 967–973 (2005)

[4] Jefferson, K.K., Cramton, S.E., Gotz, F., Pier, G.B.: Identification of a 5-nucleotide sequence that controls expression of the ica locus in *Staphylococcus aureus* and characterization of the DNA-binding properties of IcaR. Mol. Microbiol. 48, 889–899 (2003)

[5] Sarin, V.K., Kent, S.B., Tam, J.P., Merrifield, R.B.: Quantitative monitoring of solid-phase peptide synthesis by the ninhydrin reaction. Anal. Biochem. 117, 147–157 (1981)

[6] Cheng, H., Sen, T.Z., Kloczkowski, A., Margaritis, D., Jernigan, R.L.: Prediction of protein secondary structure by mining structural fragment database. Polymer (Guildf) 46, 4314–4321 (2005)

[7] Kloczkowski, A., Ting, K.L., Jernigan, R.L., Garnier, J.: Combining the GOR V algorithm with evolutionary information for protein secondary structure prediction from amino acid sequence. Proteins 49, 154–166 (2002)

[8] Lockwood, N.A., Tu, R.S., Zhang, Z., Tirrell, M.V., Thomas, D.D., Karim, C.B.: Structure and function of integral membrane protein domains resolved by peptide-amphiphiles: application to phospholamban. Biopolymers 69, 283–292 (2003)

[9] Schaumann, R., Knoop, N., Genzel, G.H., Losensky, K., Rosenkranz, C., Stîngu, C.S., Schellenberger, W., Rodloff, A.C., Eschrich, K.: A step towards the discrimination of beta-lactamase-producing clinical isolates of Enterobacteriaceae and *Pseudomonas aeruginosa* by MALDI-TOF mass spectrometry. Med. Sci. Monit. 18, MT71–MT77 (2012)

[10] Shilpakala, S.R., Ketha, M., Chintamani, D.: A Peptide Derived from Phage Display Library Exhibits Antibacterial Activity against *E. coli* and *Pseudomonas aeruginosa*. PLoS One 8, e56081 (2013)

"Head to Tail" Tool Analysis through ClustalW Alignment Algorithms and Construction of Distance Method Neighbor-Joining Trees Based on Genus *Fusarium* Genomic Distances

Juan David Henao[1], S. Melissa Rincón[2], and D. Juan Jose Filgueira[3]

[1] Student of Applied Biology of Nueva Granada Military University (UMNG), Bogotá, Colombia
[2] Research Assistant Faculty of Basic and Applied Sciences, UMNG, Bogotá, Colombia
[3] Director of Molecular phytopathology Research Group,
Faculty of Basic and Applied Sciences, UMNG, Bogotá, Colombia
Juanhenao.sanchez@gmail.com

Abstract. The genomic sequences concatenations are the most used tool for phylogenetic studies; such arrangements are characterized by eliminating all gaps arising in the process of alignment to improve phylogenetic constructions. However, no studies dedicated to the analysis of the concatenations with gaps, as these regions represent genetic transformation events that are crucial evolutionary events. For the concatenation analysis, nucleotide sequences of 11 species of the genus *Fusarium*, were experimentally obtained. For each species we used sequences of 10 amplicons, corresponding to 10 genic regions. Later several permutations were generated, concerning the order of the sequences, to observe topologies changes on the resulting trees with minimal changes in the "Head to Tail" arrangements. Multiple alignment of the DNA sequences, were performed using the ClustalW algorithm. Subsequently a feasibility analysis of sequences for phylogenetic analysis method was generated based on the likelihood-mapping tool using the Tree-puzzle-5.2 program. From this analysis, molecular inferences from trees were made using MEGA5 software, through a Neighbor-Joining distance method with 1000 bootstrap replicates, to support the resulting trees. We observed that there is variation level in the trees using "Head to Tail" arrays, which prevents showing the uniformity of the resulting cluster, keeping alignment gaps regardless of the order of the array. So far, the results obtained indicate that the "Head to Tail" arrangements are subject to the order of the genomic sequences that comprise it, and they are susceptible to possessing a sequence difference relative to another; i.e., the input of a single species whose "Head to Tail" arrangement possesses a range of major change, in comparison to the others in terms of concatenated with gaps included, generates a considerable change in the output of the resulting tree.

Keywords: Neighbor-Joining, ClustalW, *Fusarium*, Distance method, concatenations.

L.F. Castillo et al. (eds.), , *Advances in Computational Biology*,
Advances in Intelligent Systems and Computing 232,
DOI: 10.1007/978-3-319-01568-2_14, © Springer International Publishing Switzerland 2014

1 Introduction

Currently, the method to construct phylogenetic trees regardless of the construction algorithm, is based on the global alignment of sequences with slight variations over time, such generated "head to tail" arrangements produce concatenated sequences which eliminates gaps, are therefore considered null information and therefore alters the final results [1].

This concatenated method without gaps is mathematically feasible, then the resulting tree topologies, regardless of the order of "Head to Tail" arrangement tend to remain intact [2], but it should be remembered that gaps are defined as possible events of genetic transformation of the sequences sometimes in the evolutionary history of the organisms studied [3] so as to eliminate such "non-data", is to remove information from a biological evolution [4]. This paper analyzes how gaps alter the results of the final topologies of the trees changing orders in various "Head to Tail" arrangements to determine how much this type of biological information comes in contrast to the mathematical analysis and bioinformatics.

2 Materials and Methods

Obtaining of sequences. Gene sequences were obtained from 11 species belonging to the genus *Fusarium* in Plant Biotechnology Laboratory of Nueva Granada Military University, the species used as a model for phylogenetic analysis were: *Fusarium avenaceum, Fusarium culmorum, Fusarium equiseti, Fusarium foetens, Fusarium graminearum, Fusarium oxysporum, Fusarium proliferatum, Fusarium solani, Fusarium sporotrichioides, Fusarium subglutinans and Fusarium verticillioides*. The amplified regions belong to the next genes: β-tubulin (Bt), Elongation Factor 1α (EF), Cytochrome Oxidase (AHy), Ribosomal Intergenic Spacer region (ITS), Histone 3 and Histone 4 (H3 and H4), and genic regions were used: (Bt1a [5] 5' TTC CCC CGT CTC CAC TTC TTC ATG3'/Bt1b[5] 5'GAC GAG ATC GTT CAT GTT GAA CTC3'), (Bt2a [5] 5'GGT AAC CAA ATC GGT GCT GCT TTC3'/Bt2b [5] 5'ACC CTC AGT GTA GTG ACC CTT GGC3'), (EF-1H [6] 5'ATG GGT AAG GAA GAC AAG AC3'/EF-2T [6] 5'GGA AGT ACC AGT GAT CAT GTT3'), (AHyFuF [7] 5'CTT AGT GGG CCA GGA GTT CAA TA3'/AHyFuR [7] 5'ACC TCA GGG TGT CCG AAG AAT3'), (ITSFuF [8] 5'CAA CTC CCA AAC CCC TGT GA3'/ITSFuR [8] 5'GCG ACG ATT ACC AGT AAC GA3'; ITS-1 [8] 5'TCC GTA GGT GAA CCT GCG G3'/ITS-2 [8] 5'GCT GCG TTC TTC ATC GAT GC3'; ITS-1 [8] 5'TCC GTA GGT GAA CCT GCG G3'/ITS-4 [8] 5'TCC TCC GCT TAT TGA TAT GC3'; ITS-4 [8] 5'TCC TCC GCT TAT TGA TAT GC3'/ITS-5 [8] 5'GGA AGT AAA AGT CGT AAC AAG G3'), (H3-1a [5] 5'ACT AAG CAG ACC GCC CGC AGG3'/H3-1b [5] 5'GCG GGC GAG CTG GAT GTC CTT3'), (H4-1a [5] 5'GCT ATC CGC CGT CTC GCT3'/H4-1b [5] 5'GGT ACG GCC CTG GCG CTT3'). The alignment was done using the ClustalW algorithm [9].

***A priori* and *a posteriori* analysis.** With the aligned sequences, several "head to tail" arrangements were performed, in such way that each resulting concatenated possessed the gaps corresponding to the alignment carried out individually, did not align the new concatenated sequences because of lost information about gene transformation events [4].

The *a priori* analysis of the sequences was performed using concatenated likelihood-mapping method analysis [10]. To do this we used the Tree-puzzle-5.2 program [11], this analysis was performed for sequences whose "Head to Tail" arrangement varied less than two changes in the order of the sequences to establish the behavior of the results with the minimal changes.

A posteriori analysis was based on the distance tree construction with Neighbor-Joining algorithm [12] using the MEGA5 software [13] with 1000 bootstrap replicates for statistical support [14]. Each tree was constructed using the concatenated obtained on *a priori* analysis based on the values given by the Bootstrap process.

3 Results and Discussion

Phylogenetic analysis *a priori*. The arrangements "Head to Tail" used for *a priori* and *a posteriori* analysis consisted of an arrangement whose order of the sequences produced by the pooling of amplicons belonging to the same gene, such that the resulting concatenated outside binding sequences of the same gene and not amplicons dispersed in the array. A second arrangement "head to tail" consisted in the change of the order of amplicons corresponding to sequences of β-tubulin (Bt1a/Bt1b and Bt2a/Bt2b) which determined how the results varied with the introduction of variations like other sequences.

Analyzing the sequences using the Likelihood-mapping analysis, showed that even with gaps the trees possess a prediction percentage sustainable for predicting a tree-type topology and not for inconsistent as intermediate topology or star topology-type [15]. The first order of arrangement "Head to Tail" corresponded to the amplicons Bt1a/Bt1b, Bt2a/Bt2b, AhyFuF/AHyFuR, EF-1H/EF-2T, H3-1a/H3-1b, H4-1a/H4-1b, ITS FuR/ITS FuF, ITS-1/ITS-2, ITS-1/ITS-4, ITS-4/ITS-5 according to the analysis reflected in the form of the triangle diagram based on probabilities for 7 basins [10], [15] (Fig. 1), shows that the sum of the probabilities of obtaining a topology defined is 97.6%, indicating that the alignment of the sequences having the characteristics necessary for phylogenetic analysis [15] (Fig. 1a).

The second order of arrangement "Head to Tail" corresponded to the amplicons Bt2a/Bt2b, Bt1a/Bt1b, AhyFuF/AHyFuR, EF-1H/EF-2T, H3-1a/H3-1b, H4-1a/H4-1b, ITS FuR/ITS FuF, ITS-1/ITS-2, ITS-1/ITS-4, ITS-4/ITS-5 (the difference with the first one is the order of Bt1 and Bt2 amplicons) and formulating the same analysis to the previous concatenated sequence, the result of the sum of the probabilities of obtaining one of the three topologies based in quartets analysis (n/4) [9], [14] (Fig. 1b), was 97.9%, which indicates that both arrangements "head to tail" are feasible to generate phylogenetic analyzes.

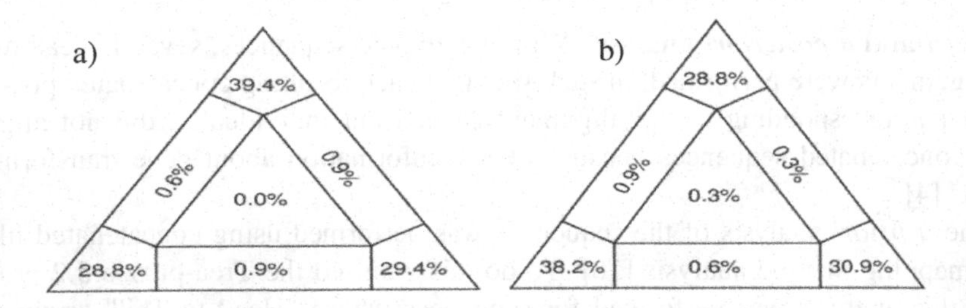

Fig. 1. scheme of the likelihood-mapping analysis obtained with the software Tree-puzzle-5.2, a) Analysis of concatenated comprising the sequences Bt1a/Bt1b, Bt2a/Bt2b, AhyFuF/AHyFuR, EF-1H/EF-2T, H3-1a/H3-1b, H4-1a/H4-1b, ITS FuR/ITS FuF, ITS-1/ITS-2, ITS-1/ITS-4, ITS-4/ITS-5; b) Analysis of concatenated comprising the sequences Bt2a/Bt2b, Bt1a/Bt1b, AhyFuF/AHyFuR, EF-1H/EF-2T, H3-1a/H3-1b, H4-1a/H4-1b, ITS FuR/ITS FuF, ITS-1/ITS-2, ITS-1/ITS-4, ITS-4/ITS-5. The central triangle is the probability of obtaining a star-type topology, the rectangles correspond to the probability for intermediate topology of the corners that are fully resolved topologies for analyzing quartets (n/4), where n is the number of sequences aligned.

In Fig.1. can be seen, as despite getting a sum of probabilities, that favors fully resolved topologies for both sequences do not exist a probability that favors one of the topologies given by the study of quartets that defines the topology most likely [10], [15].

Phylogenetic analysis *a posteriori*. Building distance trees was done using MEGA5, to use the neighbor-joining algorithm with a Bootstap of 1000 replicates. The resulting trees were completely different from each other. The tree corresponding to the arrangement Bt1a/Bt1b, Bt2a/Bt2b, AhyFuF/AHyFuR, EF-1H/EF-2T, H3-1a/H3-1b, H4-1a/H4-1b, ITS FuR/ITS FuF, ITS-1/ITS-2, ITS-1/ITS-4, ITS-4/ITS-5 has a topology more robust, sustained by Bootstrap values (Fig. 2a), in relation to distance tree generated using the second arrangement "Head to Tail" to the order of the sequences Bt2a/Bt2b, Bt1a/Bt1b, AhyFuF/AHyFuR, EF-1H/EF-2T, H3-1a/H3-1b, H4-1a/H4-1b, ITS FuR/ITS FuF, ITS-1/ITS-2, ITS-1/ITS-4, ITS-4/ITS-5 (Fig. 2b), indicating that the order according to which it generates a concatenated sequence containing gaps, is considerably susceptible to change in the order of the sequences that make up an arrangement preventing the formation of reported clusters [16], [17].

As can be seen in Fig. 2, the trees do not have a topology supported by bootstrap values and the clusters were inconsistent regarding previous reports for evolutionary relationships of the genus *Fusarium* species [14], [16], [17] . This is because the likelihood-mapping analysis was performed to validate the use of the sequences in a phylogenetic analysis, but not to give carriers to the internal branches of the trees [10], [15], therefore considered valid sequences for phylogenetic analyzes, although the concatenated sequences are susceptible to generate different results depending on the arrangement that compose.

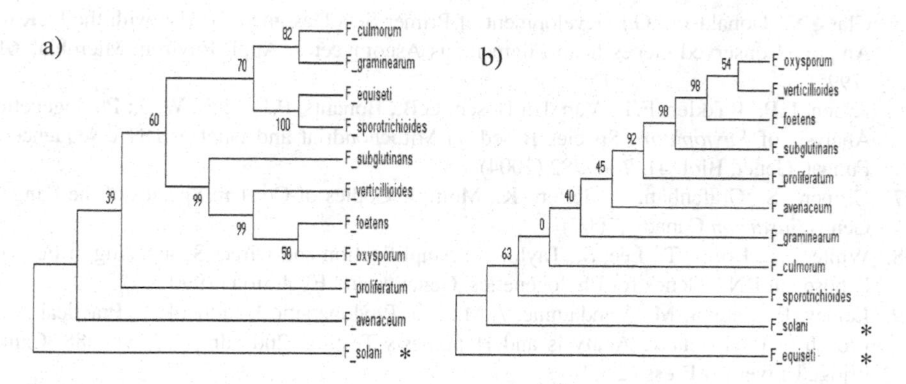

Fig. 2. Evolutionary histories inferred from Neighbor-Joining algorithm, a) tree generated using the arrangement mentioned in Fig. 1a, b) tree generated using the arrangement mentioned in Fig. 1b, Bootstrap consensus tree from 1000 replicates analyzing 3274 positions, reference species (*F. solani*and *F. equiseti)* shown with an asterisk (*).

4 Conclusions

The aligned sequences containing gaps, biologically related to the hypothesis of genetic transformation events, are crucial in the evolutionary process [4], but these gaps, are excluded in phylogenetic analyzes altering the possible biological interpretations [1]. With the results is established that concatenated sequences that maintain alignment gaps are valid for phylogenetic analyzes, but do not possess the necessary mathematical support to generate specific topologies and well established to support evolutionary relationships of the genus *Fusarium* reported in previous studies [16], [17].

References

1. Yli-Mattila, Y., Mach, R.L., Alekhina, I.A., Bulat, S.A., Koskinen, S., Kulling-Gtadigner, C.M., Kubicek, C.P., Klemsdal, S.S.: Phylogenetic Relationship of *Fusarium langsethiae* to *Fusarium poae* and *Fusarium sporotrichioides* as Inferred by IGS, ITS, β-tubulin Sequences and UP-PCR Hybridization Analysis. International Journal of Food Microbiology 95, 267–285 (2004)
2. Stenglein, S.A., Rodriguero, M.S., Chandler, E., Jennings, P., Salerno, G.L., Nicholson, P.: Phylogenetic Relationship of *Fusarium poae* Based on EF-1α and mtSSu Sequences. Fungal Biology 114, 96–106 (2010)
3. Shi, J., Blundell, T.M., Mizuguchi, K.: FUGUE: Sequence-structure Homology Recognition Using Environment-specific Substitution Tables and Structure-dependent Gap Penalties. J. Mol. Biol. 310, 243–257 (2001)
4. Scott, J., Chakraborty, S.: Multilocus Sequence Analysis of *Fusarium pseudograminearum* Reveals a Single Phylogenetic Species. Mycological Research 110, 1413–1425 (2006)

5. Glass, N., Donaldson, G.: Development of Primer Sets Designed for Use with the PCR to Amplify Conserved Genes from Filamentous Ascomycetes. Appl. Environ. Microbial. 61 (1995)
6. Kroon, L.P., Bakkler, F.T., Van den Bosch, G.B., Bonants, P.J., Flier, W.G.: Phylogenetic Analysis of *Phytophtora* Species Based on Mitochondrial and Nuclear DNA Sequences. Fungal. Genet. Biol. 41, 766–782 (2004)
7. Gilmore, S., Grafenhan, T., Seifert, K.: Multiple Copies of Cox1 in Species of the Fungal Genus *Fusarium* Canada (2007)
8. White, T.J., Bruns, T., Lee, S., Taylor, J.: Amplification and Direct Sequencing of Fungal Ribosomal RNA Genes for Phylogenetics. Genetics and Evolution (1990)
9. Lemey, P., Salemi, M., Vandamme, A.M.: The Phylogenetic Handbook, A Practical Approach to Phylogenetic Analysis and Hypothesis Testing, 2nd edn., pp. 196–388. Cambridge University Press (2009)
10. Thompson, J.D., Higgins, D.G., Gibson, T.: CLUSTALW: Improving the Sensitivity of Progressive Multiple Sequence Aligment through Sequence Weighting, Position-specific Gap Penalties and Weight Matrix Choice. Nucleic Acid Research 22, 4673–4680 (1994)
11. Schmidt, H.A., Strimmer, K., Vingron, M., Von Haeseler, A.: TREE-PUZZLE: Maximum Likelihood Phylogenetic Analysis Using Quaetets and Parallel Computing. Bioinformatics 18, 502–504 (2002)
12. Saitou, N., Nei, M.: The Neighbor-Joining Method: A New Method for Reconstructing Phylogenetic Trees. Molecular Biology and Evolution 4, 406–425 (1987)
13. Tamura, K., Peterson, D., Peterson, N., Stecher, G., Nei, M., Kumar, S.: MEGA5: Molecular Evolutionary Genetics Analysis Using Maximum Likelihood, Evolutionary Distance, and Parsimony Methods. Molecular Biology and Evolution 28, 2731–2739 (2011)
14. Felsenstein, J.: Confidence Limits on Phylogenies: An Approach Using the Bootstrap. Evolution 39, 783–791 (1985)
15. Strimmer, K., Von Haeseler, A.: Likelihood-mapping: A Simple Method to Visualize Phylogenetic Contents of a Sequence Aligment. Proc. Nat. Acad. Sci. 94, 6815–6819 (1997)
16. Watanabe, M., Yonezawa, T., Lee, K., Kumagai, S., Sugita-Konishi, Y., Goto, K., Hara, K.: Molecular Phylogeny of the Higher and Lower Taxonomy of the *Fusarium* Genus and Differences in the Evolutionary Histories of Multiple Genes. BMC Evolutionary Biology 11, 322–338 (2011)
17. Sarver, B.A., Ward, T.J., Gale, L.R., Broz, K., Kistler, C., Aoki, T., Nicholson, P., Carter, J., O'Donnell, K.: Novel *Fusarium* Head Blight Pathogens from Nepal and Louisiana Reveled by Multilocus Genealogical Concordance. Fungal Genetics and Biology 48, 1096–1107 (2011)

False Positive Reduction in Automatic Segmentation System

Jheyson Vargas[1], Jairo Andres Velasco[1], Gloria Ines Alvarez[1],
Diego Luis Linares[1], and Enrique Bravo[2]

Pontificia Universidad Javeriana Cali, Universidad del Valle, Cali, Colombia
{jairov,jheysonv,galvarez,dlinares}@javerianacali.edu.co,
enrique.bravo@correounivalle.edu.co

Abstract. An application has been developed for automatic segmentation of Potyvirus polyproteins through stochastic models of Pattern Recognition. These models usually find the correct location of the cleavage site but also suggest other possible locations called false positives. For reducing the number of false positives, we evaluated three methods. The first is to shrink the search range skipping portions of polyprotein with low probability of containing the cleavage site. In the second and third approach, we use a measure to rank candidate locations in order to maximize the ranking of the correct cleavage site. Here we evaluate probability emitted by Hidden Markov Models (HMM) and Minimum Editing Distance (MED) as measure alternatives. Our results indicate that HMM probability is a better quality measure of a candidate location than MED. This probability is useful to eliminate most of false positive. Besides, it allows to quantify the quality of an automatic segmentation.

1 Introduction

The viruses of the family Potyviridae are single stranded RNA having a sequence that contains a single ORF that encodes a polyprotein. This polyprotein is processed by three virus-encoded proteinases to form ten mature proteins [1] called: P1, HC-Pro, P3, 6K1, CI, 6K2, VPg, NIa, NIb, CP. The places where a polyprotein is cut to originate mature proteins are called cleavage sites.

The prediction of cleavage sites allows the isolation of specific segments, the annotation of new sequences, and the comparison of all these findings with existing databases such as GenBank and PDB. Our previus work has focused on the evaluation of multiple modeling techniques to build an automatic segmentation tool that can predict cleavage site using information from primary structure only.

We have developed an automatic segmentation system based on *Hidden Markov Models* (HMM) [4], which commonly suggests several possible options for the location of a cleavage site. The purpose of this article is to show how to decrease the number of false positives, maximizing the ability to select the right location for each site. First, we briefly describe the segmentation system. Then, three independent approaches to reduce false positives and distinguish them of

L.F. Castillo et al. (eds.), *Advances in Computational Biology*,
Advances in Intelligent Systems and Computing 232,
DOI: 10.1007/978-3-319-01568-2_15, © Springer International Publishing Switzerland 2014

the real site are presented. Afterwards, we introduce experimental results of applying each method and an analysis of them. Finally, we propose the conclusions of this work.

2 Segmentation System

The proposed segmentation system takes as input an unsegmented polyprotein of Potyvirus for which we want to find their cleavage sites. The expected output of the system is a collection of tuples showing the locations of cleavage sites and their names.

There are two stages for the segmentation system: *training-time* and *runtime*. Training-time corresponds to the stage where the system is still under construction, the parameters of their internal models are being estimated. Runtime corresponds to the stage when the segmentation system has been assembled and is ready for use segmenting polyproteins.

Our automatic segmentation software is built using multiple classifiers based on HMM. We assemble a classifier for each of the nine cleavage sites recognized in Potyviruses. Each classifier is used to determine the sites in amino acid sequence with high probability to be cleavage sites. Each classifier consists of several HMM and a voting system to decide if a given location is a suitable cleavage site.

We construct two datasets: the training dataset: which is used to train the HMM, and testing dataset: which is used to measure the generalization ability achieved by the trained models. The HMM are trained using the parameter estimation algorithm Baum-Welch [5].

In order to segment a complete polyprotein at runtime, we must slide a window across the sequence to get all samples from it. Then, samples are evaluated one by one through the classifiers and thus we obtain samples having high probability of containing the cleavage site. The location of the samples in the original sequence indicates the location of potential cleavage sites predicted by the classifier.

3 Methods

Because the prediction system can propose multiple locations for the same cleavage site, we propose three heuristics to discard false positive keeping the most probable locations of the actual site. We try several approaches using available information about tagged sequences: statistical information, distance measurement and probability. The rest of the section describes the most promising approaches.

3.1 Search Range Stretching

Since the order of appearance of the segments in the polyprotein is usually the same, each cleavage site typically appears in a specific region of the sequence. Therefore, it is possible to restrict the search range for each site saving computing time and removing some false positives.

Table 1. Minimum and maximum location for each cleavage site found into tagged sequences. Computed initial and final location of the resulting search range.

Cleavage Site	Minimun	Maximum	Initial Position	Final Position
P1—HC-Pro	211	745	137	819
HC-Pro—P3	652	1147	621	1178
P3—6K1	704	1499	648	1555
6K1—CI	1065	1551	965	1651
CI—6K2	1386	2194	1248	2296
6K2—VPg	1572	2247	1549	2270
VPg—NIa-Pro	1806	2439	1783	2462
NIa-Pro—NIb	2147	2682	2124	2732
NIb—CP	2534	3203	2482	END

We extract the real location of cleavage sites for 445 tagged sequences. Then, we calculate the minimum and maximum location for each cleavage site (see Table 1) and the length of each segment. We calculate the average length and standard deviation of the length for each segment. Based on these informations we establish a search range using the minimum and maximum locations. To reduce the probability that the correct cleavage site gets excluded, we broaden the search range subtracting the standard deviation from minimum location to obtain a new minimum value. Similarly, we add the standard deviation to maximum location to obtain a new maximum value. Resulting search ranges are presented in Table 1.

3.2 Output Sorting by Minimum Editing Distance

The *Minimum Editing Distance* (MED) [6] algorithm measures a distance between the candidate sample and the most common pattern for the cleavage site called the reference sample. There is a reference sample for each cleavage site.

For example, if we seek the cleavage site P1—HC-Pro, we must compare all candidate windows with the reference sample. Because the samples compared have the same length, the penalty in the MED algorithm comes mainly from the substitution component. However, there are substitutions between amino acids that can be penalized less severely than others. The amino acids are classified in different groups according to the properties of their side chain. One amino acid may belong to several amino acid groups. Therefore, the criteria for determining the penalty of the distance measure is based on the number of common groups between candidate sequence and the corresponding reference sample. If the number common groups is 0, the penalty is 1. If there is 1 group in common, the penalty is 0.5. Otherwise, if number common groups is greater than 1, then penalty is found as $0.5 - \frac{0.5n}{9}$ where n indicates the amount of common groups.

This criterion is used to sort candidates with respect to their distance to the reference. A short distance is understood as a high confidence on the candidate.

So we can sort candidates in ascendent order and the first candidate is assumed to be the best.

3.3 Output Sorting by HMM Probability

This approach is based in a feature of our classifiers based on HMM, they compute the probability of the existence of a cleavage site in a given sample. So, this probability may be used to compare candidates. A high probability of acceptance means a high confidence on the candidate. Sorting candidates in descending order, the first one should be the best.

4 Results

To compare the effectiveness of sorting orders, we applied both measures: distance and probability to the segmentation of 445 complete polyproteins. The segmentation system produces a list of candidate locations for each cleavage sites. Those lists are ordered with both criteria. Then, we search the real location for each site in each sequence, since the actual location appears in the candidates list almost always. Figure 1 and figure 2 shows the result of this task for one sequence. In figure 1 distance is the ordering criterion, and probability in figure 2. We can see that the length of candidates lists is variable. Besides, according to the ordering criterion the real location may be next to the top or far of it.

Sorting by HMM probability

Position	P1/HC-PRO	HC-PRO/P3	P3/6K1	6K1/CI	CI/6K2	6K2/VPG	VPG/Nia-Pro	Nia-Pro/Nib	Nib/CP
1.	308	766	1116	1168	1803	1856	2049	2289	2807
2.	309	453	876	1164	1330	1513	2044	2250	2803
3.	288	456	1117	1166	1401	1517	2045	2253	2810
4.	302	642	1140	1167	1715	1519	2047	2292	2805
5.	383	767		1169	1774	1853	2048	2255	2806
6.				1165	1436	1854	2043	2290	2808
7.						1855	2050	2291	2809
8.						1518	2046	2288	2804
9.						1516		2293	2811
									2812

Real Location

Candidates List

Fig. 1. Results of candidate list sorting by HMM probability

Once a real location is found, we consider its position in the ordered list as a measure of the effectiveness of the ordering criteria and as a bound to the number of false positive that we must maintain to avoid discarding the real location.

Given a cleavage site, we calculate the fairest position of the real location in the ordered candidates list for each sequence. This number is an upper bound of the number of candidates we should not eliminate in order to guarantee that we will not eliminate the real location. Table 2 shows these bounds calculated from

Sorting by MED

Position	P1/HC-PRO	HC-PRO/P3	P3/6K1	6K1/CI	CI/6K2	6K2/VPG	VPG/Nia-Pro	Nia-Pro/Nib	Nib/CP
1.	309	453	876	1164	1330	1513	2044	2289	2807
2.	288	456	1116	1166	1401	1517	2045	2250	2803
3.	302	642	1117	1167	1715	1519	2047	2253	2805
4.	383	767	1140	1168	1803	1853	2048	2255	2806
5.	308	766		1169	1774	1854	2043	2290	2808
6.				1165	1436	1855	2050	2292	2810
7.						1856	2049	2291	2809
8.						1518	2046	2288	2804
9.						1516		2293	2811
									2812

Real Location

Candidates List

Fig. 2. Results of candidate list sorting by Minimum Editing Distance

lists ordered with each criterion. Notice that probability ordering leads to shorter candidate lists which containing the actual location. It means that probability is a good measure to distinguish a good candidate from a false positive. In order to evaluate more tightly this hypothesis, we count how many times the actual location equals a given element into sorted candidates list using our sequence set. Table 3 shows the results obtained.

Notice that more than 90% of the cleavage sites are predicted into the three first options in candidates list, it allows to present to biologists a few possibilities highly probable for the location of the cleavage site.

Table 2. Maximum position for each actual cleavage site and criterion into sorted list of candidates. **S1**:P1—HC-Pro, **S2**:HC-Pro—P3, **S3**:P3—6K1, **S4**:6K1—CI, **S5**:CI—6K2, **S6**:6K2—VPg, **S7**:VPg—NIa-Pro, **S8**:NIa-Pro—NIb, **S9**:NIb—CP.

Criteria	S1	S2	S3	S4	S5	S6	S7	S8	S9
Minimum Editing Distance	19	6	8	8	7	19	15	41	43
Probability	6	5	3	5	5	5	10	16	27

Table 3. Match count for each position of candidate list using average probability from HMM classifiers. Columns means cleavage sites. Rows represent index into candidate list. **S1**:P1—HC-Pro, **S2**:HC-Pro—P3, **S3**:P3—6K1, **S4**:6K1—CI, **S5**:CI—6K2, **S6**:6K2—VPg, **S7**:VPg—NIa-Pro, **S8**:NIa-Pro—NIb, **S9**:NIb—CP.

Positions	S1	S2	S3	S4	S5	S6	S7	S8	S9
1	358	416	420	410	419	404	221	324	208
2	16	10	4	7	4	11	149	80	185
3	22		3	1	4	1	26	3	5
4	2	1		3	1	1	4	3	6
5		1		2	1	1	2	7	8
6	1						6	1	9

The tables 2 and 3 show that probability is not only a good criterion for ordering of candidates but also a measure that allows to discover the real location among other candidates. Even in cases where the best probability candidate does not match with the real location, candidate usually is near to the actual position. We use these results to suggest to biologists the most confident segmentation, choosing the first element of the candidate list for each site and putting them together.

5 Conclusions

Bounding of searching ranges saves computational cost and limits the number of false positives. Besides, using a measure of the quality of candidate locations allows to trim most of false positive without eliminate the real location from the list. Probability of acceptance is a very accurate criterion and it allows not only to purge candidates lists but also to propose a high quality automatic segmentation for a polyprotein.

Finally, the product of the probabilities of each prediction in the segmentation is a quantitative measure of the quality of the segmentation as a whole. This additional information may be useful to biologists when using the segmentation system.

Acknowledgements. This work is funded by Departamento Administrativo de Ciencia, Tecnología e Innovación de Colombia (COLCIENCIAS) under grant project code 1251-521-28290.

References

1. Nicolas, O., Laliberté, J.F.: The complete nucleotide sequence of turnip mosaic potyvirus rna. Journal of General Virology 73(Pt 11), 2785–2793 (1992)
2. von Heijne, G.: Patterns of amino acids near signal-sequence cleavage sites. Eur. J. Biochem. 133(1), 17–21 (1983); PubMed: 6852022
3. Li, B.Q., Cai, Y.D., Feng, K.Y., Zhao, G.J.: Prediction of protein cleavage site with feature selection by random forest. PLoS ONE 7(9), e45854 (2012) doi:10.1371/journal.pone.0045854, PubMed Central:PMC3445488, PubMed:23029276
4. Rabiner, L.: A tutorial on hidden markov models and selected applications in speech recognition. Proceedings of the IEEE 77(2), 257–286 (1989)
5. Baum, L.E., Petrie, T., Soules, G., Weiss, N.: A Maximization Technique Occurring in the Statistical Analysis of Probabilistic Functions of Markov Chains. The Annals of Mathematical Statistics 41(1), 164–171 (1970)
6. Wagner, R.A., Fischer, M.J.: The string-to-string correction problem. J. ACM 21(1), 168–173 (1974)

Photosynthesis Thermodynamic Efficiency Facing Climate Change

Víctor Alonso López-Agudelo, Julián Cerón-Figueroa, and Daniel Barragán

Escuela de Química, Facultad de Ciencias, Universidad Nacional de Colombia,
Bloque 16 Oficina 413, Calle 59A No 63-20, Medellín, Colombia
viclopezag@gmail.com, dalbarraganr@unal.edu.co

Abstract. A mathematical model that describes the oscillatory dynamic experimentally observed in the volumetric flows of CO_2/O_2 during photosynthesis is used in order to study the response of the photosynthetic process to changes in the external temperature. The model allows modeling steady, oscillatory and damped transitions between states, in relation the flows of matter and the substrate concentrations, but in order to study the effect of temperature, we added the energy balance equation to the model and we took the entire photosynthetic process to the scale of a reactor chloroplast. Variation in external temperature is carried out in different ways and. in order to analyze the photosynthetic model's response to thermal changes; we choose the variation in the generation of entropy as the second law criteria. Results show that entropy generated during the heating process is specific to the way it's carried out and that the system reacts more efficiently in response to a Fourier heating.

Keywords: photosynthesis, entropy production, climate change.

1 Introduction

Plants can carry out photosynthesis at wide temperature intervals, for example between 0°C and 30°C for plants adapted to cold temperatures [1] and between 15°C and 45°C for plants in arid environments such as deserts [2]. However, for most of the biomass available in our planet, optimal temperature for the photosynthetic process is between 20 and 35 Celsius. The photosynthesis efficiency is affected by the saturation of sunlight, the composition of atmospheric gases and temperature [3]. Climate change and gases from greenhouse effect are considered destabilizing factors for ecosystems with adverse future consequences over diversity and the available biomass [4-5]. Facing the complexity of photosynthesis, an experimental work is carrying out with the purpose of understanding the main regulatory steps of this process in C3 and C4 type of plants at a molecular level, in order to be able to design alternatives that counter adverse external processes that affect the plant's photosynthetic performance, as has been suggested for the catalytic activity of RuBisCO enzyme through genetic modifications [6]. Temperature is a control parameter of productivity and distribution of the photosynthetic organisms in our planet, which affects the assimilation capacity

L.F. Castillo et al. (eds.), *Advances in Computational Biology*,
Advances in Intelligent Systems and Computing 232,
DOI: 10.1007/978-3-319-01568-2_16,

of RuBisCO in plants, electron transport chain and the metabolic capacity of the Calvin's cycle, and regeneration of inorganic phosphate [7], these steps are determinant for the distribution, quantity and quality of the biomass. Parallel to the experimental work carried out in the study of the effect of temperature in photosynthesis [8-9], it is important to work with mathematical models in order to contribute to the understanding of kinetic, dynamic and energetic aspects of the process [10-13].

In this work, we used a simplified model proposed by Dubinsky et al. [14] to study the photosynthesis' response to external changes in temperature in an interval between 298,15 K and 313,15 K carried out in different ways: isothermal change , linear heating, Fourier heating and periodical variation. In order to analyze the processes' response, we used as second law criteria the rate of generation of entropy due to irreversible processes that take place during photosynthesis [15-16], for this purpose, we added the energy balance equation to the mathematical model and scaled the entire process to the level of a reactor chloroplast in which processes of mass transfer, flow of heat and enzymatic reactions take place.

2 Model of Photosynthetic Oscillations

The model used is proposed by Dubinsky (2010) [14], which is mainly based on Roussel et al's [17] the experimental results, which prove the real existence of an oscillatory regime in *Nicotiana Tabacuma* leaves in dark phase of photosynthesis, specifically in the coupling between the process of CO_2 assimilation and photorespiration where RuBisCO enzyme acts as a switch between the two processes. This model allows studying different dynamic states of photosynthesis and includes only the step of CO_2 assimilation of the enzyme RuBisCO, an outflow and the subsequent consumption of sugars. The model equations are (Eq. 1-2),

$$dX/dt = (1/5)k_cXC - V_{out}(X/(X + K_{out})) \tag{1}$$

$$dC/dt = -k_cXC + k_d(C_i - C) \tag{2}$$

where X and C represent the molar concentration of carbohydrates and CO_2 respectively, V_{out} is the maximum rate of consumption of sugars, K_{out} is Michaelis-Menten's constant of the pseudo enzyme that models the total consumption of sugars, k_d is CO_2 diffusion coefficient from the external medium to the chloroplast, C_i is CO_2 concentration in the external medium and k_c is the kinetic constant of the CO_2 assimilation reaction (carboxylation of ribulose 1,5-biphosphate - RuBP -).

3 Methods

To include the effect of temperature on the model, the chloroplast is considered as an idealized open reactor, at constant volume, wherein the photosynthesis processes takes place as a mixture of homogenous reaction in liquid phase and with constant agitation, exchange of matter and energy through the walls, as shown in Fig.1. In order to include the energy balance in the reactor chloroplast, we considered enthalpy changes due to flow of CO_2, exergonic changes of RuBisCO reaction enthalpy ($\Delta_r H = -21.3kJ/mol$) enthalpy of enzyme-substrate binding ($\Delta_b H = -56.5\ kJ/mol$) and protonation enthalpy ($\Delta_p H = -30\ kJ/mol$); whereas for transfer of energy through the walls, Newton's law of cooling was used. The energy balance that involves all terms previously mentioned is shown in (Eq. 3),

$$\rho V c_p (dT/dt) = -F^{\alpha}(\sum C_i c_{pi})(T - T^{\alpha}) - \dot{Q} + (-\sum_i^N (\Delta_r H_i + \Delta_b H_i + \Delta_p H_i) r_i)V \quad (3)$$

The heat transfer coefficient, U, is taken as control parameter of the dynamic of the process. Temperature modifies the reaction kinetics by means of the kinetic constants k_c and V_{out}, which we consider to follow Arrhenius law as follows: $k_c(T) = k_{c,0}\, exp((Ea/R)(1/T^o - 1/T))$, $V_{out}(T) = V_{out,0} exp((Ea/R)(1/T^o - 1/T))$, where T^o is the reference temperature for the initial value of constant k_c y V_{out}. The effect of the variation of the external temperature on the photosynthesis model is evaluated by means of different temperature profiles in variable T^{α} of (Eq. 3), these variations are expressed as temperature functions, with lineal, exponential or Fourier's profiles, and periodic temperature values between 298 and 313 K. Functions used in order to model heat profiles or external temperature are: (a) linear: $T^{\propto}(t) = 0.1136\, t + 276.075$, (b) Fourier:$T^{\propto}(t) = 313 + (209.65 - 313) exp(-0.01t)$ and (c) sinusoidal: $T^{\propto}(t) = -7.5COS(\pi/12(t-1)) + 305.65$. To study the model's response to changes in the external temperature through the second law of thermodynamics, it is necessary to use the formalism of thermodynamics of non-equilibrium processes in order to calculate the rate of generation of entropy of the different stages of photosynthesis [15,16]: heat transfer (dS_{heat}/dt) and mass transport (dS_{mass}/dt) (Eq. 4-5), for Dubinsky et al's model,

$$dS_{heat}/dt = UA((T - T^{\alpha})^2/TT^{\alpha}) \quad (4)$$

$$dS_{mass}/dt = R[k_d(Ci - C)]\ln(C) \quad (5)$$

Global generation of entropy in the photosynthesis model is the sum of the contributions of each irreversible process, $\sigma = dS_{heat}/dt + dS_{mass}/dt$, and it is evaluated in relation to the external temperature, and taking the coefficient of heat transfer, U, as the control reference will allow us to analyze the answer of photosynthesis with thermodynamic efficiency criteria. Average generation of global enthalpy and by heat transport for oscillatory states is calculated using the equation: $< \sigma > = (1/\tau) \int_0^t T(dS_i/dt)dt$, with τ being oscillation period. Values of the equation's parameters (Eq. 1-2) were taken from [14] for Dubinsky's model, and the rest of the parameters of (Eq. 3) were estimated and adjusted to the scale of a reactor chloroplast, as follows: $C_i = 10\mu M$, $k_d = 2.5s^{-1}$, $K_{out} = 0.25\mu M$, $V_{out\_0} = 5\mu Ms^{-1*}$, $k_{c\_0} = 15\mu Ms^{-1*}$, $Ea = \frac{79.43kJ}{mol}$, $A = 1.88x10^{-13}m^2$, $V = 4.18x10^{-15}m^3$, $C_p = 7.5kJ/molK$, $C_{pi} = 0.037kJ/molK$, $\rho = 6.5x10^4 mol/m^3$. The initial temperature of the system is 302 K.

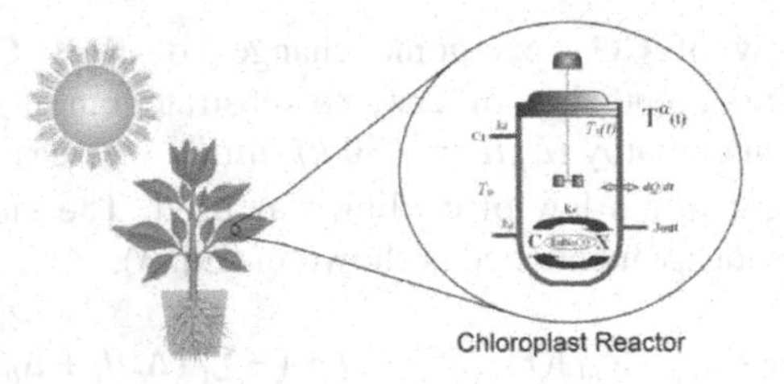

Fig. 1. Schematic representation of photosynthesis in an open reactor chloroplast

4 Results

Fig 2 shows the variation of the entropy generated in relation to the external temperature for two values of the coefficient of heat transfer U. These values of U were chosen based on a previous analysis of the model's time series, and they are characterized by taking the system to dynamic states of sustained (U=10 W/m^{2}K) and damped (U=40 W/m^{2}K) oscillations. Fig 2. shows the existence of a maximum that concurs with the transition of the dynamic range from states of damped to sustained oscillations which indicates in both cases that the external temperature favors the system's sustained oscillatory dynamic of the photosynthetic system, furthermore, it can be noted that this oscillatory dynamic is characterized by presenting low levels of global generation of entropy, which indicates a better use in photosynthesis's energetic processes.

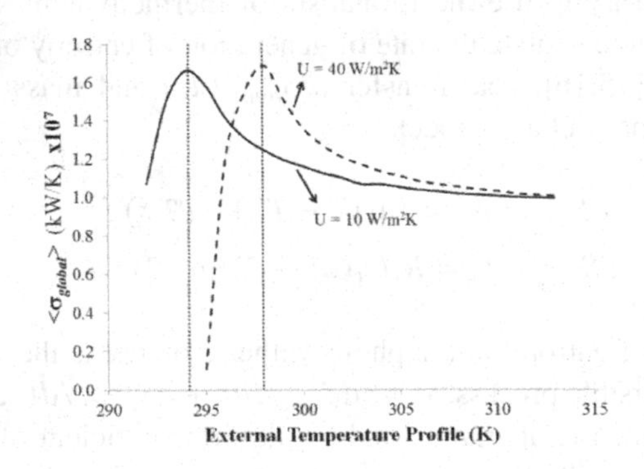

Fig. 2. Global generation of entropy in relation to isothermal increase of external temperature evaluated to two values of the parameter U: 10 W/m^{2}K and 40 W/m^{2}K

Fig. 3(a-c), show the effect of different heating profiles over the system, starting from a value of constant external temperature of 298K, a state characterized by the presence of a dynamic of damped oscillations for the flows of CO_2/O_2, with a U=38 W/m^{2}K. The evaluation interval of the linear and periodic heating profile was the same (193 – 325 seconds), whereas for Fourier's heating profile it was (193-1223.9 seconds), both intervals allow a variation of the external temperature from 298 K to 313 K. A transition or change of dynamic regime, from steady states to sustained oscillations in the point where external temperature is approximately 301K always takes place in these evaluation intervals of each heating profile. For the generation of entropy due to heat transfer (grey line Fig. 3(a-c)), we can see how its average starts decreasing as sustained oscillations begin and they increase the system's temperature. This effect is observed for the evaluation of the three heating profiles (Fig. 3(d)).

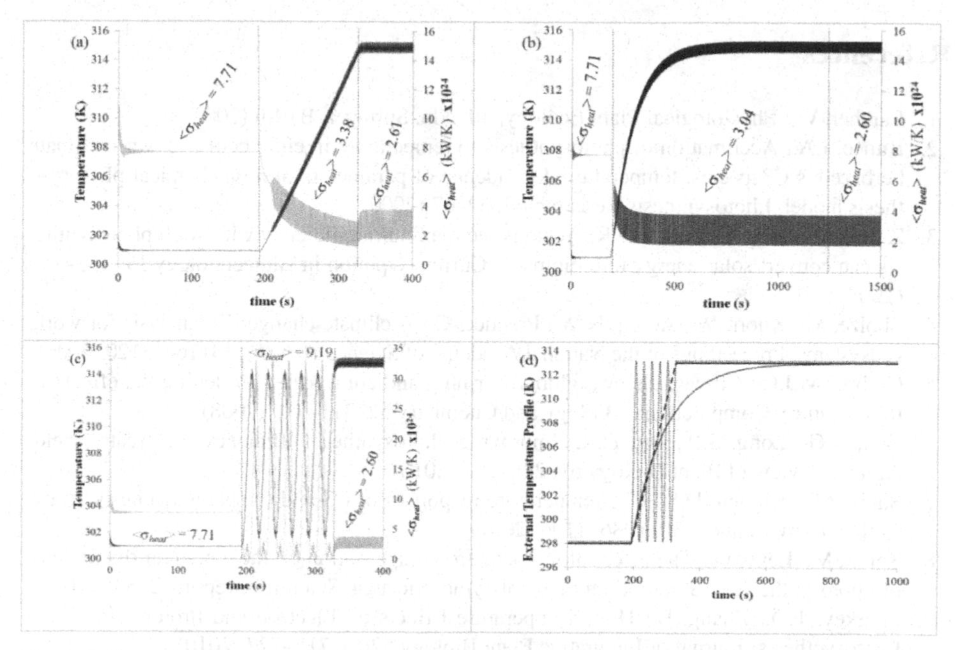

Fig. 3. Effect of different heating profiles (linear, Fourier and periodic) over the efficiency of the photosynthetic system (a), (b), (c), the dark line indicates the system's temperature, the grey line indicates the effect on the generation of entropy due to heat transport, the area between the dotted vertical lines indicates the time of the profile disturbance. (d) Linear, Fourier and sinusoidal periodic heating.

In Fig 3 we see important differences in the corresponding value of average generation of entropy due to transfer of heat in the heating interval, with the values: $<\sigma_{heat}>$ =3.38 kW/K (Fig. 3(a)) for linear heating, $<\sigma_{heat}>$ =3.04 kW/K (Fig. 3(b)) for Fourier heating, and $<\sigma_{heat}>$ =9.19 kW/K (Fig. 3(c)) for periodic heating. These values show that facing a natural heating, such as the one described by the phenomenological processes of heat transfer, such as Fourier's type, the system responds with a minor in the generation of entropy.

5 Conclusions

After using the model proposed by Dubinsky et. al. (2010) for the study of the effect of changes in the external temperature on photosynthesis at the scale of a chloroplast, with the purpose of modeling the global warming phenomenon, results indicate that in relation to the increase in the external temperature, transition towards dynamic oscillatory states helps upholding the performance of the photosynthetic process. In relation to fluctuating variations in the external temperature, this work indicates that, compared to other possible heating profiles, Fourier-type heating is the one that allows a better adaptation of the plant and therefore a better response of photosynthesis.

References

1. Larcher, W.: Physiological Plant Ecology, 4th edn. Springer, Berlin (2003)
2. Bunce, J.A.: Acclimation of photosynthesis to temperature in eight cool and warm climate herbaceous C3species: temperature dependence of parameters of a biochemical photosynthesis model. Photosynthesis Research 63, 59–67 (2000)
3. Zhu, X.G., Long, S.P., Ort, D.R.: What is the maximum efficiency with wich photosynthesis can convert solar energy into biomass? Current Opinion in Biotechnology 19, 153–159 (2008)
4. Sholze, M., Knorr, W., Arnell, N.W., Prentice, C.: A climate-change risk analysis for world ecosystems. Proceedings of the National Academy of Science 103(35), 13116–13120 (2006)
5. Crabbe, M.J.C.: Climate change, global warming and coral reefs: Modelling the effects of ttemperature. Computational Biology and Chemistry 32, 311–314 (2008)
6. Zhu, X.G., Long, S.P., Ort, D.R.: Improving photosynthetic efficiency for greater yield. Annual Review of Plant Biology 61, 235–261 (2010)
7. Sage, R.F., Kubien, D.S.: The temperature responses of C3 and C4 photosynthesis. Plant, Cell and Environment 30, 1086–1106 (2007)
8. Sorek, M., Levy, O.: The effect of temperature compensation on the circadian rhythmicity of photosynthesis in Symbiodinium, coral-symbiotic alga. Scientific Reports 2, 536 (2012)
9. Sharkey, T.D., Zhang, R.: High Temperature Effects on Electron and Proton Circuits of Photosynthesis. Journal of Integrative Plant Biology 52(8), 712–722 (2010)
10. Parent, B., Turc, O., Gibon, Y., Stitt, M., Tardieu, F.: Modelling temperature-compensated physiological rates, based on the co-ordination of responses to temperature of developmental processes. Journal of Experimental Botany 61(8), 2057–2069 (2010)
11. Riznichenko, G., Lebedeva, G., Demin, O., Rubin, A.: Kinetic mechanisms of biological regulation in photosynthetic organisms. Journal of Biological Physics 25(2), 177–192 (1999)
12. Vershubskii, A.V., Priklonskii, V.I., Tikhonov, A.N.: A mathematical model of electron and proton transport in oxygenic photosynthetic systems. Russian Journal of General Chemistry 77(11), 2027–2039 (2007)
13. Juretic, D.: Photosynthetic models with maximum entropy production in irreversible charge transfer steps. Computational Biology and Chemistry 27(6), 541–553 (2003)
14. Dubinsky, A.Y., Ivlev, A.A., Igamberdiev, A.U.: Theoretical Analysis of the Possibility of Existence of Oscillations in Photosynthesis. Biophysics 55, 55–58 (2010)
15. Kjelstrup, S., Bedeaux, D., Johannessen, E., Gross, J.: Non-equilibrium Thermodynamics for Engineers. World Scientific, Singapore (2010)
16. Kondepudi, D.: Introduction to Modern Thermodynamics. John Wiley, England (2008)
17. Roussel, M.R., Ivlev, A.A., Igamberdiev, A.U.: Oscillations of the internal CO_2 concentration in tobacco leaves transferred to low CO2. Journal of Plant Physiology 164(9), 1188–1196 (2007)

Thermogenesis Driven by ATP Hydrolysis in a Model with Cubic Autocatalysis

Julián Cerón-Figueroa, Víctor Alonso López-Agudelo, and Daniel Barragán

Escuela de Química, Facultad de Ciencias, Universidad Nacional de Colombia,
Bloque 16 Oficina 413, Calle 59A No 63-20, Medellín, Colombia
ejuliansjpd@gmail.com, dalbarraganr@unal.edu.co

Abstract. A cubic autocatalytic model is used in order to study thermogenesis of a metabolic process driven by hydrolysis of ATP, with the purpose of modeling temperature gradients measured experimentally in living cells that carry out the active transport of the Ca^{2+} ion. The model was taken to the scale of a living cell and the equation of energy balance was added in order to incorporate the effect of temperature in the process dynamic. A second law analysis was applied in order to determine the dynamic state and the value of the bifurcation parameters that favor efficiency of the system's thermogenic activity. Heat pulses generated with the model were studied in a 2-D array of 101*x*101 cells with radii of 50nm each. Results show that at distances inferior to the 300 nm of the cell with thermogenic activity, temperature gradients that range between 0.3K and 1K can be achieved, depending on the values of the bifurcation parameters, gradients that are in accordance with those measured experimentally.

Keywords: thermogenesis, calcium transport, ATP hydrolysis, entropy production.

1 Introduction

Metabolism of the skeletal muscle has a determining role in the regulation of the body's energetic consumption when in resting state and the homeostasis of the body temperature. There's still uncertainty regarding the mechanism responsible of the metabolic regulation and the thermogenic activity of the skeletal muscle, but experimental evidences indicate that the catalytic cycle of *Ca^{2+}-ATPases -SERCA-* plays an important role in these processes. Skeletal muscle *Ca^{2+}-ATPases* modulate free energy obtained from the hydrolysis of ATP, whether they dissipate it completely as heat, they use it for the active transport of ions or they store it as osmotic work in the Ca^{2+} gradients [1,2]. Calorimetric experiments show that the amount of heat dissipated through the SERCA *Ca^{2+}-ATPases* of the skeletal muscle is function of some membrane proteins and the ion gradients [2]. This thermodynamic ability of *Ca^{2+}-ATPases* -SERCA- has been with classic thermodynamics in order to estimate the temperature gradients that can be generated during the active transport [3] and to develop a non-equilibrium thermodynamic theory that allows understanding the phenomenological coupling between the different processes involved: hydrolysis-synthesis of ATP,

L.F. Castillo et al. (eds.), *Advances in Computational Biology*,
Advances in Intelligent Systems and Computing 232,
DOI: 10.1007/978-3-319-01568-2_17,

active transport of ions and flow of heat [4-5]. In this work we will use a dynamic model of enzymatic hydrolysis in order to model temperature gradients that have been measured experimentally.

2 Model with Cubic Autocatalysis

The enzymatic cubic autocatalytic model is based on Selkov's scheme proposed to explain oscillations during glycolysis, which supposes that enzyme *PFK* hydrolyzes *ATP* in the presence of λ molecules of *ADP* linked to its molecular structure and obey to the law of mass action (Eq. 1 – 3) [6],

$$S_f \leftrightarrow S \quad ; \quad v_1 = k_1 S_f \quad , v_{-1} = k_{-1} S \tag{1}$$

$$S + 2P \leftrightarrow 3P \; ; \quad v_2 = k_2 S P^2 \quad , v_{-2} = k_{-2} P^3 \tag{2}$$

$$P \leftrightarrow P_f ; \quad v_3 = k_3 P \quad , v_{-3} = k_3 P_f \tag{3}$$

where $S \equiv ATP$, $P \equiv ADP$, S_f and P_f are the concentrations of the inflow reactants. In the system that is being modeled, *the enzyme*, a cubic autocatalytic hydrolysis is carried out, $S + 2P \leftrightarrow 3P$, in a region of the intracellular space with permanent availability of reactants S_f and P_f which are administered at constant rate with kinetic constants k_1 y k_3 respectively. Species *S* and *P* exit the system towards the intracellular space at a rate determined by kinetic constants k_{-1} and k_{-3} respectively. A schematic representation of the model is shown in Fig.1 with the purpose of giving contexts to its meaning. Hydrolysis of *ATP* is exergonic with a $\Delta_r H$= -20.5 kJ/mol [3], this energy is released in the system and transferred to the surroundings as heat promoting the surge of temperature gradients. The effect of temperature in reaction kinetic is incorporated with the following Arrhenius equation, , $\frac{k_{i,T}}{k_{i,o}} = \exp\left(\frac{E_i}{R}\frac{(T-T^\circ)}{(TT^\circ)}\right)$, where T° is the reference temperature for the initial value of the constants $k_{i,o}$.The system's change in temperature is calculated from the energy balance of the interaction between the system, the enzyme, and its surroundings, the intracellular space, so the cubic autocatalytic model is described through the following equations (Eq. 4-6):

$$\frac{dS}{dt} = v_1 - v_{-1} - v_2 + v_{-2} \tag{4}$$

$$\frac{dP}{dt} = v_2 - v_{-2} - v_3 + v_{-3} \tag{5}$$

$$\frac{dT}{dt} = \left(\frac{UA}{mC_p}\right)(T^\bullet - T) + \frac{V\Delta_r H}{mC_p}[v_2 - v_{-2}] \tag{6}$$

In the energy balance (Eq. 6) only the reaction enthalpy of cubic autocatalysis is considered, for this reason, changes in enthalpy associated to flow of reactants and protonation of the pH buffer are not taken into account.

The values of the equation's parameters (Eq. 4-6) were established for a real biological system at the scale of an only living cell, as follows: k_1=8x10^{-3} ; k_{-1}=5x10^{-5} ;

k_2=2.5x10^{-2} ; k_{-2}=2x10^{-2} ; k_3=1x10^{-3} ; k_{-3}=9x10^{-3} ; E_{a1}=2x10^{4} J mol^{-1} ; E_{a-1}= 25x10^{3} J mol^{-1} ; E_{a2}=8x10^{4} J mol^{-1} ; E_{a-2}=81x10^{3} J mol^{-1} ; E_{a3}=2x10^{4} J mol^{-1} ; E_{a-3}=25x10^{3} J mol^{-1} ; m=2.908x10^{-17} moles ; A=3.141x10^{-12} m^2 ; V=5.236x10^{-16} L ; Cp=75.31 J mol^{-1} K^{-1} ; $\Delta_r H_{(S+2P \rightarrow 3P)}$=-20.5 kJ mol^{-1}. The initial temperature of the intracellular space is $T^{\bullet}$=298.15 K.

Parameters S_f and P_f, and U, are control parameters that determine the dynamic states of the process. The model with cubic autocatalysis may exhibit a wide range of dynamic states: stable steady states, sustained oscillations and damped oscillations, and in all of them enthalpy is dissipated. For this reason, in order to decide which dynamic state the system must be in in order to study thermogenesis, we choose as criteria the generation of entropy during the process of enzymatic hydrolysis. The thermodynamic analysis is carried out by calculating generation of entropy due to each of the following irreversible processes [7]: chemical reaction $\left(\frac{dS_r}{dt} = RV\ln\left(\frac{v_2}{v_{-2}}\right)(v_2 - v_{-2})\frac{dS_r}{dt}\right)$, flow of reactants $\left(\frac{dS_f}{dt} = -RV(v_1 - v_{-1})\ln S - RV(v_{-3} - v_3)\ln P\frac{dS_f}{dt}\right)$ and transfer of heat $\left(\frac{dS_q}{dt} = \mathrm{UA}\frac{(T-T^{\bullet})^2}{TT^{\bullet}}\frac{dS_q}{dt}\right)$. The second law thermodynamic efficiency of a process is maximized in the direction in which the generation of entropy is minimized because of the irreversibilities, so the behavior of the total generation of entropy, $\sigma = \frac{\mathrm{dS_t}}{\mathrm{dt}} = \frac{dS_r}{dt} + \frac{dS_f}{dt} + \frac{dS_q}{dt}$, in relation to the control parameters of the process allows to carry out a first approach in order to establish the dynamic state thorough which the system takes optimal advantage of energy [8].

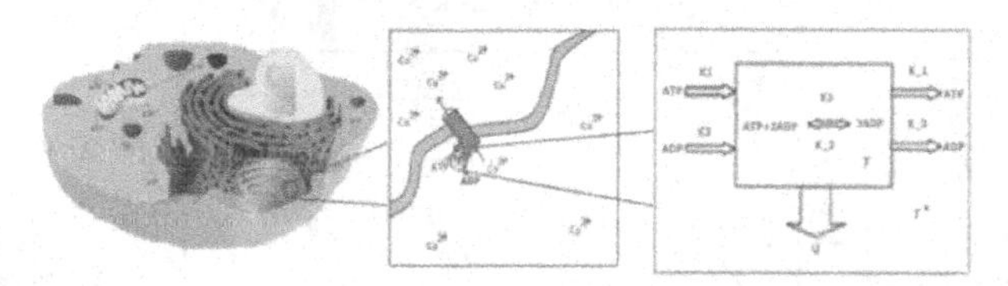

Fig. 1. Schematic representation of the cubic autocatalytic model: from molecular to mathematical cell biology

3 Methods

The system of differential equations proposed for the model with cubic autocatalysis (Eq. 1-3) was transformed to its reduced form by the following variable changes: $s^* = {}^{S}/_{S_f}$, $p^* = {}^{P}/_{S_f}$, $\tau = tk_1$, $T^* = {}^{T}/_{T^{\bullet}}$, $\sigma^* = \frac{\sigma}{k_1 RVS_f}$, the following system of dimensionless equations was obtained (Eq. 7-10):

$$\frac{ds^*}{d\tau} = 1 - \alpha s^* - \beta s^* p^{*2} + \gamma p^{*3} \tag{7}$$

$$\frac{dp^*}{d\tau} = \beta s^* p^{*2} - \gamma p^{*3} - \delta p^* + \varepsilon \tag{8}$$

$$\frac{dT^*}{d\tau} = \theta(1 - T^*) + \lambda(\beta s^* p^{*2} - \gamma p^{*3}) \tag{9}$$

$$\sigma^* = (\beta s^* p^* - \gamma p^{*3}) ln \frac{\beta s^* p^{*2}}{\gamma p^{*3}} + \theta' \frac{(T^*-1)^2}{T^*} - \{(1-\alpha s^*) ln s^* + (\varepsilon - \delta p^*) ln p^*\} \quad (10)$$

where the parameters of the previous equations (Eq. 7-10) are given by the following relations: $\alpha = \frac{k_{-1}}{k_1}$, $\beta = \frac{k_2}{k_1} S_f^2$, $\gamma = \frac{k_{-2}}{k_1} S_f^2$, $\delta = \frac{k_3}{k_1}$, $\varepsilon = \frac{k_{-3}}{k_1} \frac{S_f}{P_f}$, $\theta = \frac{UA}{mC_p k_1}$, $\theta' = \frac{UA}{VRk_1 S_f}$, $\lambda = \frac{VQ_r S_f}{mC_p T^\bullet}$. In (Eq. 10) the terms on the right corresponds to the generation of entropy by: chemical reaction, transfer of heat and the third one between brackets to the flow of reactants. All systems of differential equations were solved numerically using the *DLSODE* subroutine, in order to analyze time series and the model's bifurcation diagrams as a function of parameters S_f, P_f and U, then the generation of entropy was calculated and the modeling of the propagation of heat pulses was carried out. For the analysis of the propagation of heat through space and time, a 2-D array of 101*x*101 cells was designed, each cell having a radii of 50nm, in a space that's not continual but discreet, as the Fitzhugh-Nagumo type. Initially, the whole array is at a temperature of 298.15 K and in the cell located at the center, position (50,50), the reaction takes place over a Von Newman type of neighborhood.

4 Results

Fig. 2 shows the diagram of bifurcations for the system of equations 7-9.

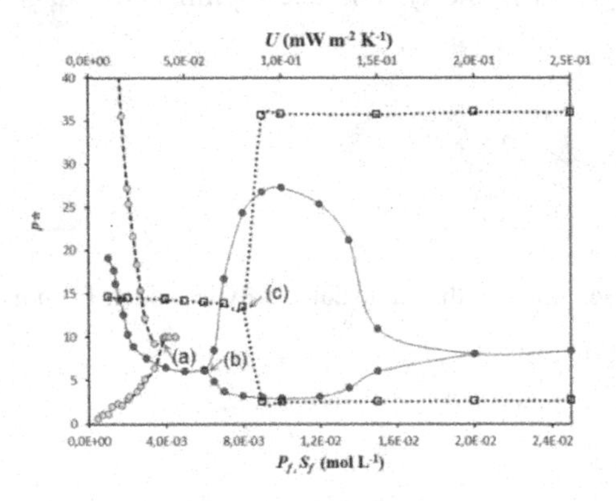

Fig. 2. Bifurcation diagrams of the cubic autocatalytic model, for parameters: (a) P_f, inflow of *ADP* ; (b) S_f , inflow of *ATP* ; (c) *U*, global heat transfer coefficient. p^* corresponds to dimensionless *ADP* concentration in the cell. The parameter values are: $S_f = 1x10^{-2}$ mol L^{-1} (for (b) and (c)); $P_f = 2x10^{-3}$ mol L^{-1} (for (a) and (c)) and $U = 0.1$ mW m^{-2} K^{-1} (for (a) and (b)).

The points of control parameters S_f, P_f and U for which the system makes the transition between stable steady states and oscillations are indicated in the curves of the figure, as follows: Fig 2(a) shows that when the parameter P_f is increased, the system switches from sustained oscillations (the two curves before the bifurcation point show the maximum and minimum values during the oscillation and they come together until they reach a stable value) to steady states, however, curves of Fig.2(b) and

2(c), show that when parameters S_f and U are increased, the system makes a transition from stable steady states to sustained oscillations. Steady states close to the bifurcation point are reached through damped oscillations, for any of the parameters. For this system model there is wide range of possible combinations between values of the bifurcation parameters, this allows having a set of steady and oscillatory states with different amplitude and frequency for each variable. With either of the two dynamic states heat waves that propagate through a medium can be generated; however, in order to choose one, the criteria used will be the generation of entropy due to irreversible processes. We compared generation of entropy between states in relation to the bifurcation parameters. Fig. 3 shows the thermodynamic transition between steady and oscillatory states for the model (Eq. 7-9) in relation to the parameter U. This result shows with second law criteria that oscillatory states are more efficient, as they generate less entropy. Based on this result, it has been decided to model the propagation of the heat generated by the system when it is in an oscillatory state.

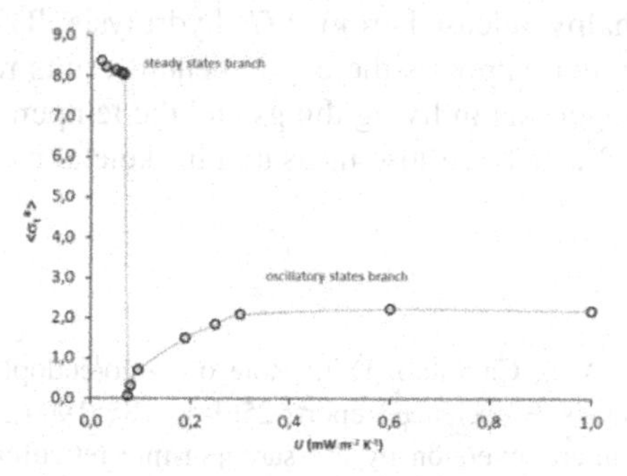

Fig. 3. Thermodynamic transition from the stable steady states to the sustained oscillatory states in the cubic autocatalytic model, with S_f=1.7x10^{-2} mol L^{-1} and P_f= 2x10^{-3} mol L^{-1}. The entropy production was computed with the equation, $<\sigma_t^* >= \frac{1}{\tau_r}\int_0^t T\left(\frac{d\sigma_t^*}{dt}\right) dt.$

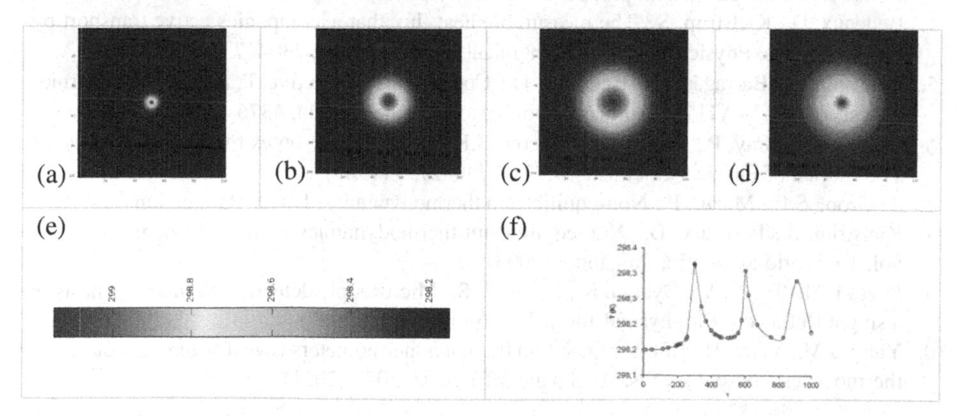

Fig. 4. Thermal wave propagation triggered by the cubic autocatalytic model. The sequence of snapshots corresponds to: (a) τ=602 ; (b) τ=674; (c)τ =755; (d) τ=827. The color temperature scale is showed in (e) and, the temperature sensed at 223.6 nm to the heat source is showed in (f). The parameter values are: S_f=1.7x10^{-2} mol L^{-1}, P_f= 2x10^{-3} mol L^{-1}; U=0.1 mW m^{-2} K^{-1}.

Fig. 4 shows different snapshots in time of the propagation of the heat wave generated by the hydrolysis model with cubic autocatalysis from a cell with radii of 50nm located at the center of a 2-D array of $101x101$ cells. In the dimensionless time scale, it can be observed that there is a difference in temperature gradients (see Figure 4 (e)) between Figure 4 (a) and Figure 4 (d) which is due to a change in the intensity of the oscillation in the model dynamics, as shown by the time series of Figure 4 (f) for a point located at 223.6 nm below the center of the array. Depending on the values used for the bifurcation parameters, temperature gradients between the cell located at the center and its neighbors at less than 1000 nm can be of up to 1K.

5 Conclusions

Based on the dynamic model with cubic autocatalysis at the scale of a real cell and with values of the parameters adjusted to physiological conditions heat waves were modeled from the enthalpy released from *ATP* hydrolysis. The result constitutes a computational evidence that supports the experimental results reported in the studies on the origins of thermogenesis in living things and the temperature gradients present at the cellular level [9,10], such as those measured in skeletal muscle cells.

References

1. de Meis, L., Arruda, A.P., Carvalho, D.P.: Role of sarco/endoplasmic reticulum Ca^{2+}-ATPase in thermogenesis. Bioscience Reports 25(3-4), 181–190 (2005)
2. Meis, L.D.: Energy interconversion by the sarcoplasmic reticulum Ca^{2+}-ATPase: ATP hydrolysis, Ca^{2+} transport, ATP synthesis and heat production. Anais da Academia Brasileira de Ciências 72(3), 365–379 (2000)
3. Kjelstrup, S., de Meis, L., Bedeaux, D., Simon, J.M.: Is the Ca^{2+}-ATPase from sarcoplasmic reticulum also a heat pump? European Biophysics Journal 38(1), 59–67 (2008)
4. Bedeaux, D., Kjelstrup, S.: The measurable heat flux that accompanies active transport by Ca^{2+}-ATPase. Physical Chemistry Chemical Physics 10(48), 7304–7317 (2008)
5. Kjelstrup, S., Barragán, D., Bedeaux, D.: Coefficients for Active Transport and Thermogenesis of Ca^{2+}-ATPase Isoforms. Biophysical Journal 96(11), 4376–4386 (2009)
6. Cook, G.B., Gray, P., Knapp, D.G., Scott, S.K.: Bimolecular routes to cubic autocatalysis. The Journal of Physical Chemistry 93(7), 2749–2755 (1989)
7. de Groot, S.R., Mazur, P.: Non-equilibrium thermodynamics. Dover Publications (1963)
8. Kjelstrup, S., Bedeaux, D.: Non-equilibrium thermodynamics of heterogeneous systems, vol. 16. World Scientific, Singapore (2008)
9. Suzuki, M., Tseeb, V., Oyama, K., Ishiwata, S.: Microscopic detection of thermogenesis in a single HeLa cell. Biophysical Journal 92(6), L46–L48 (2007)
10. Yang, J.M., Yang, H., Lin, L.: Quantum dot nano thermometers reveal heterogeneous local thermogenesis in living cells. ACS nano 5(6), 5067–5071 (2011)

Towards a Linked Open Data Model for Coffee Functional Relationships

Luis Bertel-Paternina[1], Luis F. Castillo[2], Gustavo Isaza[2], and Alvaro Gaitán-Bustamente[3]

[1] Universidad de Manizales,
Departamento de Ciencias e Ingeniería
lbertel@umanizales.edu.co
[2] Universidad de Caldas. Manizales (Colombia),
Departamento de Sistema e Informática
{luis.castillo,gustavo.isaza}@ucaldas.edu.co
[3] Centro Nacional de Investigación del Café – CENICAFÉ
alvaro.gaitan@cafedecolombia.com

Abstract. The development of technologies for massively disclose genetic information stored within cells generated the genomics revolution, where everyday exponentially growing databases with the description of the chemical structure of novel genes obtained with the automated nucleic acid sequencing. Each of these sequences is necessary to assign a function, in a process known as gene annotation comparing information of unknown sequence with sequences previously studied in the laboratory. This will identify candidate genes associated with traits of interest such as disease susceptibility or resistance, adaptation to the environment or animal or plant production. Given the volume of information, it is necessary that this be presented in a structured way so that such comparisons can be performed by computer agents quickly and it is open to the possibility of generating new knowledge by finding features and novel relationships between genes. This paper aims to present a bioinformatic application for functional relations search of transcriptome for coffee genetic material provided by the Coffee Investigation Center (CENICAFE), using the Bio2RDF biological database, applying new standards for LOD (Linked Open data) necessary to communicate effectively with other reference databases already operating under the scheme or Semantic Web.

Keywords: functional gene networks, Semantic Web bioinformatics, Linked Data, Bio2RDF.

1 Introduction

The cultivation of coffee is the source of income of 20 million people worldwide and generates a million direct jobs in Colombia [1]. The country is cultivated Coffea called Arabica that produces mild coffees, which is threatened by disease and pest problems, such as rust and bit, climate change, loss of biodiversity, and high

L.F. Castillo et al. (eds.), *Advances in Computational Biology*,
Advances in Intelligent Systems and Computing 232,
DOI: 10.1007/978-3-319-01568-2_18,

production costs. One response to these challenges is the generation of new coffee varieties that retain the best agronomic characteristics of traditional varieties, but which incorporate genetic advantages that allow them to overcome the drawbacks without further intervention from you growers. These advantages are represented physically genetic genes, which are the information units that are inherited between generations, and which chemically corresponds to nucleic acid strands that are stored in the cell, which are known as DNA.

For the process we have used, as reference annotation databases cured, among which stand *GenBank*, the database of genetic sequences, *Uniprot* (Universal Protein Resource) and *PDB* (Protein Data Bank), which is a central repository of sequences and functions of proteins, KEGG (Kyoto Encyclopedia of Genes and Genomes), which assigns a metabolic pathway genes, and InterPro, a meta-search protein structures to predict these function.

The goal of Linked data on the web is that you can share data easily structured in the same way they can share documents today. This term was coined in 2006 by Tim Berners Lee in his research on Linked Data Web architecture. The two basic elements that should be addressed are: using the RDF data model to publish structured data on the web and using RDF links to reference data from different origins. Applying these two principles can share data on the web, the common data are called Data Web or Semantic Web. Knowledge is used to infer the SPARQL query language which is applied to a set of triplets RDL to infer knowledge and in this case the different datasets used in bioinformatics reference to be found published in the WEB under the standard LOD.

2 Linked Open Data

The Linked Open Data (LOD) is a community project in the World Wide Web Consortium (W3C), which aims to expand "the network with a common data by publishing various open data sets as RDF on the Web and by establishing links between data items from different RDF data sources "[2], [3]. In this context, many biomedical databases have been made available (an open diagram based on cloud data is available online [4]). Many of these data sets are derived deBio2RDF, but there are also some that were built independently. Approaches based on Semantic Web for biomedical data integration have been proposed in some instances in recent years [5], [6], [7], [8]. In the biomedical field, a resource issue has been represented by Bio2RDF [9], a system for integrated access to a large number of biomedical databases through Semantic Web technologies RDF, ie for data representation and SPARQL (SPARQL Protocol and RDF Query Language) for queries. For this purpose, many databases have been converted to RDF by special scripts, called RDFizers, while some systems offer a viable format information and interfaces directly related to the system. An extension of this proposal is demonstrated in [10] and [11] which presents the portal Chem2Bio2RDF Linked Open Data (LOD) portal for systems chemical biology aiming for facilitating drug discovery. It converts about 25 different datasets on genes, compounds, drugs, pathways, side effects, diseases and RDF triples that links to other LOD bubbles, such Bio2RDF, LODD and DBPedia. The portal is based on D2R server and provides a SPARQL endpoint, but adds a few unique features such as RDF faceted, easy to use from SPARQL query generator, MEDLINE / PubMed cross validation service, and Cytoscape visualization track. More recent efforts like [12] present a portal that is aimed at developing community around linked biological data (LOD). This public space offers several services and collaboration infrastructure in order to stimulate the generation of activity in biological data consumption linked and therefore contributes to the implementation of the benefits of the Web of data in this area.

3 Materials and Methods

To design and implement the prototype we have used UP Open methodology as agile approach to software development, with only fundamental content provides a simplified set of artifacts, roles, tasks and guide work from an iterative process Software development is minimal, complete, and extensible.

The evolution of information representation methods to ontological models and semantic followed, in many cases, an empirical process and a process lacking Formal which evidences its life cycle, however, in recent years have been maturing some initiatives that tends to define methodologies and specifications in the context of engineering processes and software engineering knowledge to address the methodological component of the analysis and design of ontologies. For representing the ontology, we decided to use METHONTOLOGY [13], [14]. With the data of the coffee transcriptome annotations found by the Coffee Research Center of Colombia (Cenicafe) where to have used bioinformatics tools such as BLAST and Interproscan, the data is processed and taken three main components in order to perform searches through functional relationships linking through open linked data to biological databases publicly available. The results of the relationships between Cenicafé annotations and the data set of proteins Protein Data Bank (PDB), is one of the objectives of this study. The components used in the studio are detailed in table 1.

Table 1. Data used in functional search through open linked data

Type	Description
cen	Index into Cenicafe used for the classification of the coffee transcriptome sequences, an example would be: CEN396590
protein	Canonical name of the protein. The name is obtained by *Interproscan* eg: HYDROXYMETHYLTRANSFERASE
pdb	Link to resource Protein Data Bank, in this case used the link with http://bio2rdf.org/pdb:1fp2 Bio2RDF through, if used directly Protein Data Bank can access the resource http:/ / rdf.wwpdb.org/pdb/1fp2.

The information is organized in data triplets RDF (Resource Description Framework) and is stored in the Sesame repository, which is a database to store content of linked open data. The Sesame RDF repository provides an endpoint to query web requests through GET type and / or POST, the repositories are used by sparql query language or web services through REST-Full. The visualization of the results of queries to the Sesame repository tool was used *sgvizler* (https://code.google.com/p/

sgvizler/). Sgvizler is a tool created in javascript which in turn consists of other tools oriented data representation sparql query results. *Sgvizler* is very flexible when display sparql query results, the types of graphs that can be generated are the location coordinates maps, charts, bubble diagram, time lines, tables, lists.

4 Results and Discussion

The query results unfold figures through interconnected graph of the information defined in the RDF triples stored in *sesame*.

One of the basic query is the search for annotations that have relation with a specific protein, in this case the search is performed with the protein *aquaporin*. The query is performed in the repository of the resources associated Cenicafé through Pdb Bio2RDF.

```
PREFIX rdf: <http://www.w3.org/1999/02/22-rdf-syntax-ns#>
PREFIX rdfs: <http://www.w3.org/2000/01/rdf-schema#>
PREFIX owl: <http://www.w3.org/2002/07/owl#>
PREFIX xsd: <http://www.w3.org/2001/XMLSchema#>
PREFIX dc: <http://purl.org/dc/terms/>
PREFIX link: <http://bio2rdf.org/bio2rdf_resource:>
PREFIX cenicafe: <http://cenicafe.org/anotacion/cen/>
PREFIX recurso: http://bio2rdf.org/pdb:

SELECT ?cen ?proteina
where
{
?cen rdfs:label "aquaporin";
       rdfs:label ?proteina .
}
```

The results of the query are shown in Figure 1.

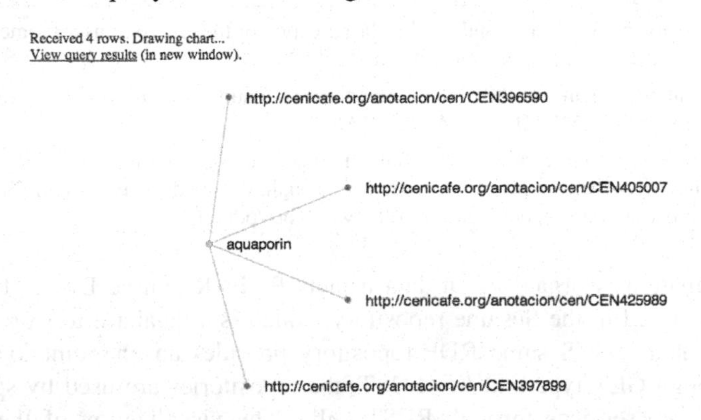

Fig. 1. Query result of the annotations associated with aquaporin protein

The following result is the search for proteins containing the term *kina*. The result can be seen in Figure 2.

```
select ?cen ?biologico ?pdb
where
{
        ?s rdfs:label ?biologico ;
        link:linkedToFrom ?pdb;
        dc:title ?cen .
        filter( regex(?biologico, "kina" , "i"))
}
```

Fig. 2. Kina Term Query

5 Conclusions

Using bioinformatics' processes semantic standards has reached a level of maturity that makes possible the discovery of genetic information derived from techniques of distributed computational intelligence. The technologies used in Linked Open Data let search for patterns, functional relationships from which the researcher will provide new data to be interpreted. Methodologies validated using semantic management contributed to develop an appropriate prototype for coffee semantic functional relationships.

References

1. Hartwick, E.: The Cultural Turn in Geography: A New Link in the Commodity Chain. In: Encounters and Engagements between Economic and Cultural Geography. GeoJournal Library, vol. 104, pp. 39–46 (2012)
2. Berners-Lee, T.: Semantic Web Road map (1998), `http://www.w3.org/DesignIssues/Semantic.html` (retrieved May 2012)
3. Linked Data-Connect Distributed Data across the web, `http://linkeddata.org/`
4. The Linking Open Data cloud diagram, `http://lod-cloud.net/`

5. Stephens, S., LaVigna, D., DiLascio, M., Luciano, J.: Aggregation of bioinformatics data using Semantic Web technology. Journal of Web Semantics 4, 216–221 (2006)
6. Dhanapalan, L., Chen, J.Y.: A case study of integrating protein interaction data using semantic web technology. Int. J. Bioinform Res. Appl. 3, 286–302 (2007)
7. Deus, H.F., Stanislaus, R., Veiga, D.F., Behrens, C., Wistuba II, Minna, J.D., Garner, H.R., Swisher, S.G., Roth, J.A., Correa, A.M., Broom, B., Coombes, K., Chang, A., Vogel, L.H., Almeida, J.S.: A Semantic Web management model for integrative biomedical informatics. 2008
8. Miles, A., Zhao, J., Klyne, G., White-Cooper, H., Shotton, D.: OpenFlyData: an exemplar data web integrating gene expression data on the fruit fly Drosophila melanogaster. J. Biomed. Inform 43, 752–761 (2010)
9. Belleau, F., Nolin, M.-A., Tourigny, N., Rigault, P., Morissette, J.: Bio2RDF: Towards a mashup to build bioinformatics knowledge systems. Journal of Biomedical Informatics 41, 706–716 (2008)
10. Chen, B., Ding, Y., Wang, H., Wild, D.J., Dong, X., Sun, Y., Zhu, Q., Sankaranarayanan, M.: Chem2Bio2RDF: A Linked Open Data Portal for Systems Chemical Biology. Web Intelligence, 232–239 (2010)
11. Chen, B., Dong, X., Jiao, D., Wang, H., Zhu, Q., Ding, Y., Wild, D.J.: Chem2Bio2RDF: a semantic framework for linking and data mining chemogenomic and systems chemical biology data. BMC Bioinformatics 11, 255 (2010)
12. García-Godoy, M.J., Navas-Delgado, I., Aldana-Montes, J.: Bioqueries: a social community sharing experiences while querying biological linked data. Published in Proceedings of the 4th International Workshop on Semantic Web Applications and Tools for the Life Sciences, SWAT4LS 2011, pp. 24–31. ACM, New York (2012) ISBN: 978-1-4503-1076-5
13. Corcho, O., Fernandez, M., Gomez-Perez, A., Lopez-Cima, Á.: Building Legal Ontologies with METHONTOLOGY and WebODE. In: Benjamins, V. (ed.) Law and the Semantic Web, pp. 142–157. Springer, Heidelberg (2005a) ISSN 0302-9743 (Print) 1611-3349 (Online)
14. Gómez-Pérez, A., Fernández-López, M., Corcho, O.: Ontological engineering, Springer Verlag as part of the Advanced Information and Knowledge Processing series (2004) ISBN 1-85233-551-3

Stability Analysis of Antimicrobial Peptides in Solvation Conditions by Molecular Dynamics

Daniel Osorio[1], Paola Rondón-Villarreal[2], and Rodrigo Torres[3]

[1] Escuela de Biología, Facultad de Ciencias, UIS
[2] Escuela de Ingeniería Eléctrica, Electrónica y Telecomunicaciones, Facultad de Ingenierías Físico-Mecánicas, UIS
[3] Escuela de Química, Facultad de Ciencias, Universidad Industrial de Santander (UIS), Cra 27 Calle 9 Bucaramanga, Colombia
daniel.osorio@correo.uis.edu.co, paitorv@gmail.com, rtorres@uis.edu.co
http://ciencias.uis.edu.co/gibim

Abstract. Cationic antimicrobial peptides are a family of highly homologous proteins conserved in all multicellular organisms with great potential as broad-spectrum antibiotics. In this paper, we analyze the stability of the 3D structure of three antimicrobial peptides reported in CAMP database, working in solvation conditions at three pHs and three different temperatures. We found that one of the tested peptides did not form stable three-dimensional structure. For this reason, it is not expected a bactericidal action *per se*. The other peptides showed an α-helix conformation under certain conditions evaluated.

Keywords: AMP, Antimicrobial peptides, Stability, 3D-Structure, Molecular Dynamics.

1 Introduction

Cationic antimicrobial peptides are a high conserved family of proteins having structural and physicochemical characteristics (charge, hydrophobicity, size, amino acid composition, etc.), which allow its antimicrobial activity [3]. These biological molecules are currently subject of research due to its potential use as a new generation of broad-spectrum antibiotics, because they show a near zero rate of resistance by bacteria [5]. These studies cover topics such as the design and processing of known peptide sequences, *in vitro* bioassay and *in vivo* analysis of the mechanism of action, 'docking' and interactions with molecular membrane by molecular dynamics. Due to the complexity in the formation and maintenance of the 3D structure of the peptides, which determines its antimicrobial activity [7], it is very important to predict conformation and folding of antimicrobial peptides. In this study, we analyzed the stability of the three dimensional structure of three antimicrobial peptides under solvation conditions. For this aim, we worked at both three pH values (5, 7 and 9) and temperatures (298K, 310K and 323K), carrying out simulations of peptides by molecular dynamics, and using as models two antimicrobial undecapeptides with sequences $KLKL_5K\text{-}NH_2$ and

L.F. Castillo et al. (eds.), *Advances in Computational Biology*, Advances in Intelligent Systems and Computing 232,
DOI: 10.1007/978-3-319-01568-2_19,

$RLKL_5RLK-NH_2$, which are derived from *sapecin β* reported by Alvares-Bravo et al. [2], and the peptide $FLPIPRPILLGLL-NH_2$ obtained by Xuequing et al. [9] from the venom of *Vespa magnifica* .

2 Materials and Methods

Three-dimensional models of peptide structures were obtained through the 3D-JIGSAW [4] website portal by selecting the model with the lowest free energy value (Fig. 1). The molecular dynamics simulations were processed in GROMACS [6] 4.6 GPU-MPI-SP package using the OPLS force field-AA/L, configured under isothermal and isobaric conditions at salt concentrations (NaCl) of 0.13 mM. Depending on the pH to be evaluated in the computer simulation, the amino acids were protonated or deprotonated using the application *pdb2gmx*, which is included in the GROMACS package. We set three stages in the simulation: energy minimization to achieve a potential energy of 1 (kJ/mol), the temperature stabilization (NVT) - pressure (NPT) and using a simulation time of 2ns. In order to assess the stability of the peptides, we evaluated for each simulation at the pH and temperature conditions studied, the root mean square deviation (RMSD) of the aligned proteins, and the progression of temperature, pressure and density in the simulated system. All simulations were render in *trjconv* program included in GROMACS package and visualized using VMD software and Radeon HD6750 graphic card.

3 Results and Discussion

In Figure 1 are shown simulations of the three peptides stabilized at expected isobaric and isothermal configurations.

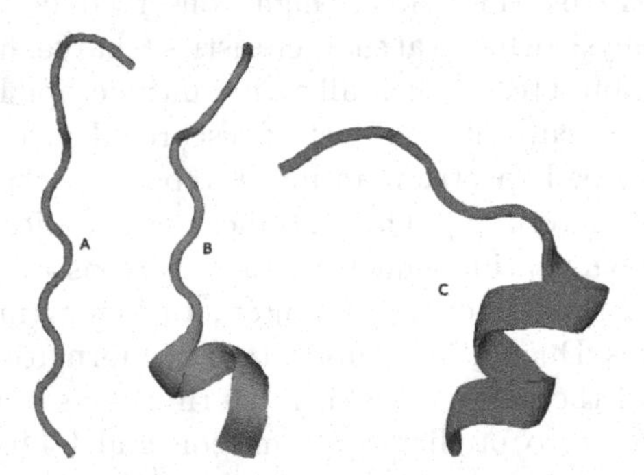

Fig. 1. Three-dimensional structures obtained by homology modeling through 3D-JIGSAW of A: Peptide with $KLKLLLLKLK-NH_2$ sequence B: Peptide with $RLKLLLLLRLK-NH_2$ sequence C: peptide derived from the venom of *Vespa magnifica* with $FLPIPRPILLGLL-NH_2$ sequence

The first peptide analyzed (KLKL$_5$KLK-NH$_2$) did not show secondary structure formation at any pH or temperature evaluated in this study (vacuum and solvated). This effect can be explained in terms of an structural unbalance, because we obtained high deviation (0.25 ± 1.5 nm) of RMSD compared with the original structure (Fig. 2A) at all pHs and temperatures evaluated.

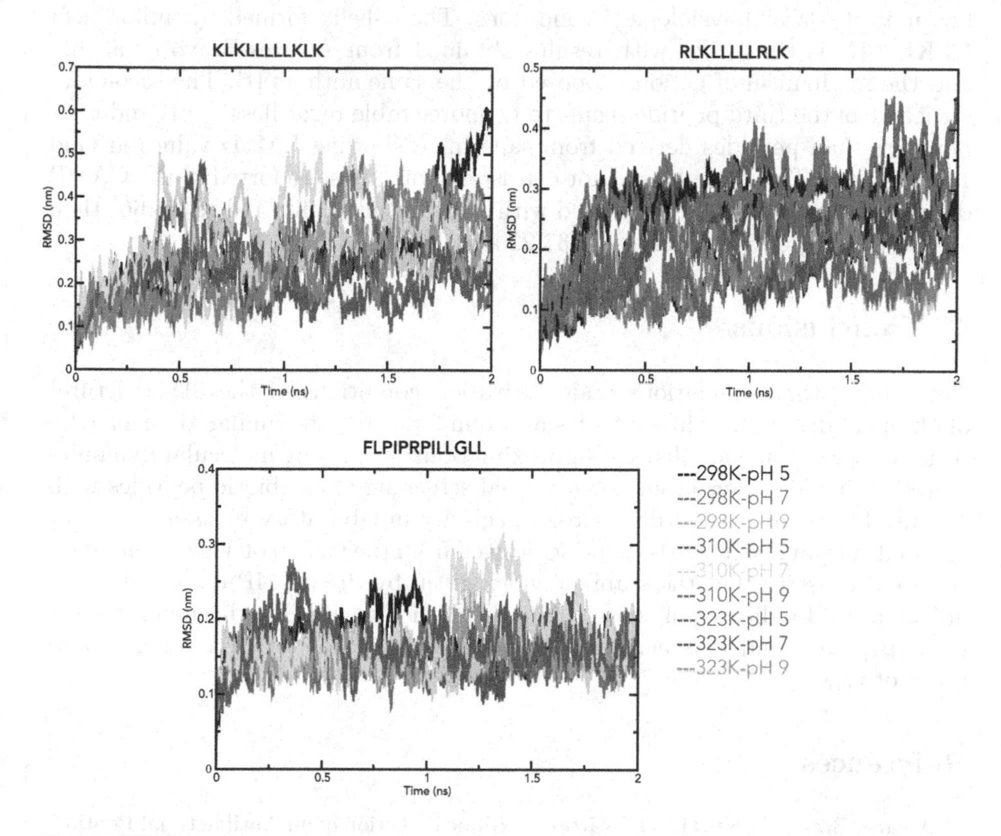

Fig. 2. RMSD of the three antimicrobial peptides simulated at three different values of pH and temperature.

Initially, all structures for the second peptide showed an alpha-helix formed by the amino acids LKL-NH$_2$ with a free energy of formation of -7.12 kcal/mol. This structure (in vacuum) was kept after the protonation or deprotonation in all pHs studied, which indicates a conformational stability of the peptide. By modeling at 298K and 310K, and at pH5 and pH7, the α-helix added the R neighbor amino acid to its three dimensional structure, while simulations with pH 9 to 298K and 323K showed a denaturation of the peptide structure. The third modeled peptide obtained and kept (in solvation and vacuum at all pH and temperature simulated) an α-helix structure, which was formed by the amino acids RPILLGL-NH$_2$ with free energy of formation of -7.87 kcal/mol.

Considering that the peptide with sequence KLKL5KLK-NH_2 does not form a defined secondary structure at any temperature and pH evaluated, it is unlikely that it can display antimicrobial activity per se. However, this depends on the conformation that this peptide acquires during their interaction with the cell membrane of microorganisms. Furthermore, the stability of the secondary structure of the peptide RLKL5RLK- NH_2 is dependent on the temperature and pH, but it is stable at physiological conditions. The α-helix formed by amino acid RLKL-NH_2 is consistent with results obtained from Alvares-Bravo et al. [2], and the mechanism of action proposed by the same authors [1]. The secondary structure of the third peptide seems to be more stable regardless of pH and temperature than peptides derived from sapecin β, showing RMSD values around 1.5 ± 0.5 nm. These results are not consistent with those reported in the CAMP database [8], where these are scored with a higher instability index (45.66) than the other two peptides analyzed (-37.22 and 3.45).

4 Conclusions

Considering that simulations under solvation conditions of the 3D-structures of the peptides under different pHs and temperatures are similar than *in vitro* results, we can conclude that the approximations *in silico* by molecular dynamics methods provide information for a virtual screening of antibiotic peptides with potential to be tested for antimicrobial activity in laboratory bioassays.

Finally, from results of the peptide derived from the venom of *Vespa magnifica*, we could consider that the stability index given by the CAMP database would not be a good indicator of the stability at physiological pH and temperature of these peptides, and it is necessary to assess this parameter in a experimental point of view.

References

1. Alvares-Bravo, J., Kurata, S., Natori, S.: Mode of Action of an Antibacterial Peptide, KLKLLLLLKLK-NH_2. J. of Biochem. 117(6), 1312–1316 (1995)
2. Alvares-Bravo, J., Kurata, S., Natori, S.: Novel synthetic antimicrobial peptides effective against methicillin-resistant Staphylococcus aureus. Biochem. J. 302, 535–538 (1994)
3. Brown, K., Hancock, R.: Cationic host defense (antimicrobial) peptides. Curr. Opin. Immunology 18, 24–30 (2006)
4. Contreras-Moreira, B., Fitzjohn, P.W., Offman, M., Smith, G.R., Bates, P.A.: Novel use of a genetic algorithm for protein structure prediction: searching template and sequence alignment space. Proteins 6, 424–429 (2003)
5. Fjell, C., Hiss, J., Hancock, R., Schneider, G.: Designing antimicrobial peptides: form follows function. Nat. rev. Drug Discov. 11, 37–51 (2012)
6. Lindahl, E., Hess, B., Spoel, D.: GROMACS 3.0: a package for molecular simulation and trajectory analysis. J. Mol. Model 7, 306–317 (2001)

7. Pikkemaat, M., Linssen, A., Berendsen, H., Janssen, D.: Molecular dynamics simulations as a tool for improving protein stability. Prot. Eng. 15, 185–192 (2002)
8. Thomas, S., Karnik, S., Barai, R., Jayaraman, V.K., Idicula-Thomas, S.: CAMP: a useful resource for research on antimicrobial peptides. Nucleic Acids Res. 38, D774–D780 (2010)
9. Xueqing, X., Jianxu, L., Qiuming, L., Hailong, Y., Yungong, Z., Ren, L.: Two families of antimicrobial peptides from wasp (*Vespa magnifica*) venom. Toxicon 47, 249–253 (2006)

Application of Genome Studies of Coffee Rust

Marco Cristancho[1,*], William Giraldo[1], David Botero[1,2], Javier Tabima[1,2], Diana Ortiz[1], Alejandro Peralta[1], Álvaro Gaitán[1], Silvia Restrepo[2], and Diego Riaño[2]

[1] CENICAFE, Chinchiná, Caldas, Colombia
{marco.cristancho,william.giraldo,david.botero,Javier.tabima, diana.ortiz,alejandro.peralta,alvaro.gaitan}@cafedecolombia.com
[2] Universidad de los Andes, Bogotá, Colombia
{srestrep,dm.riano122}@uniandes.edu.co

Abstract. The coffee rust, caused by the fungus *Hemileia vastatrix*, is the most serious disease of this crop worldwide. In Colombia the pathogen causes a reduction in production of up to 30% in susceptible varieties of *Coffea arabica*, if not controlled mild epidemics can occur and complete crop losses in strong epidemics. We applied genomics to study the evolution of the population of this pathogen due to recent outbreaks of the disease in Colombia.

Sequencing was performed using 454 and Illumina technology, using DNA and RNA of 8 and 3 isolates of *H. vastatrix*, respectively. With the software CEGMA we made a first estimate of the genome size of *H. vastatrix* resulting in an approximate size of 250 Mb. The hybrid assembly was performed with all sequenced genomes given a coverage of 92% with a GC content of 32%.

Keywords: Coffee, genomics, coffee rust.

1 Introduction

Rusts are caused by a wide group of obligate fungi belonging to the phylum Basidiomycota. These fungi cause some of the most important diseases in economical terms including the cereal rusts caused by different *formae speciales* of *Puccinia graminis* [1], the poplar rust caused by *Melampsora larici-populina* [2] and the corn smut caused by *Ustilago maydis* [3]. Coffee leaf rust, caused by the fungus *Hemileia vastatrix,* is the most limiting disease wherever coffee is cultivated. In Colombia, coffee represents 16% of the country's agricultural GDP and since 2008 high incidence of coffee leaf rust in crops established with susceptible varieties has caused significant reduction in yield, which can reach 30% [4]. Recent epidemics have been caused by increased patterns in rainfall and it is necessary to continue monitoring for the emergence of new races of the pathogen [5].

Next generation sequencing techniques like 454 and Illumina are often used to sequence the genomes of microorganisms and particularly of plant pathogens [1,2,3]. After cleaning and assembly processes, genome structural and functional annotations

* Corresponding author.

L.F. Castillo et al. (eds.), *Advances in Computational Biology*,
Advances in Intelligent Systems and Computing 232,
DOI: 10.1007/978-3-319-01568-2_20,

are performed. In addition, proteomic, transcriptomic and other information are routinely integrated into an information system. Comparisons with closely related organisms give insights into the genome organization and the functional implications of all rearrangements [6].

Plant pathologists and breeders are interested in the processes of infection by the pathogen and the plant-pathogen interactions. Several models have tried to explain the molecular interactions between the pathogen virulence proteins (effectors) and the plant resistance and defense proteins [7,8]. In some cases, effector proteins secreted by the pathogen are key factors in the infection process [9]. The products of these genes are directed through secretory pathways, so predictions of secreted proteins by software tools are very important in the search of pathogenicity factors.

In this study we tested software tools to assemble the *H. vastatrix* genome and to predict secreted proteins. We compared different fungal genomes with the assembled *H. vastatrix* genome. This is to our knowledge the first approach to sequence the *H. vastatrix* genome, which might help to unravel the infection mechanisms of coffee by this fungus.

2 Materials and Methods

Nine samples of dikaryotic urediniospores of *H. vastatrix* of different isolates were collected from infected leaves, taking care of collecting spores from lesions free of the hyperparasite *Lecanicillium lecanii*, the main biological control antagonist of *H. vastatrix*.

The nine samples of *H. vastatrix* genome were sequenced by Illumina™ and ROCHE™ 454 technologies. Reads were subjected to quality control with FastQC (Babraham Bioinformatics, Babraham Institute). Then, they were trimmed (CLCbio script), masked or filtered by low complexity end regions and reads shorter than 70 nucleotides were discarded. Mdust and SeqClean were used for the cleaning process. Several assemblers were tested with different combinations of clean reads. A hybrid assembly was performed (Illumina and 454 clean reads) using the CLC cell assembly software (http://www.clcbio.com/). The quality of the assembly was assessed with CLC tools and in-house R scripts.

The hybrid assembly was analyzed with MEGAN 4 [10] to assess the level of possible contamination and to perform a first approximation of the biological communities associated to *H. vastatrix* on the coffee leaf. Blastx was performed with the contigs from the hybrid assembly (396264 contigs) against the non-redundant protein database at NCBI. An E-value of 1E-3 was used as a cut-off. With the aim of filtering out putative contaminated sequences, contigs that presented similarities to reported fungal sequences were extracted to form a reliable set of *H. vastatrix* genome contigs.

The reliable set of *H. vastatrix* genome contigs were aligned against the genomes of related organisms by Mauve [11]. The genomes used were *P. graminis*, *M. laricis-populina* and *U. maydis.* The Low Collinear Blocks (LCB) values were set by visual inspection searching the best block size for each pair of alignments (largest coverage of both genomes). Finally, values used for LCB were: *P. graminis* 12154, *M. laricis-populina* 10409 and *U. maydis* 1203.

For RNA-seq experiments, a normalized library construction was performed at Evrogen, Moscow, Russia using Kamchatka crab duplex-specific nuclease. Illumina genome and RNA-seq sequencing was performed at BGI, China. RNA-Seq sequences were assembled with Trinity assembler [12]. Protein predictions were performed by mapping transcripts of *H. vastatrix* to the genome assembly with TopHat [13]. Then the mapped genome contigs were extracted. Predictions in MAKER were performed using the Augustus gene predictor (with the *Saccharomyces* probability matrix). Predictions used *H. vastatrix* rust transcriptome, *H. vastatrix* genome contigs selected and NCBI non-redundant protein database as evidences. Finally, the proteins predicted were polished, filtering out transposons with RepeatMasker (A.F.A. Smit, R. Hubley & P. Green, RepeatMasker at http://repeatmasker.org). A final set of 14425 proteins was obtained and used for subsequent analyses. Predicted proteins were classified into secreted or non-secreted proteins. The programs SignalP 4.0 [14] and PProwler [15] were used to predict secreted proteins. For the case of PProwler a prediction probability cut-off of 0.9 was used.

A set of secreted proteins predicted in a previous study for *H. vastatrix* by Fernández et al. (2012) [16] was compared with our predictions. Shortly, a Blastn (1E-30) was performed between our set of *H. vastatrix* transcripts and Fernández transcript contigs with secreted prediction. Then the proteins predicted based on these transcripts (and genome assembly) were extracted. Thus, a set of proteins that showed similarity with the proteins predicted by Fernández et al. (2012) as secreted proteins was obtained. Multiple comparisons of the three sets of secreted proteins were performed (SignalP, PProwler and Fernández-Blastp).

3 Results and Discussion

We obtained 412 million short-reads from Illumina and 5.8 million reads from 454. The quality analysis of the 454 reads showed low quality before base at position 200 (Table 1). The mean of the quality value was below 20 around base position 600 and a bias of nucleotide composition before this position was detected. The mean read quality was good. In the case of Illumina reads, the mean quality value was above 20 for all isolates, although some reads had low quality after base at position 95. Reads had little bias in nucleotide composition at the beginning of the read. The mean read quality was very high for all isolates.

The mean GC content for *H. vastatrix* was approximately 33%, based on the genome sequencing. The level of read duplication was low for 454 reads. Illumina reads showed some level of duplication, approximately 20% of reads had 10 or more reads duplicated, this probably was an artifact of sequencing and was took into account for the read cleaning process.

Several assemblers were tested: SOAP de novo, MIRA, Velvet and CLC. Finally, CLC assembler (CLCbio, 2012) was chosen to generate a draft assembly. We tested different combinations of assemblies with CLC (Table 1). The best assembly with

CLC was a hybrid of 454 and Illumina reads. As a result, 396264 contigs with a GC content of 32% were obtained. Table 1 depicts some statistics of this assembly; 23.2% of the paired reads mapped in the same contig. Most of unpaired reads (66.8%) matched two different contigs (useful for scaffolding, data not shown). Most of the contigs showed high coverage. Some contigs had coverage greater than 100X.

The most important clades represented in the Megan analysis were Fungi (31.376), Bacteria (8.826 contigs), Viridiplantae (13.193), Metazoa (9.100) and Stramenopiles (674) (Figure 1). Most of contigs (296.813) did not show similarity by BLAST with any organism in NCBI nr protein database. Megan results for Illumina reads obtained with Blastx showed that the most important clades were Fungi (12.210), Metazoa (6.714) and Viridiplantae (4.989) and Bacteria (2295 reads). Again, most reads (1.555.151) did not show similarity with any organism. A total of 2905 reads showed low complexity.

Megan results for Illumina reads obtained with Blastn and showed that the most important clades were Fungi (67.179), Metazoa (10.650), Bacteria (5.586), Viridiplantae (9.927) and Low complexity (2.905) reads. Again, most reads (1.493.471) did not show similarity with any organism. Comparisons made between *H. vastatrix* and *P. graminis* genomes showed several blocks of genome conservation, especially in *Puccinia*, although some of them are short in *Hemileia*. Comparison with *Melampsora* showed several blocks of conservation but less than with *Puccinia*. Comparison with *Ustilago maydis* showed that most of its genome is present in *Hemileia* but with very short blocks.

Table 1. Summary of coffee rust genome hybrid assembly. Clean reads were assembled with the CLC assembler. Then, the same reads were mapped against the contigs assembled with CLC.

Nº Sequences assembled		396264
Total residues assembled		333481311
Length	Max	85126
	Average	841.56
	N50	1590
Reads	Total	336649188
	Unassembled	19788611
	Assembled	316860577
	Multihit	37520793
	Potential pairs	
	Paired	78105740
	Not Paired	255469308

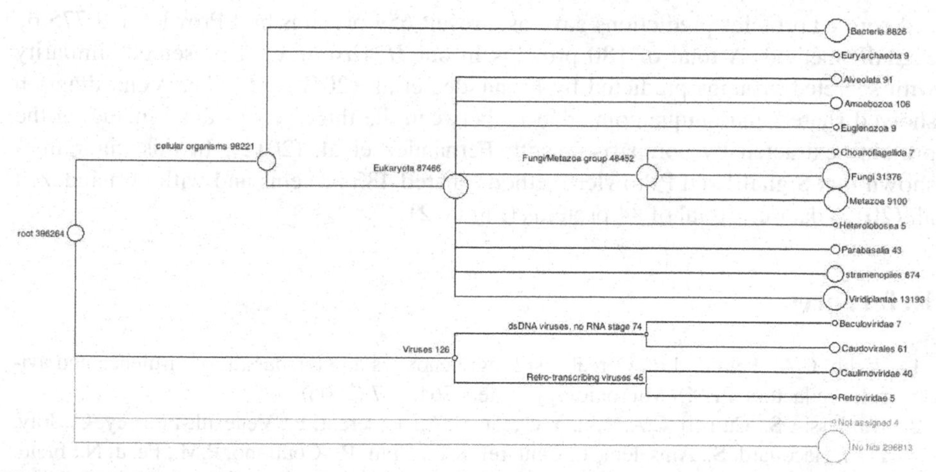

Fig. 1. Quality control of third hybrid assembly. Blastx was performed with the third hybrid assembly contigs (396264) against the non-redundant protein database of NCBI. An E-value of 1E-3 was used as a cut-off following the recommendation from the Megan developers. Megan was used to map and visualize Low Common Ancestor for every contig against NCBI tree taxonomy. Circle size shows (in logarithmic scale) the number of contigs assigned to each taxonomical category.

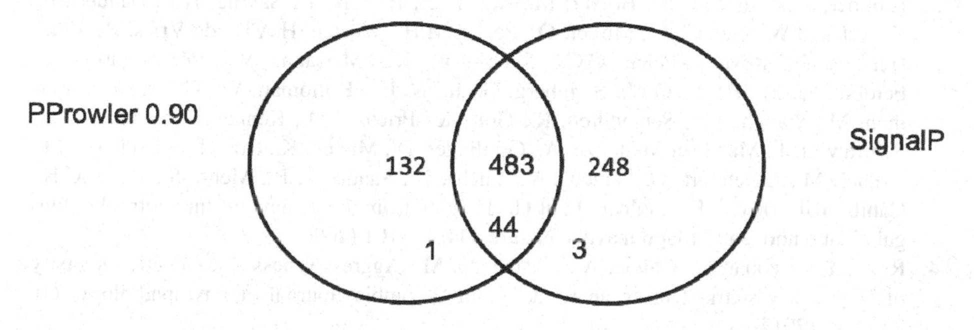

Fig. 2. Venn diagram of consensus secreted predicted proteins. SignalP and PProwler (0.9 probability cut-off) programs were used to predict which proteins are secreted. A set of 14445 putative proteins was used for classification into secreted or non-secreted proteins. The results were compared with Fernandez et al. (2012) pathogenicity predicted proteins.

When the complete *H. vastatrix* assembly (without filtering sequences with Megan) were used to Blast *Puccinia* spp. most contigs did not show similarity. Special cases were the Blast of the mitochondrial genome and transcriptome of *Puccinia* whose contigs had hits in almost all cases with the *H. vastatrix* genome. Therefore, when only fungi contigs (filtered by MEGAN) are used, enrichment in the hits is shown.

Secreted proteins predictions gave as a result 659 proteins by PProwler and 775 by SignalP method. A total of 180 proteins in our *H. vastatrix* set presented similarity with secreted proteins predicted by Fernández et al. (2012) [16]. The Venn diagram showed shared and unique coincidences between the three sets of data, including the proteins extracted by comparison with Fernández et al. (2012). In this diagram is shown that SignalP and PProwler methods shared 483 proteins and with Fernández et al. (2012) dataset a total of 44 proteins (Figure 2).

References

1. Webb, C.A., Fellers, J.P.: Cereal rust fungi genomics and the pursuit of virulence and avirulence factors. FEMS Microbiology Letters 264, 1–7 (2006)
2. Duplessis, S., Cuomo, C.A., Lin, Y.C., Aerts, A., Tisserant, E., Veneault-Fourrey, C., Joly, D.L., Hacquard, S., Amselem, J., Cantarel, B.L., Chiu, R., Coutinho, P.M., Feau, N., Field, M., Frey, P., Gelhaye, E., Goldberg, J., Grabherr, M.G., Kodira, C.D., Kohler, A., Kues, U., Lindquist, E.A., Lucas, S.M., Mago, R., Mauceli, E., Morin, E., Murat, C., Pangilinan, J.L., Park, R., Pearson, M., Quesneville, H., Rouhier, N., Sakthikumar, S., Salamov, A.A., Schmutz, J., Selles, B., Shapiro, H., Tanguay, P., Tuskan, G.A., et al.: Obligate biotrophy features unraveled by the genomic analysis of rust fungi. Proceedings of the National Academy of Sciences 108, 9166–9171 (2011)
3. Kämper, J., Kahmann, R., Bölker, M., Ma, L.-J., Brefort, T., Saville, B.J., Banuett, F., Kronstad, J.W., Gold, S.E., Müller, O., Perlin, M.H., Wösten, H.A.B., de Vries, R., Ruiz-Herrera, J., Reynaga-Peña, C.G., Snetselaar, K., McCann, M., Pérez-Martín, J., Feldbrügge, M., Basse, C.W., Steinberg, G., Ibeas, J.I., Holloman, W., Guzman, P., Farman, M., Stajich, J.E., Sentandreu, R., González-Prieto, J.M., Kennell, J.C., Molina, L., Schirawski, J., Mendoza-Mendoza, A., Greilinger, D., Münch, K., Rössel, N., Scherer, M., Vraneš, M., Ladendorf, O., Vincon, V., Fuchs, U., Sandrock, B., Meng, S., Ho, E.C.H., Cahill, M.J., Boyce, K.J., Klose, J., et al.: Insights from the genome of the biotrophic fungal plant pathogen Ustilago maydis. Nature 444, 97–101 (2006)
4. Rozo, Y., Escobar, C., Gaitán, Á., Cristancho, M.: Aggressiveness and Genetic Diversity of Hemileia vastatrix During an Epidemic in Colombia. Journal of Phytopathology 160, 732–740 (2012)
5. Cristancho, M., Rozo, Y., Escobar, C., Rivillas, C., Gaitán, Á.: Outbreak of coffee leaf rust (Hemileia vastatrix) in Colombia. New Disease Reports 25 (2012)
6. Spanu, P.D., Abbott, J.C., Amselem, J., Burgis, T.A., Soanes, D.M., Stuber, K., Loren van Themaat, E.V., Brown, J.K.M., Butcher, S.A., Gurr, S.J., Lebrun, M.H., Ridout, C.J., Schulze-Lefert, P., Talbot, N.J., Ahmadinejad, N., Ametz, C., Barton, G.R., Benjdia, M., Bidzinski, P., Bindschedler, L.V., Both, M., Brewer, M.T., Cadle-Davidson, L., Cadle-Davidson, M.M., Collemare, J., Cramer, R., Frenkel, O., Godfrey, D., Harriman, J., Hoede, C., King, B.C., Klages, S., Kleemann, J., Knoll, D., et al.: Genome Expansion and Gene Loss in Powdery Mildew Fungi Reveal Tradeoffs in Extreme Parasitism. Science 330, 1543–1546 (2010)
7. Dangl, J.L., Jones, J.D.: Plant pathogens and integrated defence responses to infection. Nature 411, 826–833 (2001)
8. Hogenhout, S.A., Van der Hoorn, R.A.L., Terauchi, R., Kamoun, S.: Emerging concepts in effector biology of plant-associated organisms. Molecular Plant-Microbe Interactions 22, 8 (2009)

9. Stajich, J.E., Saunders, D.G.O., Win, J., Cano, L.M., Szabo, L.J., Kamoun, S., Raffaele, S.: Using Hierarchical Clustering of Secreted Protein Families to Classify and Rank Candidate Effectors of Rust Fungi. PLoS ONE 7, e29847 (2012)
10. Huson, D.H., Mitra, S., Ruscheweyh, H.J., Weber, N., Schuster, S.C.: Integrative analysis of environmental sequences using MEGAN4. Genome Research 21, 1552–1560 (2011)
11. Darling, A.C.E.: Mauve: Multiple Alignment of Conserved Genomic Sequence With Rearrangements. Genome Research 14, 1394–1403 (2004)
12. Grabherr, M.G., Haas, B.J., Yassour, M., Levin, J.Z., Thompson, D.A., Amit, I., Adiconis, X., Fan, L., Raychowdhury, R., Zeng, Q., Chen, Z., Mauceli, E., Hacohen, N., Gnirke, A., Rhind, N., di Palma, F., Birren, B.W., Nusbaum, C., Lindblad-Toh, K., Friedman, N., Regev, A.: Full-length transcriptome assembly from RNA-Seq data without a reference genome. Nature Biotechnology 29, 644–652 (2011)
13. Trapnell, C., Pachter, L., Salzberg, S.L.: TopHat: discovering splice junctions with RNA-Seq. Bioinformatics 25, 1105–1111 (2009)
14. Petersen, T.N., Brunak, S., von Heijne, G., Nielsen, H.: SignalP 4.0: discriminating signal peptides from transmembrane regions. Nat Methods 8, 785–786 (2011)
15. Hawkins, J., Bodén, M.: Detecting and Sorting Targeting Peptides with Neural Networks and Support Vector Machines. Journal of Bioinformatics & Computational Biology 4, 1–18 (2006)
16. Fernandez, D., Tisserant, E., Talhinhas, P., Azinheira, H., Vieira, A., Petitot, A.S., Loureiro, A., Poulain, J., do Céu Silva, M., Duplessis, S.: 454-pyrosequencing of Coffea arabica leaves infected by the rust fungus Hemileia vastatrix reveals in planta-expressed pathogen-secreted proteins and plant functions in a late compatible plant–rust interaction. Mol. Plant Pathol. 13, 17–37 (2012)

In-silico Analysis of the Active Cavity of *N*-Acetylgalactosamine-6-Sulfate Sulfatase in Eight Species

Sergio Olarte-Avellaneda[1,2], Alexander Rodríguez-López[2], and Carlos Javier Alméciga-Díaz[2]

[1] Clinical Bacteriology Program, School of Health Sciences, Universidad Colegio Mayor de Cundinamarca, Bogotá D.C., Colombia
solarte@unicolmayor.edu.co
[2] Institute for the Study of Inborn Errors of Metabolism, School of Sciences, Pontificia Universidad Javeriana, Bogotá D.C., Colombia
{rodriguez.edwin,cjalmeciga}@javeriana.edu.co

Abstract. Mucopolysaccharidosis IV A (MPS IV A) is a lysosomal storage disease produced by the deficiency of N-acetylgalactosamine-6-sulfate sulfatase (GALNS), which is involved in the catabolism of keratan sulfate and chondroitin-6-sulfate. In the present study we performed a computational analysis of active cavity of GALNS from human and other eight species, as well as their interaction with the natural ligands. The modeled enzymes showed a highly conserved structure, although differences in the sizes of the active cavity and affinity energy for the ligands were observed among the studied GALNS. The results could be associated to the molecular evolution of the catalytic cavity and differences in the complexity of the substrate produced by the species. These results could have a significant impact towards the understanding of the molecular bases of MPS IV A and the development of efficient treatment alternatives.

Keywords: Morquio A, N-acetylgalactosamine-6-sulfate sulfatase, keratan sulfate, chondroitin-6-sulfate, molecular modeling, computational molecular docking.

1 Introduction

The Mucopolysaccharidosis IV A (MPS IV A, Morquio A disease, (MPS IV A, Morquio A disease, OMIM 253000) is a lysosomal storage disease caused by the deficiency or alteration of the human N-acetylgalactosamine-6-sulfate sulfatase (GALNS, EC 3.1.6.4). GALNS hydrolyze the sulfate group present at N-acetylgalactosamine-6-sulfate (6S-GalNAc) and galactose-6-sulfate (G6S) from the glycosaminoclycans (GAGs) chondroitin-6-sulfate (C6S) and keratan sulfate (KS), respectively. GALNS deficiency leads to the lysosomal accumulation of C6S and KS causing systemic skeletal dysplasia [1, 2]. Despite of the promising preclinical and clinical results of

L.F. Castillo et al. (eds.), *Advances in Computational Biology*,
Advances in Intelligent Systems and Computing 232,
DOI: 10.1007/978-3-319-01568-2_21, © Springer International Publishing Switzerland 2014

enzyme replacement therapy [3] it is still necessary to explore alternatives to obtain a more efficient and less immunogenic enzyme. Furthermore, it is also necessary to explore new treatment alternatives, such as the use of pharmacological chaperones that have been successfully evaluated for other lysosomal storage disorders [4]. Recently, we expanded the computational analysis of human GALNS enzyme showing the first *in-silico* assessment of ligand-GALNS interactions [5]. In this study was performed a computational analysis of active cavity of GALNS in eight species as well as their interaction with natural ligands.

2 Methods

Sequences for human GALNS (UniProtKB/Swiss-Prot P34059) and *Macaca mulatta* (rhesus macaque) H9F5L7, were retrieved from Uniptrot, while for the other species, GALNS sequences were retrieved from Genbank: *Bos taurus* (bovine) NP_001193258.1, *Mus musculus* (mouse) AAH04002.1, *Rattus norvergicus* (rat) NP_001041316.1, *Canis lupus familiaris* (dog) NP_001041585.1, *Gallus gallus* (chicken) XP_414208.1, *Oreochromis niloticus* (tilapia nilotica) XP_003445750.1, and *Sus scrofa* (pig) NP_999120.1. Prediction of signal peptide was performed with SignalP 4.1 server. Multiple alignment was carried out with MUSCLE, and phylogenetic tree was generated by using MEGA5 [6]. Prediction of tertiary structures was done with I-TASSER Server using the tertiary structure of the human Arylsulfatase A (ASA, PDB 1AUK), as template. PDBsum was used for proteins structure validation. Calcium ion was added by using YASARA View v11.4.18 (YASARA Biosciences GmbH, Vienna, Austria), constrained to Asp39, Asp40, Asp288, and Asn289 (according to human numbering) as reported for human ASA and GALNS [7, 8]. Amino acids within the active cavity and volume of cavity were predicted by using Computed Atlas of Surface Topography of Proteins (CASTp) [9]. Structural comparison was done using Swiss-PdbViewer v4.1. Partial charges, affinity energy and interactions (H-bonds, and electrostatic and steric interactions) of GALNS with the ligands 6S-GalNAc and G6S were evaluated by using Molegro Virtual Docker v5.5 (MVD, CLC bio, Aarhus N, Denmark).

3 Results and Discussion

The human GALNS enzyme has 522 amino acids including 26 amino acids of signal peptide (SP), while in the other studied GALNS the size varied from 513 to 525 amino acids with SPs between 18 to 27 residues. As show in Table 1, identity of the studied sequences against human GALNS varied between 73% to 96%. Human GALNS enzyme showed 310 completely conserved residues (63%) when compared against the other studied sequences. The phylogenetic tree showed than human enzyme was closer to *Macaca mulatta* and distant from *Gallus gallus* and *Oreochromis niloticus* (Figure 1).

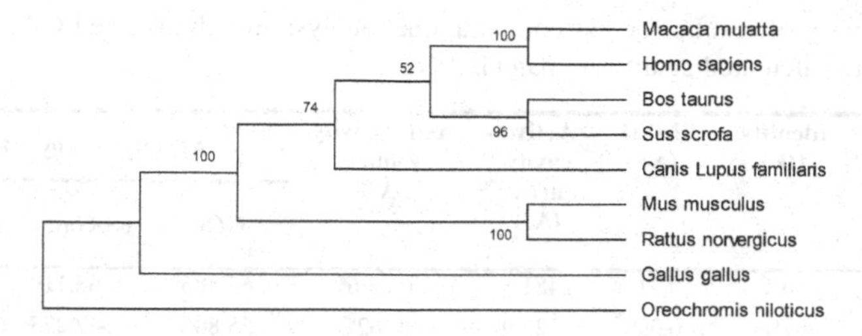

Fig. 1. Phylogenetic tree of GALNS enzyme from the studied species. GALNS sequences were aligned and Neighbor-joining phylogenetic tree was generated with a 500 bootstrap.

The active cavity of human GALNS involves 58 amino acids, with 49 residues completely conserved among the studied sequences, and I93, A102, E112, W184, I294, Q299, G300, Q311 and D388 showing different conservation profiles according to the specie (Table 2). However, from these non-completely conserved amino acids, only Ala102 in human GALNS interacts with ligands, while all the other residues are involved in cavity architecture. *Oreochromis niloticus* showed the highest difference against human sequence differing in six residues, while dog GALNS showed the lowest difference against human GALNS, and *Macaca mulatta* did not show any difference with human GALNS. These results showed the high conservation of the active cavity in GALNS enzyme, excepting *Oreochromis niloticus* GALNS, as previously reported for other sulfatases [10].

Modeled 3D structures had C-score values between 0.34 to 1.28, with over 90% of the amino acids within the two most favored regions. Structural comparison showed that all enzymes had a RMSD lower than 1Å in comparison with human GALNS (Table 1), showing the high structural conservation of this enzyme during evolution. The results of area and volume for the active cavity for the studied sequences are summarized in Table 1. Chicken showed the largest area and volume values, while pig and rhesus macaque showed the smallest ones, and human GALNS showing intermediate values. A clear correlation between these results and the phylogenetic tree was not observed.

Partial charges at the active cavity were evaluated for all the studied enzymes, showing a positive charge, which correlates with the negative charge of the native GALNS substrates [11]. In human GALNS, residues R83, H142, H236, and K310 provide a positive charge to the cavity, which might promote the interaction with the substrate, while residues D39, D40 and D288 provide a negative charge and are involved in the interaction with the cofactor (Figure 2A). Likewise, these amino acids were conserved among the other studied enzymes, but Lys140 (respect to human sequence), which was not observed in human GALNS active cavity and provided a positive charge to the cavity in these enzymes. Ghosh, D [10] proposed for Arylsulfatase C (ARSC) that Lys134, which corresponds Lys140 in human GALNS, participates in the catalytic mechanism of the ARSC, which maybe extend to other sulfatases. This difference between human ARSC and GALNS could be related with the target substrate of each enzyme.

Table 1. Summary of results for the computational analysis for the studied GALNS. Identity and RMSD were calculated against human GALNS.

Specie	Identity (%)	RMSD (Å)	Active cavity area (Å²)	Active cavity volume (Å³)	Affinity energy (kJ/mol)		
					G6S	6S-GalNAc	Σ total
Mus musculus	86%	0.71	1481.8	1663.6	-63.486	-68.119	-131.605
Canis lupus	87%	0.39	1125.9	1162	-55.806	-47.273	-103.079
R. norvergicus	85%	0.75	849.8	1526	-80.312	-94.138	-174.450
Sus scrofa	88%	0.69	522.8	732.3	-73.784	-68.318	-142.102
Bos taurus	87%	0.83	938.9	1241.9	-88.571	-108.199	-196.770
G. gallus	79%	0,38	1561.5	1699.5	-46.127	-64.603	-110.730
O. niloticus	73%	0,45	990.4	1307	-80.244	-6.730	-86.974
M. mulatta	96%	0.44	728.8	941.2	-65.635	-63.964	-129.599
Homo sapiens	---	---	1152	1244.6	-115.618	-120.104	-235.722

Table 2. Amino acids differing at the active cavity of studied GALNS enzyme. Blue and red amino acids represent conservativeness and non-conservativeness residues changes.

Homo sapiens	*Macaca mulatta*	*Canis lupus*	*Mus musculus*	*Rattus norvergicus*	*Sus scrofa*	*Bos taurus*	*Gallus gallus*	*Oreochromis niloticus*
I93	I67	I92	I93	I94	I92	I93	V83	I93
A102	A76	R101	A101	A103	G101	G102	A92	A102
E112	E86	E111	E111	E113	E111	E112	D102	E112
W184	W158	W183	W183	W185	W183	Q184	W174	S184
I294	I268	I294	I293	I296	V294	I295	I285	M295
Q299	Q273	Q299	E299	E301	Q299	Q300	Q290	E300
G300	G274	G300	G300	G302	G300	G301	G291	S301
Q311	Q285	Q311	Q286	Q313	Q311	Q312	Q302	E312
D388	D361	D385	N388	N390	N388	N389	N375	N391

The results of molecular docking for human GALNS with the monomers G6S and 6S-GalNAc showed affinity energy of -115.618 and -120.104 kJ/mol, respectively. Previously we identify that key amino acids for ligand-enzyme interactions for human GALNS are Asp39, Asp40, Ser80, Cys79, Arg83, Ala102, Tyr108, His142, Cys165, Tyr181, His236, Asp288, Asn289, Lys310 and $Ca2^+$ [5]. For all the other studied enzymes it was observed a lower affinity for G6S and 6S-GalNAc than that observed with human GALNS (Table 1). The closest affinity energies to human values were observed for *Bos taurus*, *Rattus norvergicus* and *Sus scrofa,* which are phylogenetically close to human. Although human and dog GALNS only differ in one residue at the active cavity (Ala102 and Arg102, for human and dog, respectively), it was observed a large difference in the sumatoria of affinity energies for both ligands (-235.722 vs. -103.079 kJ/mol, for human and dog, respectively). This difference could be associated to partial charges in dog GALNS in comparison to human

GALNS (Figure 2B) and to the interaction of Ala211 with the carbon chain of ligands that is not observed in human GALNS.

Currently there is not a natural animal model for MPS IV A and three genetic engineered mouse models have been developed. Although these animals share some of the histological features of MPS IVA patients, they lack of the skeletal abnormalities observed in human patients [1, 12, 13]. The absence of bone deformities observed in MPS IV A mice models could be due to differences in the complexity and distribution of KS in this animals [14]. In addition, rats seem to be have more KS than mice, while bovine and human proteoglycans are richer in KS than in mouse [12, 14]. We observed a large difference in affinity energy by G6S, a constituent of KS, for human and mouse GALNS, with values of -115.618 and -63.486 kJ/mol, respectively. However, only two amino acids differ between human and mouse GALNS (Q273/E273 and D362/N362), which were not directly involved in the ligand-enzyme interaction. These two resides also differ in rat GALNS, but this enzyme showed a higher affinity energy for G6S (-80.312 kJ/mol) than that observed for mouse GALNS. These results might suggest that in mice and rats GALNS could have an additional substrate that the observed in humans. The amino acids involved in the interaction with the sulfate group of the ligands 6S-GalNAc and G6S are highly conserved in GALNS, although in non-human GALNS new residues interact with the carbon chain of the ligands, and Lys140 interacts with ligands in all the studied enzymes excepting humans.

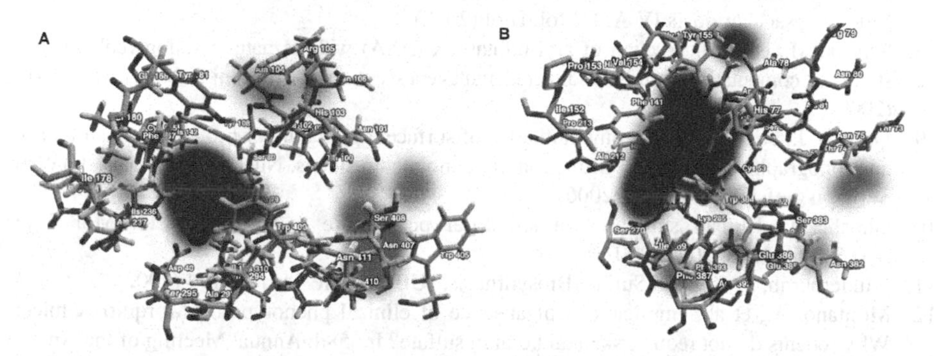

Fig. 2. Partial charges in human (A) and dog (B) GALNS. Partial charges were generated with MVD v5.5. Blue and red zones represent positive and negative partial charges, respectively.

4 Conclusions

These results confirm the importance of amino acids involve in the ligand-enzyme interaction and contribute to the knowledge of the evolution of the active cavity of the GALNS enzyme. We observed that differences in affinity energy could be associated with the evolution of the catalytic cavity as the result in changes of substrate complexity and differences in the composition of the proteoglycans produced by each species [15]. These results could serve as a starting point for the design of recombinant enzymes with higher affinity for their substrates and in the search for pharmacological chaperones, contributing to the development of alternative therapeutics for MPS IVA.

References

1. Alméciga-Díaz, C.J., et al.: Contribución colombiana al conocimiento de la enfermedad de Morquio. A Imbiomed 34(3), 221–241 (2012)
2. Tomatsu, S., et al.: Current and emerging treatments and surgical interventions for Morquio A syndrome: a review. Research and Reports in Endocrine Disorders 2, 65–77 (2012)
3. Hendriksz, C., et al.: A multi-national, randomized, double-blind, placebo-controlled study to evaluate the efficacy and safety of BMN 110 treatment for mucopolysaccharidosis IVA (Morquio syndrome type A). Molecular Genetics and Metabolism 108, S48 (2013)
4. Boyd, R.E., et al.: Pharmacological chaperones as therapeutics for lysosomal storage diseases. J. Med. Chem. 56(7), 2705–2725 (2013)
5. Olarte-Avellaneda, S., et al.: Computational analysis of human N-acetylgalactosamine-6-sulfate sulfatase enzyme. Molecular Genetics and Metabolism 108(2), 70–71 (2013)
6. Tamura, K., et al.: MEGA5: molecular evolutionary genetics analysis using maximum likelihood, evolutionary distance, and maximum parsimony methods. Mol. Biol. Evol. 28(10), 2731–2739 (2011)
7. Rivera-Colon, Y., et al.: The Structure of Human GALNS Reveals the Molecular Basis for Mucopolysaccharidosis IV A. J. Mol. Biol (2012)
8. Schenk, M., et al.: Interaction of arylsulfatase-A (ASA) with its natural sulfoglycolipid substrates: a computational and site-directed mutagenesis study. Glycoconj J. 26(8), 1029–1045 (2009)
9. Dundas, J., et al.: CASTp: computed atlas of surface topography of proteins with structural and topographical mapping of functionally annotated residues. Nucleic Acids Res. 34(Web Server issue), W116–W118 (2006)
10. Ghosh, D.: Human sulfatases: a structural perspective to catalysis. Cell Mol. Life Sci. 64(15), 2013–2022 (2007)
11. Funderburgh, J.: Keratan Sulfate Biosynthesis. IUBMB Life 54, 187–194 (2002)
12. Montano, A., et al.: Implications of absence of clinical phenotype on Morquio A mice: Why rodents do not require skeletal keratan sulfate? In: 58th Annual Meeting of the American Society of Human Genetics (ASHG) ABSTRACT 2008, Philadelphia, Pennsylvania (2008)
13. Tomatsu, S., et al.: Mouse model of N-acetylgalactosamine-6-sulfate sulfatase deficiency (Galns–/–) produced by targeted disruption of the gene defective in Morquio A disease. Hum. Mol. Genet. 12(24), 3349–3358 (2003)
14. Venn, G., Mason, R.M.: Absence of keratan sulphate from skeletal tissues of mouse and rat. Biochem J. 228(2), 443–450 (1985)
15. Barry, F.P., et al.: Length variation in the keratan sulfate domain of mammalian aggrecan. Matrix Biol. 14(4), 323–328 (1994)

Gene Predictors Ensemble for Complex Metagenomes

Nestor Díaz[1], Andres Felipe Ruiz Velazco[2], and Cristian Alberto Olaya Márquez[2]

[1] Universidad Del Cauca, Facultad de Ingeniería Electrónica y Telecomunicaciones, Departamento de Sistemas, Popayán, Colombia
nediaz@unicauca.edu.co

[2] Universidad Del Cauca, Facultad de Ingeniería Electrónica y Telecomunicaciones, Popayán, Colombia
{Andresfeliperuizv,warrdnez}@gmail.com

Abstract. Is presented an ensemble of predictors of genes that focuses on improving the performance of traditional predictors when applied to metagenomes obtained by sequencing 454 and are characterized by very short reads. The proposed ensemble is based on the use of data mining techniques, such as decision trees and k-means, complemented by structural information of the sequence provided by the fractal dimension. The assembly obtained can overcome the performance from the best ab initio predictor in a proportion of 15 to 20%.

Keywords: expert ensemble, metagenomics, gene predictor, data mining.

1 Introduction

The prediction of genes in metagenomic studies turns in a more difficult task than in traditional genomics. The reasons behind this are originated from the impossibility to obtain genetic material in enough quantity in environmental samples. The low cost sequencing technologies, such as 454, typically obtains reads between 100 and 300 bp (base pairs) of average length [1],[2]. In some cases, the complexity of the environmental samples affects negatively the sequencing process and the reads are very short.

Very short reads implies, less efficiency in assembly and less precision in gene prediction. For single species sequences, gene predictors such as Glimmer[3], Prodigal[4] and MetageneAnnotator[5], brings precision superior to 90%, in short reads the performance decays around 50%.

Each one of the gene predictors obtains different results when is used on the same sequences. This is because each predictor uses a different approach in the prediction process exploiting the sequence characteristics in diverse ways. This fact gives place to some works that try to enhance the prediction result through integrative methods, based on the preliminary results of existing predictors[6], [7].

One of the most recognized meta – tool that combines other tools is Yacop [6]. The gene prediction tools used are Glimmer, Critica[8] and ZCURVE[9]. The

L.F. Castillo et al. (eds.), *Advances in Computational Biology*,
Advances in Intelligent Systems and Computing 232,
DOI: 10.1007/978-3-319-01568-2_22, © Springer International Publishing Switzerland 2014

assembly approach is based in set theory and the results are enhanced through the union of the results of Critica with the matching results of Glimmer and ZCURVE. The results show the improvement of the specificity and the sensitivity of the individual predictors. The drawback of this work is that the results are principally based on Critica and therefore the predictions are homologous genes principally. In metagenomic projects, is usual that the objectives of gene prediction are the finding of novel genes.

Consorf[10] is another meta – tool that try to carry out an analogous approach to YACOP. The approach is two way, in the first a homology gene prediction is developed using the FASTX[11] algorithm; in the other the ab initio gene predictors GeneMark[12], Glimmer and GeneMark.hmm[13] are used and the consensus gene prediction are considered as possible genes. Finally, the representative predicted genes are determined through pair wise alignment against a protein library. This approach shows the same drawbacks of YACOP for metagenomic projects.

Expert Combination (ensemble) has often shown that it can improve the performance of the individual expert algorithms. [14],[15]. In accordance with the above, and considering the drawbacks of the existing predictors an meta – tools, we have developed an ensemble of gene predictors that enhances the results of each predictor, using an approach based on data mining and taking advantage of the fractal dimension of the sequence.

The rest of the document is organized as follows. Section 2 describes the assembly and the algorithms used and the synthetic dataset created for this work. Section 3 describes the prediction result of the assembly and those of the individual predictors. In the final section all the work is revised and discussed.

2 Materials and Methods

The proposed assembly is based on the use of data mining techniques, such as decision trees and k-means, complemented by structural information of the sequence provided by the fractal dimension.

This section describes the data used, the preliminary work and the new meta predictor of genes for short meta genomic reads.

2.1 Datasets

In order to make the training and testing process relevant for metagemomic projects, we constructed several datasets. These datasets were synthetically constructed using the Metasim tool [16], that simulates sequencing using some technologies such as 454. The reference genomes were those corresponding to the taxonomy of the Figure 1. On each dataset were simulated 400.000 reads with average length of 180 bp, belonging to 973 complete genomes available at NCBI (ftp://ftp.ncbi.nlm.nih.gov/genomes/Bacteria/all.fna.tar.gz, visited: April 2011). The total base pairs were 76.3 Mbp.

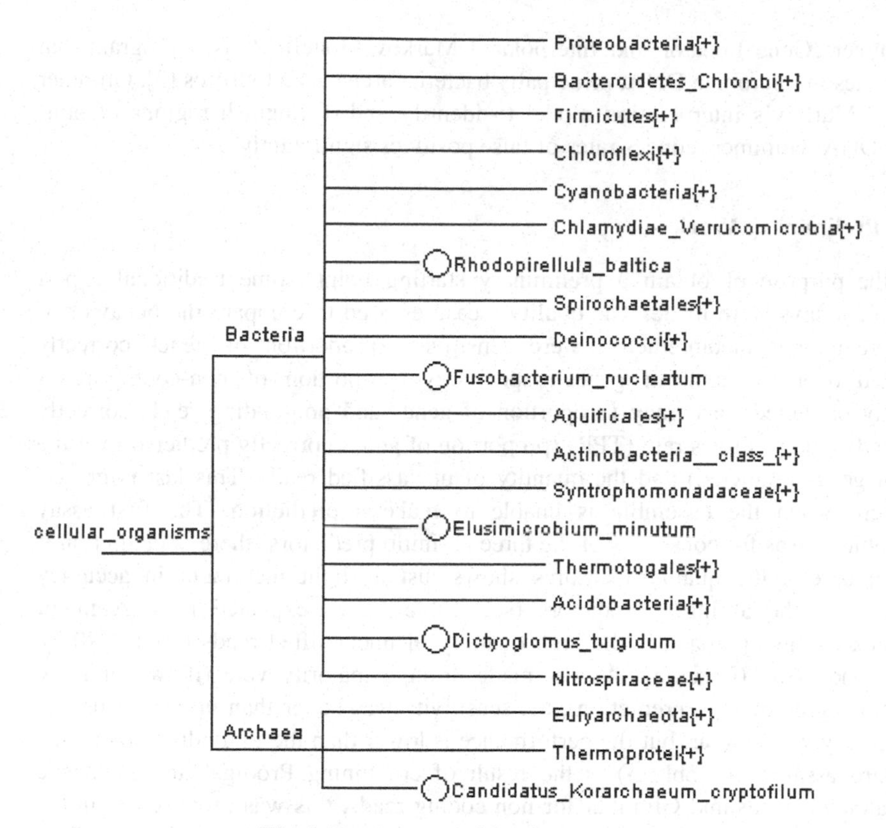

Fig. 1. Taxonomy used for dataset construction

2.2 Ab Initio Gene Predictors

Below are described three ab-initio gene predictors, selected from a more extensive and exhaustive list for its free licensing and standalone installations on local machines, besides of good performance as reported in literature.

Prodigal is a search algorithm using a “trial and error” approach[4]. With the building of a set of curated genomes, general rules were determined about the nature of the prokaryotic genes. With this information, Prodigal is able to learn all the necessary properties about input organism and build a complete training profile. Prodigal automatically determines a set of genes called "real" about the training. Analyzing content codons, and applying metrics, the algorithm builds an overall score for each gene.

Metagene is a prokaryote genes searcher[5]. It can predict a range of prokaryotic genes from fragmented genomic sequences. The prediction is made in two steps, first the possible ORFs are extracted from a sequence and these are marked for their lengths and base compositions, then an optimal combination of ORFs is calculated using the orientation markers and the lengths nearness, in addition to the markers that every ORF have.

Glimmer (Gene Locator and Interpolated Markov ModellER) is a program that seeks genes in microbian DNA, principally bacteria, archaea and viruses [3]. Glimmer uses the Markov's interpolation model to identify and distinguish regions of non-coding DNA. Glimmer reduces rates of false positives significantly.

2.3 Preliminary Work

With the purpose of obtain a preliminary starting point, some traditional expert ensemble assays were made. The quality measures used to compare the behavior of each predictor / metapredictor, were sensitivity (proportion of genes correctly predicted over the total of genes), specificity (proportion of non-coding reads correctly predicted), accuracy (proportion of genes and non-coding reads correctly predicted), true positives rate (TPR) (proportion of genes correctly predicted over the total of genes predicted) and the quantity of unclassified reads. This last is greater than zero when the ensemble is unable to make a prediction. The first assay (Ensemble 1) was for consensus of the three ab-initio predictors (there is prediction if they all agree), the quality measures shows just a slight increment in accuracy compared to the ab-initio predictors (see Table 1), as expected [17]. A major drawback of this approach is the high quantity of unclassified reads (around 70%). The second assay (Ensemble 2) was made through majority vote (if two or more predictors agree there is prediction), the sensitivity was better than in consensus and the accuracy was similar, but the performance is lower than the individual predictors. The third assay (Ensemble 3) is the result of combining Prodigal and Metagene Annotator for genes and Glimmer for non-coding reads, this was proposed from the better sensitivity performance of the first two, and the specificity of Glimmer. This approach outperforms in sensitivity the individual predictors with the drawbacks of lowering the specificity and left 36000 reads unclassified. The last assay (Ensemble 4) combines the majority vote for genes and Glimmer for non-coding reads, this approach outperforms all other.

It's remarkable the very similar performance of the TPR for all predictors, the values around 0.5 shows a tendency to do incorrect prediction of genes in half of the cases. This is an issue to consider, as the number of false predicted genes affects the development of post gene prediction work.

2.4 Assembly Using Data Mining

The best ensemble (Table 1) in sensitivity is the worst in TPR and specificity, this shows tendency to over predict non-coding reads as genes. A significant improvement in prediction should avoid this, while improving sensitivity and accuracy. The little success of the preliminary work indicates the need for a less heuristic, which try to exploit the characteristics of the individual predictors and this is provided by the machine learning. Data mining techniques, such clustering and classification, brings learning machine capabilities. Through experimentation it was found that techniques such as k-means and decision trees, by themselves failed to obtain a significant improvement in the behavior of the ensemble. For this, additional structural features, such as fractal dimension and length of the sequence of nucleotides were used.

Table 1. Quality measures for single predictors and traditional ensembles

Predictor	Sensitivity	Specificity	Accuracy	TPR	Unclassified
Glimmer	0.38699	0.6043	0.50069	0.47123	0
Metagene Annotator	0.67571	0.34359	0.50194	0.48402	0
Prodigal	0.69599	0.32581	0.50231	0.48473	0
Ensemble 1	0.65655	0.36401	0.50341	0.48445	274353
Ensemble 2	0.66039	0.35916	0.50278	0.48429	1
Ensemble 3	0.72118	0.29169	0.49776	0.48429	36012
Ensemble 4	0.74468	0.26391	0.49313	0.47968	0

The training and testing sets for machine learning are constructed from the predictions results of Glimmer, Metagene and Prodigal, for the dataset described above, and then the fractal dimension [18] and the length of each read is added. This minable view is used for clustering with the k-means algorithm, the learned groups (number of clusters were obtained by successive refinements) shows some interesting characteristics (Table 2). Two of the five groups (cluster 0 and 2) are free of grouping error, i.e. only were grouped together genes or non-coding reads. The other three groups have mixed readings, and for this reason the ambiguity is resolved using a decision tree that refines prediction results for these groups.

Table 2. Optimal clusters algorithm k-means

Attribute	Cluster 0	Cluster 1	Cluster 2	Cluster 3	Cluster 4
METAGENE	True	False	True	False	False
PRODIGAL	True	False	True	False	False
GLIMMER	False	True	False	True	False
LENGTH	0.4004	0.1677	0.4228	0.5494	0.3222
HFD	0.3605	0.3597	0.3585	0.3387	0.3657
REALGENE	False	False	True	False	True

As result of the refinement through the decision tree, if a read is placed at cluster 1 and Glimmer classified it as non-coding the tree keeps this prediction, in other case the length acts as decision variable. In the cluster 3 the majority of Metagene and Prodigal define the rule of the decision tree. The cluster 4, results in the more complex grouping and the difficult of prediction is improved through combination of the normalized length and fractal dimension of the read.

3 Results

The assembly obtained can overcome the performance from the best single in a proportion of 15 to 20%. The training and testing of the proposed strategy was performed with a synthetic dataset that simulates the sequencing 454 and considers about 1,000 genomes of organisms. The assembly is deployed on the Web portal interface provided by the Django Framework. The tool was named Fractal PGMG Assembly.

The sensitivity of the tool (0.809941), measured on a different dataset of that used for training, was higher than that reported in Table 1 for other tools. This means that this approach is better recognizing reads with real genes in it. The drawback was the TPR (0.483312), similar to the other tools, therefore the tool was unable to reduce the proportion of reads with non-coding sequences erroneously identified.

4 Discussion

The accuracy of ab initio gene predictors traditionally used in metagenomic sequences is negatively affected when these are very short. The reason behind this behavior lies in a higher probability of obtain incomplete genes in the sequencing process, making the prediction more difficult than in single organism sequences.

For complete genomes most of the predictors can reach accuracy near to 100%, but the same predictors decay in behavior when the metagenome is more complex and the sequencing process just obtains short reads that are difficult to assembly in more large sequences.

The gene predictors selected for this work show different behavior. While Glimmer is more accurate identifying non-coding sequences, Metagene and Prodigal are for identify sequences with genes in it, but with many inconsistent predictions. In front of the disparity of results in the set of predictors, traditional techniques for expert ensemble like majority vote and consensus are inefficient to improve the gene prediction process.

For the short reads in the datasets constructed the combination of data mining techniques and fractal dimension achieve an improvement on the gene prediction, reducing the quantity of sequences with genes no identified for individual predictors.

5 Conclusions

In gene prediction of metagenomic sequences, the systematic combination of ab initio techniques enhances the identification of reads with genes in it. Our approach shows that the ensemble of predictors and structural measures of the sequence can be raised through the combination of data mining techniques, such as clustering and classification.

On the other hand, the fractal dimension was decisive for the predictors ensemble. The decision tree implemented has rules that make use of this measure as decision criteria and classify a sequence as coding or non-coding.

The future work of our group is focused in the enhancement of the TPR of the ensemble, emphasizing in the cluster assignment of the sequence. Additionally, we are planning to develop a generic tool that make easy to implement meta-predictors using the datasets and predictors that the researcher consider more convenient in any metagenomic project to develop.

Acknowledgements. We are grateful to the Colombian Center for Genomics and Bioinformatics of Extreme Environments (GeBiX) and the Universidad del Cauca for the support on the development of this work.

References

1. Metzker, M.L.: Sequencing technologies - the next generation. Nature Reviews Genetics 11(1), 31–46 (2010)
2. Chaisson, M., Pevzner, P.: Short read fragment assembly of bacterial genomes. Genome Research 18(2), 324–330 (2008)
3. Delcher, A., Bratke, K., Powers, E., Salzberg, S.: Identifying bacterial genes and endosymbiont DNA with Glimmer. Bioinformatics 1, 1–7 (2007)
4. Hyatt, D., Chen, G.-L., Locascio, P.F., Land, M.L., Larimer, F.W., Hauser, L.J.: Prodigal: prokaryotic gene recognition and translation initiation site identification. BMC Bioinformatics 11, 119 (2010)
5. Noguchi, H., Taniguchi, T., Itoh, T.: MetaGeneAnnotator: detecting species-specific patterns of ribosomal binding site for precise gene prediction in anonymous prokaryotic and phage genomes. DNA Research: An International Journal for Rapid Publication of Reports on Genes and Genomes 15(6), 387–396 (2008)
6. Tech, M., Merkl, R.: YACOP: Enhanced gene prediction obtained by a combination of existing methods. In Silico Biology 3(4), 441–451 (2003)
7. Kislyuk, A., Katz, L., Agrawal, S.: A computational genomics pipeline for prokaryotic sequencing projects. Bioinformatics 26(15), 1819–1826 (2010)
8. Badger, J.H., Olsen, G.J.: CRITICA: coding region identification tool invoking comparative analysis. Molecular Biology and Evolution 16(4), 512–524 (1999)
9. Guo, F.-B.: ZCURVE: a new system for recognizing protein-coding genes in bacterial and archaeal genomes. Nucleic Acids Research 31(6), 1780–1789 (2003)
10. Kang, S., Yang, S.-J., Kim, S., Bhak, J.: CONSORF: a consensus prediction system for prokaryotic coding sequences. Bioinformatics (Oxford, England) 23(22), 3088–3090 (2007)
11. Pearson, W.R., Wood, T., Zhang, Z., Miller, W.: Comparison of DNA sequences with protein sequences. Genomics 46(1), 24–36 (1997)
12. Borodovsky, M., McIninch, J.: GENMARK: parallel gene recognition for both DNA strands. Computers & Chemistry 17(2), 123–133 (1993)
13. Lukashin, A.V., Borodovsky, M.: GeneMark.hmm: new solutions for gene finding. Nucleic Acids Research 26(4), 1107–1115 (1998)

14. Hulth, A.: Reducing false positives by expert combination in automatic keyword indexing. Recent Advances in Natural Language Processing III :.., 367–373 (2004)
15. Dietterichl, T.: Ensemble learning. In: Arbib, M.A. (ed.) The Handbook of Brain Theory and Neural Networks, 2nd edn., Cambridge, MA, pp. 1–8 (2002)
16. Richter, D.C., Ott, F., Auch, A.F., Schmid, R., Huson, D.H.: MetaSim: a sequencing simulator for genomics and metagenomics. PloS One 3(10), e3373 (2008)
17. Dietterich, T.G.: Machine-Learning Research. AI Magazine 18(4), 97–136 (1997)
18. Higuchi, T.: Relationship between the fractal dimension and the power law index for a time series: a numerical investigation. Physica D: Nonlinear Phenomena 46(2), 254–264 (1990)

Classification of Antimicrobial Peptides by Using the *p*-spectrum Kernel and Support Vector Machines

Paola Rondón-Villarreal, Daniel A. Sierra, and Rodrigo Torres

Universidad Industrial de Santander,
Carrera 27 calle 9, Bucaramanga, Colombia
paitorv@gmail.com, dasierra@uis.edu.co, rtorres@uis.edu.co
http://www.uis.edu.co

Abstract. In the last decades, antibiotic resistance of pathogenic microorganisms constitutes a great problem of public health at global level. Multidrug-resistant bacteria cannot be controlled with the existing medications causing thousands of deaths every year. In the fight against these bacteria, antimicrobial peptides have appeared as a promising solution as therapeutic agents against pathogens. For this reason, rational design of these chemical compounds have been explored by the scientific community in order to achieve significant improvements that could lead to the discovery of new antibacterial medicine. In this sense, the present work proposes the use of the *p*-spectrum kernel with support vector machines to classify antimicrobial peptides, thus considering only the information of the order of the amino acids inside the peptide sequences. The results were satisfactory and suggest that this information should be considered in the rational design of antimicrobial peptides.

Keywords: antimicrobial peptides, kernel methods, support vector machines.

1 Introduction

Nowadays, there are multiple microorganisms that are becoming resistant to the existing medications. Among them, the multidrug-resistant bacteria kill thousands of people across the globe every year, which represents an extreme risk for the humanity. Only in the United States the situation is becoming critical. More than 40 states have been affected with at least one patient infected with CRE (carbapenem-resistant Enterobacteriaceae) bacteria. The picture is disturbing considering that current medications can no control the infections caused by these super bacteria.

On the other hand, each year in the United Kingdom die about 2,500 patients by bloodstream infections caused by multidrug resistant bacteria. The major concern is that the available medication can no kill these organisms and the pharmaceutical industry is not so much interested in antimicrobial medicine developments.

L.F. Castillo et al. (eds.), *Advances in Computational Biology*,
Advances in Intelligent Systems and Computing 232,
DOI: 10.1007/978-3-319-01568-2_23,

The situation is even worse in developing countries where the majority of bacterial infections are treated empirically, because it difficult to identify the pathogen, and even more, the lack of methods to determine the susceptibility of the bacteria causing the infection [1].

In the fight against antimicrobial resistance, specially by bacteria, antimicrobial peptides (AMPs) seems to be a promising solution due to the broad spectrum of biological activity, the high mortality rate of pathogens and the low propensity to produce resistance in bacteria [2, 3]. These chemical compounds also present interesting biological activities of great interest in medicine, such as immunoregulatory [4], antiinflamatory [3, 5], antitumoral and anticancer activity [4]. However, these peptides present some drawbacks that make difficult their commercial use, such as liability to proteases, possible toxicity at systemic level, high cost of production, possibility to develop allergies, among others [2,3,5].

For this reason, the scientific community has been using computational tools to design new AMPs that present enhanced antimicrobial activity, lower toxicity to human cells and a small number of amino acids in their sequences. In general, these *in silico* models have been developed in three scenarios: improvement of the existing peptides, prediction of important peptide characteristics and classification processes of these chemical compounds

The majority of studies related to the classification of AMPs have used QSAR techniques and learning machines [6,7]. In this work, we proposed that the order of the amino acids in peptides gives enough information to classify antimicrobial peptides. In this sense, the present work proposes the use of the *p*-spectrum kernel with support vector machines (SVM) to classify these peptides. The obtained results showed that the order of the amino acids in a peptide is an important feature to take into account in the classification processes of antimicrobial peptides.

2 Materials and Methods

2.1 Kernel Methods

In multiple situations, the classification problems present data that cannot be differentiated with linear relations. In these cases, the usage of kernel methods allows to perform a linear classification process by the mapping of the data into an N-dimensional space of order N bigger than the order of the initial space. This can be done because usually the data in this new space, called feature space, is linearly separable [8].

In the kernel methods, there are two important elements: the kernel function and the kernel matrix. The kernel function is defined by (1) [9]

$$k(\mathbf{x}, \mathbf{z}) = \langle \phi(\mathbf{x}), \phi(\mathbf{z}) \rangle . \tag{1}$$

where $\mathbf{x}, \mathbf{z}$ are elements of any set and their image $\phi(\mathbf{x})$ is a vector in $\mathbf{R}^N$. This kernel function is used to obtain the kernel matrix, defined in (2), that contains the inner product of all pairs of data points in the feature space [9].

$$\mathbf{K}_{i,j} = \langle \phi(\mathbf{x}_i), \phi(\boldsymbol{x}_j) \rangle = k(\mathbf{x}_i, \mathbf{x}_j). \tag{2}$$

where $\mathbf{x}_i$, $\mathbf{x}_j$ are elements of any set and their image $\phi(\mathbf{x}_i)$, is a vector in $\mathbf{R}^N$ and $\mathbf{K}_{i,j}$ is the element in the row i and column j of the kernel matrix. It is important to mention that the kernel function calculates the inner product of the images of two elements in the feature space without explicitly computing the mapping of these elements.

The selection of the kernel function should consider the type of the input data, and the selection of the learning algorithm depends of the process that is required: classification, prediction or clustering. One of the algorithms used in classification processes is support vector machines [9], which was selected in this study as the learning algorithm used in the classification of antimicrobial peptides.

Moreover, the *p*-spectrum kernel was used in order to determine if a classification process of the different types of peptides could be performed considering only the order of the amino acids in the peptide without any physicochemical information.

2.2 *p*-spectrum Kernel

For a sequence S, its spectrum of order p corresponds to the histogram of all their contiguous substrings of length p. The kernel based on this spectrum, allows the comparison between two sequences, calculating the number of substrings of length p they have in common [9].

In this work the *p*-spectrum was calculated using an adaptation of the *p*-spectrum recursion algorithm defined in [9]. The algorithm was modified with the function *isEqual* and in the superior limit of the sum as shown in (3).

$$k_p(s,t) = \sum_{i=1}^{|s|-p+1} \sum_{j=1}^{|t|-p+1} isEqual(s(i:i+p-1), t(j:j+p-1)) \tag{3}$$

where $k_p(s,t)$ is the *p*-spectrum for sequences s and t, $|s|$ is the length of the sequence s, $s(i:i+p-1)$ is the subsequence of s that starts in position i and ends in position $i+p-1$ and $isEqual(a,b)$ is the function defined by (4)

$$isEqual(a,b) = \begin{cases} 1\ if & a = b \\ 0 & otherwise \end{cases} \tag{4}$$

A graphical representation of the creation of one antimicrobial peptides classifier using *p*-spectrum kernel and support vector machines is shown in Fig. 1. The first step consists in the creation of the kernel matrix, followed by the calculation

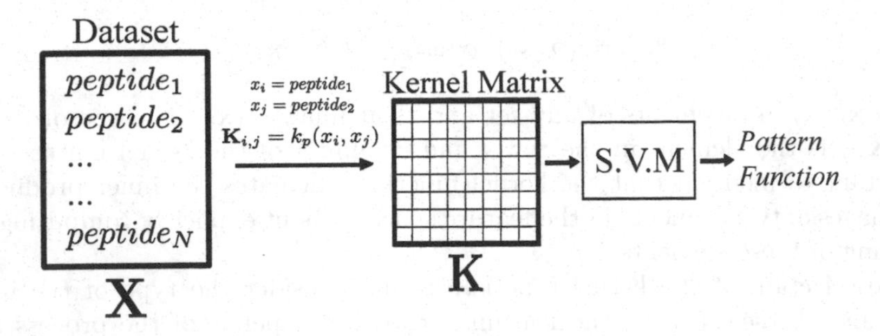

Fig. 1. General Diagram for the creation of the antimicrobial peptides classifier using the p-spectrum kernel

of the pattern function through the application of the learning algorithm, which in this work is support vector machines.

3 Results and Discussion

In this work, an antimicrobial peptide classifier was designed and created using only the information given by the p-spectrum kernel with $p = 3$, which it is different to majority of studies used in the classification of antimicrobial peptides, where QSAR techniques have been used for this aim [6,7,10]. Additionally, most of these methodologies have used an unbalanced dataset, while in this work we applied a random subsampling in order to obtain a balanced dataset.

The first step comprises the random subsampling of 600 antimicrobial peptides from the Antimicrobial Peptides Database (APD) [11] and 600 non-antimicrobial peptides from the negative dataset created by Wang *et al.* in [10]. In this sense, the initial dataset is composed by 1200 peptides sequences that will be used in the creation of 10 antimicrobial peptides classifiers using the 10-fold cross-validation technique. For each one of the 10 classifiers should be calculated a kernel matrix and then the libraries of LIBSVM [12] are used in order to obtain the pattern function using support vector machines, with the parameters values: $c = 1$ and $\gamma = 0.07$.

The mean values obtained in the 10-fold cross-validation process were: sensitivity $90.67 \pm 5.62\%$, specificity $83.50 \pm 4.19\%$, false positive rate $16.50 \pm 4.19\%$, false negative rate $9.33 \pm 5.62\%$ and precision $87.08 \pm 0.312\%$.

The final antimicrobial peptides classifier is created using the 70% of the initial data for the training process, i.e. 840 peptide sequences, and the 30% for testing (360 peptide sequences). The final antimicrobial peptides classifier created using the 70% of the data presents the following values: sensitivity 92.78%, specificity 83.89%, false positive rate 16.11%, false negative rate 7.22% and precision 88.33%.

For the sake of comparison a summary of the most relevant works in the classification of antimicrobial peptides is shown in Table 1.

Table 1. Summary of the most relevant works in classification of Antimicrobial Peptides

Tech. - Work [a]	Input dataset [b]	Val. Tech. [c]	Sn % [d]	Sp % [e]	Acc % [f]	Mcc [g]
DA [7]	2578/4011	2	-	-	87.5	0.74
NNA [10]	2752/10014	1	80.23	94.59	93.31	0.7312
RF [7]	2578/4011	2	-	-	93.2	0.86
SVM [13]	146/146	2	75.36	97.3	83.02	-
SVM [7]	2578/4011	2	-	-	93.2	0.86
This work	600/600	2	92.78	83.89	88.33	-

[a] Technique: DA=Discriminant Analysis, NNA=Nearest neighbor algorithm, RF=Random Forests, SVM=Support vector machines. [b] Input dataset: Number of positive samples/Number of negative samples. [c] Validation Technique: 1=Jackknife test, 2= 10-fold cross-validation [d] Sn=sensitivity. [e] Sp=specificity. [f] Acc=accuracy. [g] Mcc=Matthew's correlation coefficient.

4 Conclusions

From the obtained results it can be appreciated that the use of string kernels allows performing a satisfactory classification process. For this reason, it is feasible that the order of the amino acids inside the peptide sequences gives enough information to determine the presence or absence of any antimicrobial activity.

In addition, the obtained results using the 10-fold cross-validation technique show that the classifiers created using the *p*-spectrum kernel are statistically stable due to the small variation that they present in the performance measures when the training dataset is changed.

Finally, it can be observed that the performance values obtained for the antimicrobial peptides classifier created using the 70% of the initial data are closed to those values obtained using the 10-fold cross validation, suggesting that the final antimicrobial peptides classifier is statistically stable as well.

This work provides some evidence that the order of the amino acids provides enough information to classify peptides as AMPs.

References

1. Bennish, M.L., Khan, W.A.: What the Future Holds for Resistance in Developing Countries. In: Sosa, A.D.J., Byarugaba, D.K., Amábile-Cuevas, C.F., Hsueh, P.R., Kariuki, S., Okeke, I.N. (eds.) Antimicrobial Resistance in Developing Countries, Springer, New York (2010)
2. Baltzer, S.A., Brown, M.H.: Antimicrobial Peptides - Promising Alternatives to Conventional Antibiotics. J. Mol. Microbiol. Biotechnol. 20 (2011)
3. Pasupuleti, M., Schmidtchen, A., Malmsten, M.: Antimicrobial peptides: key components of the innate immune system. Crit. Rev. Biotechnol. (2011)

4. Wu, W.K.K., Wang, G., Coffelt, S.B., Betancourt, A.M., Lee, C.W., Fan, D., Wu, K., Yu, J., Sung, J.J.Y., Cho, C.H.: Emerging roles of the host defense peptideLL-37 in human cancer and its potential therapeutic applications. Int. J. Cancer 127 (2010)
5. Giuliani, A., Pirri, G., Nicoletto, S.F.: Antimicrobial peptides: an overview of a promising class of therapeutics. Cent. Eur. J. Biol. 2 (2007)
6. Torrent, M., Andreu, D., Nogués, V.M., Boix, E.: Connecting Peptide Physicochemical and Antimicrobial Properties by a Rational Prediction Model. PLoS ONE 6 (2011)
7. Thomas, S., Karnik, S., Barai, R.S., Jayaraman, V.K., Idicula-Thomas, S.: CAMP:a useful resource for research on antimicrobial peptides. Nucleic Acids Res. 38 (2010)
8. Cristianini, N., Shawe-Taylor, J., Saunders, C.: Kernel Methods: A Paradigm for Pattern Analysis. In: Camps-Valls, G., Rojo-Álvarez, J.L., Martínez-Ramón, M. (eds.) Kernel Methods in Bioengineering, Signal and Image Processing. Idea Group Publishing, London (2007)
9. Shawe-Taylor, J., Cristianini, N.: Kernel Methods for Pattern Analysis. Cambridge University Press, New York (2004)
10. Wang, P., Hu, L., Liu, G., Jiang, N., Chen, X., Xu, J., Zheng, W., Li, L., Tan, M., Chen, Z., Song, H., Cai, Y.D., Chou, K.C.: Prediction of antimicrobial peptidesbased on sequence alignment and feature selection methods. PloS one 6 (2011)
11. Wang, Z., Wang, G.: APD: the Antimicrobial Peptide Database. Nucleic Acids Res. 32 (2004)
12. Chang, C.C., Lin, C.J.: LIBSVM: A library for support vector machines. ACM TIST 2 (2011), Software available at http://www.csie.ntu.edu.tw/~cjlin/libsvm
13. Porto, W.F., Fernandes, F.C., Franco, O.L.: An SVM Model Based on Physicochemical Properties to Predict Antimicrobial Activity from Protein Sequences with Cysteine Knot Motifs. In: Ferreira, C.E., Miyano, S., Stadler, P.F. (eds.) BSB 2010. LNCS, vol. 6268, pp. 59–62. Springer, Heidelberg (2010)

Genomic Relationships among Different Timor Hybrid (*Coffea* L.) Accessions as Revealed by SNP Identification and RNA-Seq Analysis

Juan Carlos Herrera[1,*], Andrés Mauricio Villegas[1], Fernando A. Garcia[1], Alexis Dereeper[2], Marie-Christine Combes[2], Huver E. Posada[1], and Philippe Lashermes[2]

[1] Coffee Breeding Program. Centro Nacional de Investigaciones de Café, CENICAFE. Chinchiná, Caldas. Colombia
{juanc.herrera,andres.villegas,fernando.garcia, huver.posada}@cafedecolombia.com
www.cenicafé.org

[2] IRD, UMR RPB (IRD, CIRAD, Université Montpellier II), 911 avenue Agrópolis, BP 64501, 34394 Montpellier Cédex 5, France
{alexis.dereeper,marie-christine.combes, philippe.lashermes}@ird.fr
www.ird.fr

Abstract. The Timor hybrid (TH), a natural hybrid between *C. arabica* and *C. canephora*, is the main source of resistance to coffee leaf rust disease that has been used as progenitor during development of most of the modern *Coffea arabica* L. varieties. In this work a comparison of the introgression level of three accessions of the TH was conducted using massive RNAseq data analysis. To investigate the number of unigenes possibly impacted by introgression, the characterization and quantification of genome-derived SNPs were carried out on almost 20,000 unigenes. Overall results confirmed the CIFC1343 as the most introgressed accession of the TH when compared to either CIFC832-1 or CIFC832-2. Although less introgressed, the CIFC832-2 seems to be an interesting alternative for coffee breeders because it carries additional genome introgressions than observed for CIFC832-1. Our findings illustrate an alternative approach that use RNAseq data for SNP identification and interpretation in a polyploid species.

Keywords: Gene introgression, genetic resources, disease resistance, gene expression.

1 Introduction

Coffee is considered as one of the world's favorite beverages, the second most traded commodity after oil, and one crucial to the economies of several countries particularly in Latin-America. Only two species are responsible for commercial production: *Coffea arabica* (Arabica coffees) and *C. canephora* (Robusta coffees). All coffee

* Corresponding author.

L.F. Castillo et al. (eds.), *Advances in Computational Biology*,
Advances in Intelligent Systems and Computing 232,
DOI: 10.1007/978-3-319-01568-2_24,

species are diploid, except *C. arabica,* which is allotetraploid (2n = 4 x = 44) and derived from a recent (less than 50000 years ago) interspecific hybridization between two diploid species: *C. eugenioides* and *C. canephora*. Homoeologous genomes in *C. arabica* have been designated as E^a and C^a according to their parental origin (Lashermes et al. 1999; Cenci et al. 2012).

Plantations of *C. arabica* around the world are affected by several diseases, among them the most important is the coffee leaf rust (CLR), caused by the biotrophic fungus *Hemileia vastatrix* Berk & Br, also considered as the most devastating disease for this culture. In order to prevent spread of the disease, different breeding programs for rust resistance were initiated in many countries since 1970. To date, the Timor hybrid (TH), which is the result of a spontaneous cross between *C. arabica* and *C. canephora* species, is the main source of resistance to CLR which has been used as a progenitor during the production of most of the modern improved varieties in Latin America but also in Africa, Asia and Oceania (Bettencourt 1981; Rodriguez et al. 2000).

Timor hybrid presents an arabica phenotype, is self-fertile, and bears a tetraploid number of chromosomes (2n=4x=44) as *C. arabica.* At least five major resistance genes (S_H5 to S_H9) have been identified in the TH accessions, all of them supposedly coming from the Robusta side of the hybrid. From 1960, several clones and offspring of HT accessions CIFC832-1, CIFC 832-2 and CIFC1343, among others, were distributed to several coffee-producing countries, including Colombia, by the *Centro de Investigaçao das Ferrugens do Cafeeiro* (CIFC). As result, a number of commercial varieties were produced at the end of 80's and planted in extensive areas of Central and South America (Rodrigues Jr et al. 2000).

Recent advances in DNA depth sequencing have made it possible to sequence cDNA derived from cellular RNA by massively parallel sequencing technologies, a process currently termed as RNA-seq, making possible to identify not only transcriptome expression variations but also single nucleotide polymorphism (SNP) in a broad range of species including coffee (Mammadov et al. 2012). SNP-based genotypic data has been used to investigate numerous questions of evolutionary, ecological, and conservation significance in model and non-model organisms. Thanks to their broad genomic distribution and direct association with functional implications, SNPs represent today an improvement over conventional markers. Therefore, the aim of this report was to investigate the genomic relationship among three of the most important TH accessions from the Colombian coffee germplasm, throughout a RNAseq-based strategy. To do that we propose a new approach for detecting genomic differences in introgression by identification (through their SNP-content comparative analysis) of genes possibly impacted by this process in the Timor Hybrid resource.

2 Material and Methods

2.1 Rnaseq Library Preparation and Sequence Generation

Total RNA from TH accessions: CIFC1343, CIFC 832-1 and CIFC 832-2 were extracted from four different tissues (leaf, flower, young berry, old berry) using the mRNA plant mini kit (Quiagen ®) according to manufacturer's instructions. Total RNA quality and concentration were determined using the Eukaryote Total RNA Nano Assay (Agilent, Santa Clara, CA) on a 2100 Bioanalyzer (Agilent). RNA library preparation was performed using the TrueSeq RNA sample kit (Illumina, San Diego, CA) according to manufacturer's instructions. Prior to cluster generation, library concentration and size were assayed using the Agilent DNA1000 kit ®. Libraries from all samples were sequenced in a single flow-cell (3 libraries per lane) on the Illumina HiSeq 2000 using the sequence by synthesis (SBS) technology, at the MGX platform (Montpellier Genomix, *Institut de Génomique Fonctionnelle*, Montpellier France). Final reads were single-end 75 nt, with a separate read to sequence the sample index. Image analysis, base-calling and quality filtering were processed by Illumina software.

2.2 Data Analysis and Detection of Differential SNPs

The overall reads were aligned using BWA (Li & Durbin 2009) against a *C. canephora* expressed sequence tag (EST) assembly (with 56, 216 unigenes) as reference transcriptome. This reference base was built using transcripts from various tissues and Sanger as well as deep Illumina sequencing as described in Combes et al. (2013). A maximum of four mismatched nucleotides (including gaps) between the read and the reference transcriptome sequence was allowed. For each accession, the unambiguously aligned sequences were then analysed for SNP discovery with the GATK toolkit (McKenna et al. 2010; http://www.broadinstitute.org/gatk/) using the Unified Genotyper module with default parameters to obtain SNP list and allelic data, and the Depth Of Coverage module to get depth coverage information. Based on the GATK outputs, quantification and comparison of SNPs were conducted using SNiploid a dedicated web-based tool (Dereeper et al. 2011; http://sniplay.cirad.fr/cgi-bin/sniploid.cgi). SNiPloid was used to compare inter and intra-genotypic SNPs and to classify the unigenes carrying SNPs into different hypothetical evolution-based categories by comparison with the observed situation in the non-introgressed allopolyploid *C. arabica* (Figure 1).

In order to infer and quantify those unigenes with SNPs possibly impacted by the introgression process, both the relative difference, RD (i.e. difference in SNP content) and the intrinsic difference, ID (i.e. difference in SNP number) index between Caturra (P1) and each of the TH accessions (P2), were calculated per unigene as follows: RD = [Number of SNPs between P1 and P2/ Total number of positions exhibiting SNPs in either P1 or P2] * 100; ID = [Number of SNPs in P1/Number of SNPs in P2] * 100. Overall comparisons of RD and ID index between *C. arabica* and the TH accessions were interpreted as changes at the nucleotide level, as result of the introgression process into the TH genome (Figure 2). As result of this analysis, four categories of unigenes were considered as follows: Type 0: those being considered as non-introgressed; Type 1: unigenes with SNPs derived from introgression into E^a subgenome; Type 2: unigenes with SNPs derived from introgression into C^a subgenome; and Type 3: unigenes carrying uninterpretable SNPs, possibly associated with local genome homogenization.

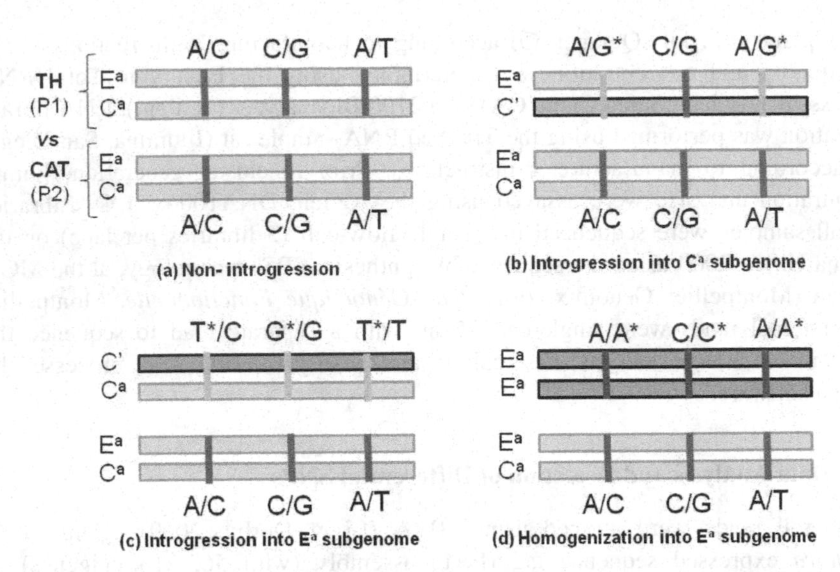

Fig. 1. Possible introgression scenarios as expected to occur during natural hybridization of the Timor hybrid resource. (a) scenario of no-introgression, as observed in *C. arabica*, (b) introgression into the C^a subgenome; (c) introgression into the E^a subgenome; (d) genome homogenization into the E^a subgenome, as result of introgression process. For each scenario single nucleotide substitutions are indicated by asterisk.

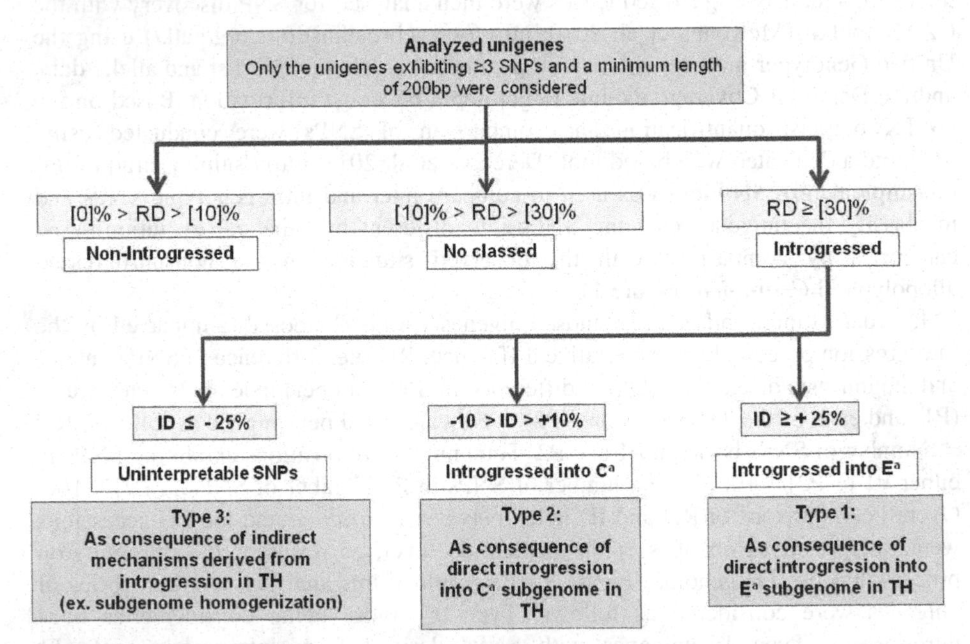

Fig. 2. Pipeline showing discrimination of unigenes by the presence of SNP possibly derived from the introgression process. Categorization of introgressed *vs.* non-introgressed unigenes was carried out by analyses of RD and ID indexes calculated for each unigene (comparison between *C. arabica* var. Caturra and each of the TH accessions).

Furthermore, mapped sequence counts were used to estimate the expression level of the different unigenes (non-introgressed as well as putatively introgressed unigenes). Differences in gene expression between the different TH accessions were tested using the feature-assigned fragment counts for each replicate as input to the DESeq package (Anders and Huber, 2010). Only genes with adjusted p-values below 0.05 were considered as differentially expressed.

3 Results and Discussion

In this report we sought to estimate the number of unigenes possibly impacted by introgression among the TH accessions throughout SNP categorization. To assess the differences of possibly introgressed-unigenes, the transcriptome of different tissues of the three accessions of the TH was sequenced. The mRNA-seq data allowed on one side to detect SNP within each sample (i.e. homoeologous SNPs) and between each TH accession and a variety of *C. arabica* used as standard and on the other side to quantify transcript (either overall or homoeologous) abundance for a large number of genes. Using this information it was possible on the one hand, to investigate the frequency of unigenes putatively impacted by introgression and another hand, to characterize the differentially expressed unigenes among the TH accessions.

Results showed that the total number of SNPs ranged from 175,419 in CIFC832-1 to 243,801 for CIFC832-2. These values corresponded to 93% of the total single nucleotide polymorphisms detected in the TH (i.e. the 3 TH accessions considered altogether). For further quantification of the introgression, only unigenes classified as Type 1, Type 2 and Type 3 were considered. As showed in Table 1 the CIFC1343 was the most introgressed accession with 16.8 % (3438) of unigenes having SNPs possibly originated from introgression process, while the CIFC832-1 was the less introgressed with only 10.6 % (1486). Among the three TH accessions, around 63 % of unigenes appeared not introgressed and for 23% of them it was not possible to infer the origin of the observed SNPs.The Venn diagram in Figure 3 showed the unique unigenes but also the impacted unigenes shared by the different TH accessions. The CIFC 1343 for example, exhibited the highest number (1670) of unique unigenes than the others (799 and 233 for CIFC832-2 and CIFC832-1, respectively).

Table 1. Comparison of type and frequency of analyzed unigenes relative to the introgression process when comparing *C. arabica* var. Caturra (P1) to the different TH accessions (P2).

Comparison P1 vs P2	Total analyzed unigenes	Total informative unigenes	Freq (%) of non-introgressed unigenes	Freq (%) of uninterpretable unigenes	Freq (%) of introgressed unigenes
CAT vs CIFC1343	30,170	20,175 (66.9 %)	60.1	22.9	16.8
CAT vs CIFC832-1	21,044	14,110 (67.0 %)	67.6	21.8	10.6
CAT vs CIFC832-2	28,339	19,055 (67.2 %)	62.2	23.7	14.2

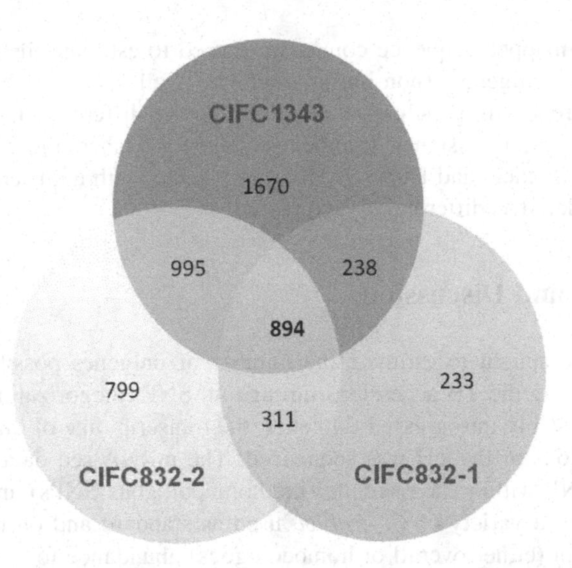

Fig. 3. Venn diagram showing the number of unique and shared unigenes (carrying SNPs possibly as product of the introgression process) between the 3 TH accessions

This accession also shared most impacted unigenes with CIFC832-2 than it does with CIFC832-1. It was interesting to note that 894 unigenes were common among the different TH accessions. Although all three TH accessions were sampled under similar field conditions, quite differences in gene expression were detected. Indeed, final DESeq analysis showed that most of the differentially expressed unigenes (419) were found when compared between CIFC1343 and CIFC832-2 (Figure 4). At the contrary, less number of differentially expressed unigenes was detected between CIFC832-1 and CIFC832-2 (78).

The CIFC1343 accession has been used intensively in development of breed derived progenies with rust resistance in Colombia. Similarly, the CIFC832-1 and CIFC832-2 represent the main sources for development of cultivars like IAPAR59 and CR95 in Brazil and Costa Rica, respectively (Alvarado and Castillo 1996; Bertrand et al. 1999; Rodrigues Jr et al. 2000). All of these resources supposedly contain different resistance genes not only for rust but for other important diseases like the coffee berry disease, CBD (Rodrigues Jr et al. 2000). Despite its importance for the *C. arabica* breeding programs, very few studies have been undertaken to compare the genetic origin and nature of these genes. In this scenario, the RNAseq approach as used here represents an important step toward knowledge of the nature of the gene introgression issued from the Timor hybrid.

Overall our results pointed the fact that CICF1343 remains the most important resource for cultivar development in Colombia because its level of gene introgression compared to other accessions. In practical terms, this suggests that CIFC1343 would to maintain additional introgressed genes of putative interest for future Arabica breeding. Nevertheless, the long exposure (more than 25 years) of CIFC1343-derived progenies to field CLR races in Colombia, obligate to think about the use of

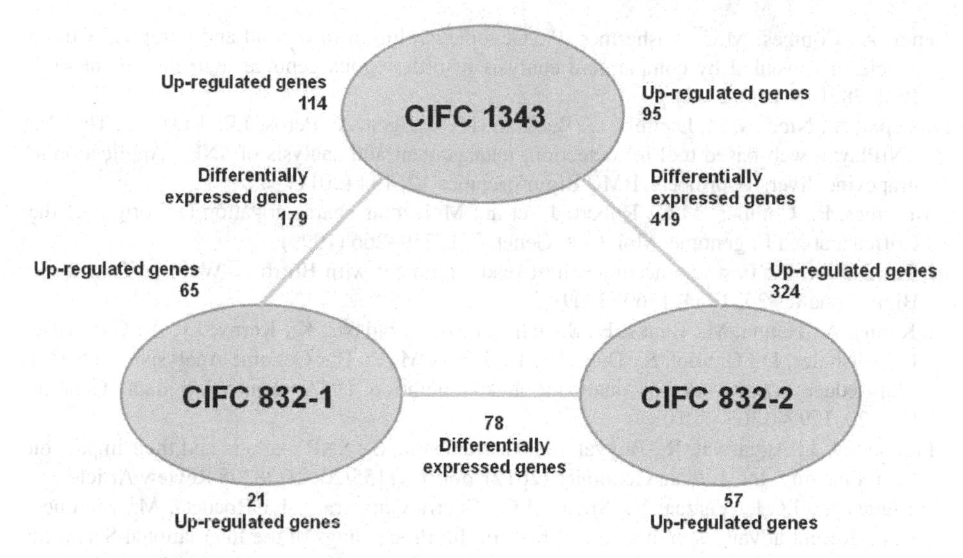

Fig. 4. Comparison between Timor hybrid accessions relative to the number of up-regulated unigenes as detected by the DESeq analysis

alternative sources for rust resistance. In this sense, the CIFC832-2 accession becomes an interesting option as source of new genes. This accession although less introgressed, seems to involve additional (new?) introgressed fragments than CIFC832-1. Further work is in progress to carry out precise annotation of introgression-related genes into the TH accessions in order to identify those involved not only on disease resistance (i.e. rust, CBD), but also on additional traits of interest for the coffee breeders (e.g. adaptation to environmental variation, cup quality).

Acknowledgements. This work is part of a co-funded joint project between CENICAFE (Colombia) and IRD (France). JCH was supported by a grant from IRD (BEST, program grants for technological and scientific exchanges).

References

Alvarado, G., Castillo, J.: Progreso de la roya del cafeto sobre genotipos resistentes y susceptibles a Hemileia vastatrix. Cenicafé 47, 42–52 (1996)

Anders, S., Huber, W.: Differential expression analysis for sequence count data. Genome Biology 11, R106 (2010)

Bertrand, B., Aguilar, G., Santacreo, R., Anzueto, F.: El mejoramiento genético en America Central. In: Bertrand, B., Rapidel, B. (eds.) Desafíos de la caficultura centroamericana. IICA Publishers, Coronado (1999)

Bettencourt, A.J.: Melhoramento genético do cafeeiro. Tranferência de factores de resistência â H. vastatrix para as principais cultivares de Coffea arabica. Centro de Investigaçao das Ferrugens do Cafeeiro, CIFC. Lisboa, Portugal. 93 p (1981)

Cenci, A., Combes, M.C., Lashermes, P.: Genome evolution in diploid and tetraploid Coffea species as revealed by comparative analysis of orthologous genome segments. Plant Mol. Biol. 78, 135–145 (2012)

Dereeper, A., Nicolas, S., Lecunff, L., Bacilieri, R., Doligez, A., Peros, J.P., Ruiz, M., This, P.: SNiPlay: a web-based tool for detection, management and analysis of SNPs. Application to grapevine diversity projects. BMC Bioinformatics 12, 134 (2011)

Lashermes, P., Combes, M.C., Robert, J., et al.: Molecular characterization and origin of the Coffea arabica L. genome. Mol. Gen. Genet. 261, 259–266 (1999)

Li, H., Durbin, R.: Fast and accurate short read alignment with Burrows–Wheeler Transform. Bioinformatics 25, 1754–1760 (2009)

McKenna, A., Hanna, M., Banks, E., Sivachenko, A., Cibulskis, K., Kernytsky, A., Garimella, K., Altshuler, D., Gabriel, S., Daly, M., De Pristo, M.A.: The Genome Analysis ToolKit: A MapReduce framework for analyzing next-generation DNA sequencing data. Genome Res. 20, 1297–1303 (2010)

Mammadov, J., Aggarwal, R., Buyyarapu, R., Kumpatla, S.: SNP markers and their impact on plant breeding. Int. J. Plant Genomics (2012), doi: 10.1155/2012/728398 Review Article

Rodrigues Jr., J.C.J., Varzea, V., Silva, M.C., Guerra-Guimaraes, L., Rocheta, M., Marques, D.V.: Recent advances on coffee leaf rust. In: In: Proceedings of the International Scientific Symposium on Coffee, CCRI. Coffee Board, Bangalore, India, pp. 179–193 (2000)

A Combined Sensitivity and Metabolic Flux Analysis Unravel the Importance of Amino Acid Feeding Strategies in Clavulanic Acid Biosynthesis

Claudia Sánchez[1], Natalia Gómez[2], Juan Carlos Quintero[3], Silvia Ochoa[3], and Rigoberto Rios[3]

[1] University of Antioquia, Food Eng. Dept. Medellin – Colombia
clapatriciasa@gmail.com
[2] University of Antioquia, Env. Academic Corp. Medellin - Colombia
natgogri@gmail.com
[3] University of Antioquia, Chemical Eng. Dept. Medellin – Colombia
{jcquinte,sochoa,rrios}@udea.edu.co

Abstract. Metabolic flux analysis (MFA) is a computationally-dependent mathematical framework which can be used for determining the more representative variables of a biological system; MFA can be greatly improved by means of a sensitivity analysis (SA). In this work, we used a SA and a MFA to study the *Streptomyces clavuligerus* metabolism for clavulanic acid (CA) production. It was observed that the measured metabolic fluxes that significantly affected the cellular system were phenylalanine, isoleucine, tyrosine and lysine. According to the flux distribution at two dilution rates (D), the biosynthesis of CA was favored at low D, leading to higher values in the precursor fluxes of the tricarboxylic acid and urea cycles. Moreover, it was found that the flux of ornithine, rather than the flux of arginine, affects the biosynthesis of CA, which evidences the importance of controlling the flux of carbon through the urea cycle.

Keywords: Clavulanic acid, MFA, *Streptomyces clavuligerus*, β-lactamase.

1 Introduction

Streptomyces *clavuligerus (S. clavuligerus)* can produce a large number of compounds among which Cephamycins and Clavulanic Acid (CA), are the most actively studied; CA is a potent β-lactamase inhibitor used to counteract the microbial resistance created by some pathogenic microorganisms, capable of producing β-lactamase enzymes for destroying the antibiotic [1]. By combining genetic and Metabolic Engineering (ME) techniques, it has been possible to improve the carbon flux distribution leading to larger CA production. Metabolic Flux Analysis (MFA) is a tool aiming at discerning and optimizing the Metabolic Flux Distribution (MFD) in target organisms [2]. Some of the more relevant works applying MFA in *Streptomyces* have been carried out on species such as *tenebrarius*, *lividans*, *coelicolor*, *clavuligerus* and *avermitilis* [3, 4]. Concerning CA, there are relevant studies which have reported the applicability of MFA. Kirk et al. (2000) used MFA for determining the MFD in the Central carbon

L.F. Castillo et al. (eds.), *Advances in Computational Biology*,
Advances in Intelligent Systems and Computing 232,
DOI: 10.1007/978-3-319-01568-2_25,

metabolism (CCM) of *S. clavuligerus,* in nutrient-limited culture media [7]. By calculating the MFD, the authors proposed a strategy for feeding amino acids which promoted the availability of arginine (Arg), the C5 precursor for CA biosynthesis. In a related work, Bushell et al. (2006) found that, feeding a mixture of aspartate (Asp), asparagine (Asn) and threonine (Thr), the fluxes toward the Arg synthesis pathway were increased, producing a CA yield 18-fold higher [4]. Meanwhile, Daae and Ison (1999) developed a SA method for studying metabolic responses when the cell faces changes in environmental conditions; by using this approach, the authors were able to analyze the effect that disturbances on measured fluxes ($\mathbf{v_m}$) such as the oxygen flux, exert on the system's MFD [5].

In this work, the batch and continuous production of CA is studied. Also, the effects of direct amino acid addition and the variation of the dilution rate (D), on the MFD, are investigated by means of a combined MFA and SA. Finally, a depiction of the effect of individual flux variations on CA accumulation is presented.

2 Materials and Methods

2.1 Microbiological Methods and Inoculum Preparation

A lyophilizate of *S. clavuligerus ATCC 27064* was activated in TSB® broth (28 ºC, 36 h). The strain was kept at -80ºC in Eppendorff tubes with a TSB® - 40% glycerol medium [6]. Pre-inoculums were prepared in TSB® medium and incubated (28°C, 220 rpm) during 36 h for ensuring exponential growth (biomass content: 8 - 9 gL^{-1}). Fermentation assays used a Production Media (PM) denoted as "*medio 2"* in [6].

2.2 Bioreactor Cultures

Cultures were performed in an instrumented 3L BIOFLO 110 bioreactor (New Brunswick), (1 L of PM, 1 vvm, 500 rpm, 28°C and pH at 6.8 ±0.2). Foaming was prevented with the early addition of $2mLL^{-1}$ of *Antifoam 143.* Batch experiments were conducted during 100 h. Samples were taken periodically for the determination of biomass (X), glycerol (GLC), amino acids, phosphate, O_2 and CA. For starting up the continuous mode it was first necessary to fulfill a 36h-batch culture. Two dilution rates (D) (0.02 and 0.03 h^{-1}) were evaluated. Steady state operation was verified by measuring X and CA concentration at each residence time.

2.3 Analytical Methods

Samples were centrifuged (14000 rpm, 4ºC, 10 min); supernatant was filtered using 0.2 μm PTFE membranes and stored at -20°C until HPLC analysis (Agilent Tech. Series 1200); recovered pellets were dried at 105°C for 4h in pre-dried Eppendorff tubes for biomass determination. CA, amino acids and glycerol were quantified by HPLC as in [6]. Phosphate content was determined by a colorimetric method (the molybdenum blue method). The dynamic technique was used for calculating oxygen consumption rate.

2.4 Metabolic Network Abstraction

The proposed stoichiometric metabolic model comprised 60 reactions and 47 metabolites present in the CCM, biosynthesis of biomass, urea cycle and CA biosynthesis. The partially known clavams' pathway as well as the Entner-Doudoroff and glyoxylate pathways (uncommon in most *Streptomyces* strains [4]) were not included. Model building was based on scientific literature and on-line databases e.g. KEGG.

2.5 Metabolic Flux and Sensitivity Analysis

MFA relies on stoichiometries which impose constraints on the carbon flux through the pathways [2]. A proper mathematical representation of the biochemical reactions is a numerical matrix, in the form of $\boldsymbol{E} \cdot \boldsymbol{v} = \boldsymbol{0}$ (eq. 1), (**E:** stoichiometric matrix; **v:** vector of fluxes). This case study has an **E** matrix 60 x 47. Equation (1) can be re-written considering that $\boldsymbol{v}$ is composed by measured fluxes ($\mathbf{v_m}$) and calculated fluxes ($\mathbf{v_c}$); re-organizing, $\mathbf{v_C} = -[\mathbf{E_C}]^{-1}\mathbf{E_m}\,\mathbf{v_m}$ (eq. 2) is acquired as the base equation for MFA. When solving eq. 2, lower error propagation is attained by formulating a well-conditioned problem. If the problem is ill-conditioned, it is necessary to choose another vector $\mathbf{v_m}$. The Condition Number (CN) is an index that determines whether the problem is well-posed or not; $\mathbf{CN} = \|\mathbf{E_C}\|\|\mathbf{E^{\#}_C}\|$ (eq. 3), ($\mathbf{E^{\#}_C}$ is the pseudo-inverse matrix). Well-conditioned problems have CNs lower than 1000. The SA is fulfilled making a perturbation in the measured fluxes (one at a time), to assess how much the vector of calculated fluxes varies. From (eq. 2), this perturbation can be re-written as $\mathbf{v_{C2}} - \mathbf{v_{C1}} = -[\mathbf{E_C}]^{-1}\mathbf{E_m}(\mathbf{v_{m2}} - \mathbf{v_{m1}})$ (eq. 4) [5]. Rearranging, (eq. 4) becomes $[d\mathbf{v_C}/d\mathbf{v_m}] = -[\mathbf{E_C}]^{-1}\mathbf{E_m}$ (eq. 5), which represents the sensitivity of $\mathbf{v_c}$ to changes in $\mathbf{v_m}$. For the current case study, 28 fluxes were proposed as measured fluxes for fulfilling the SA (GLC, O_2, AC, NH_4, (X), oxaloacetate (OAA), pyruvate (PYR), alpha-ketoglutarate (aKG) and glyceraldehyde-3-phosphate (GAP), Arg, Glu, Ser, Phe, Tyr, Gln, Trp, Val, Ala, Leu, Orn, Asn, Cys, Met, Thr, Lys, Pro, Ile and Gly). The procedure for choosing the $\mathbf{v_m}$ vector was: 1) several vectors $\mathbf{v_m}$ were proposed, 2) CNs were calculated for each proposed vector, 3) SA for each $\mathbf{v_m}$ vector with CN lower than 1000 was evaluated, 4) the $\mathbf{v_m}$ vector with lower sensitivity index was used for MFA. The SA and MFA were performed using CellNetAnalyzer [7].

3 Results and Discussion

3.1 Batch and Chemostat Production of Clavulanic Acid

CA biosynthesis (batch) reached 62.3 mgL^{-1} at 72h (Figure 1). Afterward, its concentration decreased as a result of nutrient depletion, e.g. the nitrogen source; the microorganism overcomes such nutrient depletion by consuming CA, previously synthesized. (Production/degradation of CA by *S. clavuligerus* take place simultaneously, and degradation becomes dominant when CA production stops). According to Fig. 1, at 36-42 h of cultivation the system was under Asn-limited conditions; however, there was no evidence of phosphate or GLC limitation. The DO was kept above 50% saturation (data not shown). The maximum specific growth rate (μ_{max}) was 0.0685 h^{-1}.

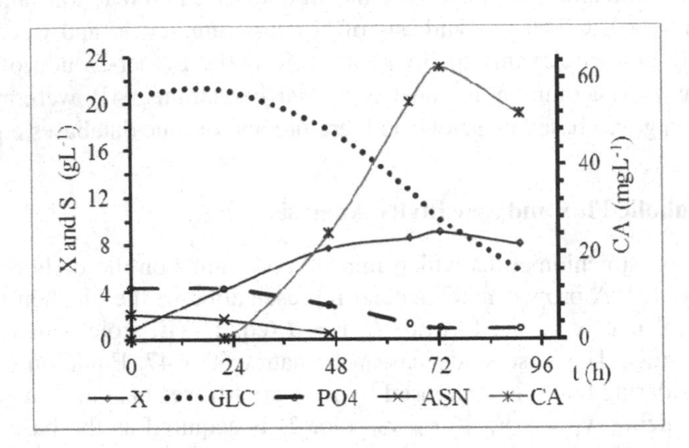

Fig. 1. Batch production of clavulanic acid. (1L, 500 rpm, 28°C, 1vvm, pH 7).

For the continuous culture, data were attained (data no shown) at two D, after verifying steady state operation. The productivity of CA was higher at lower D (0.02 h^{-1}), a fact already reported [4]. It was also observed that decreasing D, the demand for GLC, O_2 and Asn, increases. In addition, the amino acid synthesis rate (Phe, Trp, Ile, Asn, Gly, Ala), in most cases, was higher at a lower D.

3.2 Sensitivity Analysis for *Streptomyces clavuligerus*

The sensitivity values of calculated fluxes to variations in measured fluxes (estimated from equation 5), and the selected vector $\mathbf{v_m}$ that resulted in the lowest CN (CN=138) were calculated (data no shown). Considering the *accumulated sensitivity*, which represents the total impact of each element of $\boldsymbol{v_m}$ on $\boldsymbol{v_c}$, it was observed that the measured fluxes of Phe, Ile and Tyr had the largest effect on the metabolic system, accounting for about 55% of the global sensitivity. The CA flux, in turn, had the highest sensitivity with respect to these amino acids. Similarly, when CA was considered as a measured flux, it had the major influence on the system (accumulated sensitivity index: 77.64). This high sensitivity is related to its stoichiometry, involving around 22% of metabolites in the whole pathway. When Asp, Glu and Asn were considered as measured fluxes, no effect on CA production was observed.

Analyzing the individual effect of each v_m on the MFD, for perturbations in the fluxes of Phe, Ile and Tyr, the more sensitive calculated fluxes were v_{45} (transhydrogenation reaction), v_{34}, v_{35}, and v_{36} (oxidative PP pathway). Flux v_{34} corresponds to the formation of glucose-6phosphate (G6P) from fructose-6-phosphate (F6P). G6P was proposed by Orduña (2000), as a sensitive indicator of nutrient limitation, even before the microbial growth starts ceasing [8]. The sensitivity values of these fluxes with respect to Phe, Ile and Tyr, were positive, implying that increasing Phe, Ile and Tyr ends up raising the global flux in the oxidative PP pathway; consequently, depletion of carbon, available for product generation, is promoted. For perturbations in the measured fluxes of Val, Ala, GLC and O_2, the more sensitive calculated fluxes are the same fluxes found for the case in which the flux of Phe was considered as measured. The sensitivity for the flux of CA (v_{AC}), when Val was used as a measured flux, is negative, whereas for changes in the flux of Ala the v_{AC} had a positive sensitivity. Likewise, the calculated fluxes that showed to be more sensitive to disturbances of the measured flux of biomass were v_{34}, v_{35}, v_{36} and v_{45}. By increasing the flux of biomass generation, it was possible to reach an increase in the fluxes of the PP pathway and in the trans-hydrogenation flux. Correspondingly, the biomass flux had an inverse correlation with respect to the calculated product flux, also found in [5].

3.3 Metabolic Flux Analysis in *Streptomyces clavuligerus*

For computing the pathway MFD, the metabolic model was solved using equation 2, at two different D, see Figure 2. The SA results were taken into account for MFA calculations, so the vector $\mathbf{v_m}$ comprised the fluxes of Phe, Ile and Tyr. The most significant fluxes obtained from the MFA, at both dilution rates, correspond to: 1) the oxidative phosphorylation (v_{41}), 2) trans-hydrogenation (v_{45}), 3) oxidative PP pathway (v_{35}), 4) the glycolytic pathway (EMP) (v_{33}), 5) TCA cycle (v_{15}), 6) Biomass (v_{41}) and 7) CA (v_3). Results shown in Figure 2 indicate that lower D favors the production of CA, thus reducing biomass growth rate (v_{32}). Besides, by reducing D, higher fluxes in the glycolytic pathway were observed, e.g. v_{3PG}, v_{PEP} and v_{PYR}. The anaplerotic flux (v_6) promotes the metabolic activity in the TCA cycle along with the flux v_{11}. In addition, there is a higher PYR consumption at higher D, which is destined for the production of X, Lys, Ile, Leu, Ala and Val. This leads to a depletion of carbon flux towards the TCA cycle and a subsequent limitation on the availability of OAA and aKG.

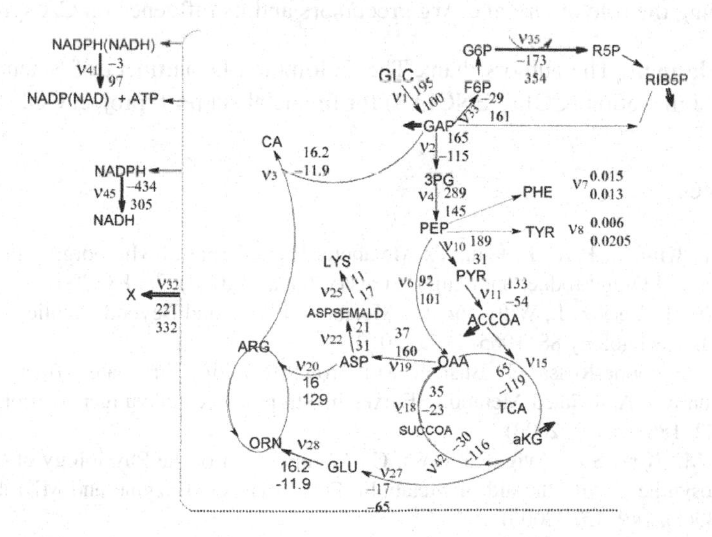

Fig. 2. Metabolic flux distribution in *Streptomyces clavuligerus* (Sc). The upper values correspond to the $0.02h^{-1}$ dilution rate; the lower ones correspond to the $0.03h^{-1}$

The metabolite aKG is the Glu precursor which favors the synthesis of Orn; the fluxes in this direction increased at high D. Interestingly, the calculated flux obtained for Orn did match that for CA, at both dilution rates; therefore, for CA biosynthesis, the Orn flux is limiting rather than Arg, since, despite of having availability of Arg and limitation of the Orn flux (at the higher D), the CA biosynthesis was not favored.

4 Conclusions

By combining a SA and a MFA it was possible to establish how the formation rate of different amino acids and/or other metabolic products affect the MFD for CA production in *S. clavuligerus*. The SA showed that compounds such as OAA, aKG, GAP, Arg, Orn and Asp do not affect significantly the MFD during CA production, despite the fact of being nodal points with high connectivity in the metabolic network. For the proposed metabolic pathway, it was found that the fluxes of Phe, Tyr and Ile are the measured fluxes with the highest impact on the cellular system. Moreover, the calculated fluxes which are highly affected by those measured, are the fluxes involved in NADPH generation in both, the oxidative pentose phosphate pathway and in the trans-hydrogenation reaction. Overall, the MFA showed that for reaching high yield in CA production, it is necessary to assure a good availability of the GAP, PEP, PYR, OAA and Orn fluxes. By observing the complete MFD, one can infer that the CA precursor, Orn, increased at lower dilution rate, thus improving CA biosynthesis. High D values did also favor Asp and TCA precursors' consumption, but not CA production. Therefore, a feasible strategy for increasing CA biosynthesis might be Orn supplementation. It is also crucial to explore the urea cycle, for the purpose of understanding the role of Orn and Arg precursors and its influence on CA synthesis.

Acknowledgment. The authors thank The Colombian Department of Science, Technology and Innovation (COLCIENCIAS) for financial support; project COL08-1-06.

References

1. Lee, S.H., Kim, J., Park, J., Kim, T.: Metabolic Engineering of Microorganisms: General Strategies and Drug Production. Drug Discovery Today 14(1-2), 78–88 (2009)
2. Kohlstedt, M., Becker, J., Wittmann, C.: Metabolic Fluxes and Beyond. Applied Microbiology and Biotechnology 88, 1065–1075 (2010)
3. Kirk, S., Avignone-Rossa, C., Bushell, M.: Growth Limiting Substrate Affects Antibiotic production and Associated Metabolic Fluxes in Streptomyces clavuligerus. Biotechnology Letters 22, 1803–1809 (2000)
4. Bushell, M., Kirk, S.Z., Avignone-Rosa, C.: Manipulation of the Physiology of Clavulanic Acid Biosynthesis with the aid of Metabolic Flux Analysis. Enzyme and Microbial Technology 39(1), 149–157 (2006)

5. Daae, E., Ison, A.: Classification and Sensitivity Analysis of a Proposed Primary Metabolic Network for Streptomyces Lividans. Metabolic Eng. 1(2), 53–165 (1999)
6. Sánchez, C., Gómez, N., Quintero, J.: Producción de Ácido clavulánico por fermentación de Streptomyces Clavuligerus. Dyna 175, 158–165 (2012)
7. Klamt, S.: CellNet Analyzer, Structural Analysis of Cellular Networks, Version 9.6, Max-Planck-Institute for Dynamics of Complex Technical Systems, Magdeburg, Germany (2011)
8. Naeimpoor, F., Mavituna, F.: Metabolic Flux Analysis in Streptomyces coelicolor under Various Nutrient Limitations. Metabolic Engineering 2(2), 140–148 (2000)

Flux Balance Analysis and Strain Optimization for Ethanol Production in *Saccharomyces cerevisiae*

León Toro[1], Laura Pinilla[1], Juan Carlos Quintero[2], and Rigoberto Rios[2]

[1] University of Antioquia, Biology Institute, Medellin – Colombia
{leonftn,laurapinilla2209}@gmail.com
[2] University of Antioquia, Chemical Eng. Dept. Medellin – Colombia
{jcquinte,rrios}@udea.edu.co

Abstract. Metabolic flux analysis provides a quantitative description of the degree of involvement of different pathways in cellular functionality. In this work, the metabolic flux distribution in *Saccharomyces cerevisiae* was evaluated to explaining low-yield of biomass and ethanol production. Furthermore, strain improvement alternatives for higher ethanol yield were proposed by means of an elementary mode analysis. Besides stoichiometric restrictions, experimental data were used as constraints for the solution of the LP problem. The corresponding flux distribution brought about indications on the fate of carbon through the entire metabolic system, after glucose uptake. The higher yield for product biosynthesis was acquired when the carbon flux was attenuated in the pentose phosphate pathway, thus compromising important biomass precursors. Reactions for glucose uptake and the synthesis of glycolytic intermediate metabolites were essential for both, biomass and product biosynthesis. This combined computational and experimental approach rendered reliable hypothesis prone to be experimentally tested.

Keywords: Ethanol, metabolic flux analysis, *Saccharomyces cerevisiae*, elementary mode analysis, Cell Net Analyzer.

1 Introduction

Cellular metabolism comprises all anabolic and catabolic reactions, responsible for substrate conversion and energy production. Despite its complexity, the diverse metabolic networks have certain characteristics making them subject of a systemic analysis [1]. Conversely, the metabolic fluxes provide a reliable quantitative description of the degree of involvement of different pathways in cellular functionality and metabolic processes [2]. Flux Balance Analysis (FBA) allows for the determination of the metabolic flux distribution (MFD) throughout the entire metabolic network of an organism, by means of a stoichiometric model that describes its metabolic capabilities [3]. For the special case of fermentative processes, the stoichiometric balances have great importance for its understanding and improvement studies. Flux Balance Analysis is particularly useful in connection with production studies of important metabolites, e.g. ethanol; these studies, usually aim at directing

L.F. Castillo et al. (eds.), *Advances in Computational Biology*,
Advances in Intelligent Systems and Computing 232,
DOI: 10.1007/978-3-319-01568-2_26,

as much carbon as possible towards a specific metabolic product, are importantly favored by FBA for estimating non-measurable intra and extracellular metabolic fluxes, maximum theoretical yields, alternatives metabolic pathways and metabolic flux bifurcation at a metabolic node [4]. Conversely, Elementary Modes (EMs) are considered as minimal functional units of metabolic networks that allow a balanced operation of the network at steady-state [6]. Each stationary flux distribution of the network can be described as a weighted combination of EMs; therefore, an optimal flux distribution will be a linear combination of optimal EMs [7].

Ethanol is traditionally produced by anaerobic fermentation of glucose using mainly *Saccharomyces cerevisiae* (*S. cerevisiae*). The metabolic process renders glycerol, CO_2 and biomass as by-products. This work is aim at evaluating the metabolic flux distribution in *S. cerevisiae* for ethanol production, to explaining low-yield of biomass and product biosynthesis. Furthermore, strain improvement alternatives for higher ethanol yield are proposed by means of an elementary mode analysis (EMA). For this purpose, the Cell Net Analyzer tool, CASOP, was used [7].

2 Materials and Methods

2.1 Organism

The yeast *Saccharomyces cerevisiae*, Ethanol Red®, was used for this study. The activated strain was stored at -80ºC in Eppendorff tubes with 40% glycerol medium. These tubes were used as seed for later experiments

2.2 Bioreactor Cultures

Anaerobic cultures were performed in a 5 L bioreactor, with a working volume of 2.7 L (Applikon® Technologies). Batch ethanol production was achieved at 35 °C, 150 rpm, and 5.46 pH units. Fermentation time reached 9 hours. Inoculum comprised 4 g/L activated yeast in a chemically defined medium (in [g/L]: glucose 50, peptone 20 and yeast extract 10).

2.3 Analytical Methods

Samples for extracellular metabolite determination were passed through 0.45 μm cellulose filters. Glucose, ethanol, acetic acid and glycerol concentrations were quantified by HPLC (Agilent 1200), as described in [8]. Biomass was assessed by the dry weight method and Absorbance at 600 nm.

2.4 Metabolic Model, Elementary Modes and Flux Balance Analysis

The metabolic model accounted for 33 reactions and 24 metabolites, comprising mainly the central carbon metabolism. The solution for the subsequent system was acquired by solving the LP problem with ethanol flux as objective function.

Experimentally measured glycerol, ethanol, acetate and glucose concentrations were used as model constraints. The optimization problem was solved by means of the software CellNetAnalyzer [4]. For strain improvement the CellNetAnalyzer tool, CASOP (Computational Approach for Strain Optimization aiming at high Productivity), was used; this computational tool relies on a stoichiometric framework associated with the elementary mode analysis to determine the importance (ω) of individual reactions in the attainment of the highest ethanol yield, without compromising the carbon availability for biomass precursors. The evaluated metabolic events obey a mass proportion (γ) between biomass (γ=0) and ethanol (γ=1) [6].

3 Results and Discussion

3.1 Batch Production of Ethanol

Figure 1 shows the experimental data for *S. cerevisiae* fermentation in a time course of 9 hours. Ethanol Red® is an industrial strain with high ethanol tolerance and easy adaption to diverse environmental conditions. Due to its physiological characteristics, Ethanol Red® produced low levels of biomass and high ethanol titers; substrate consumption profile was also apposite for Ethanol Red®. The largest ethanol accumulation was reached at the 6th hour of batch cultures, which matched the highest biomass production.

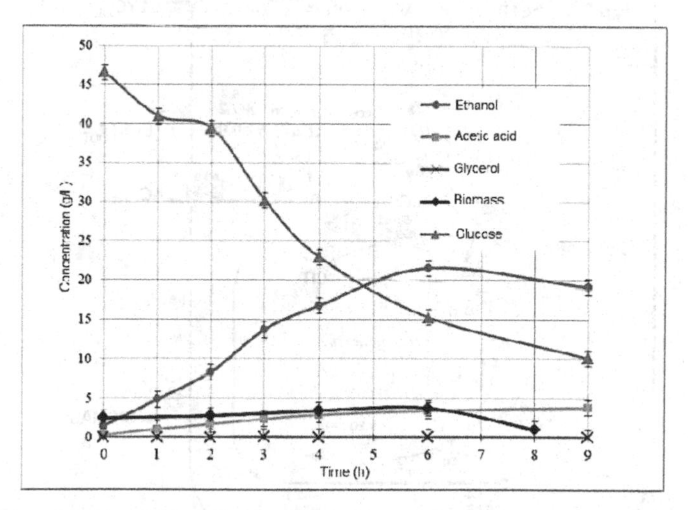

Fig. 1. Batch production of ethanol using *S cerevisiae.* (2.71L, 150 rpm, 35°C, 1vvm, pH 5.46)

3.2 Flux Balance Analysis of Ethanol Production by *S. cerevisiae*

Metabolic flux distributions were attained from the solution of the LP problem, (with ethanol flux as the objective function) using experimental constraints. For this purpose, experimental data were taken from the exponential phase of growth, where all cells are dividing at a constant rate, and the most favorable environmental conditions are reachable for the cells to execute its metabolic activity. Metabolic fluxes, in [*mmol/h*], were calculated for glucose (308.86), ethanol (266.75), glycerol (0.26) and acetic acid (40.86), and used as experimental constraints. The cellular activity was represented by the metabolic flux distribution, constrained by experimental data. The flux distribution yielding the largest product biosynthesis was achieved when the carbon flux was constrained in the oxidative pentose phosphate pathway by lessening expression levels of glucose-6-phosphate dehydrogenase; this action would have also compromised the availability of important biomass precursors such as ribulose-5-phosphate (R5P) and erythrose-4-phosphate (E4P). Ethanol yield (Y_{sp}), for this flux distribution, increased 56% compared to that from experimental data. The larger ethanol production was constrained by the carbon fate at the pyruvate branch point. The greater Y_{sp} was attained when the carbon flux going to Acetyl Coenzyme A (AcCoA) signified about 5% of the total carbon flux coming from glycolysis (See Figure 3. Though the figure shows cytosolic and mitochondrial compartments, the metabolic model did not account for compartmentalization).

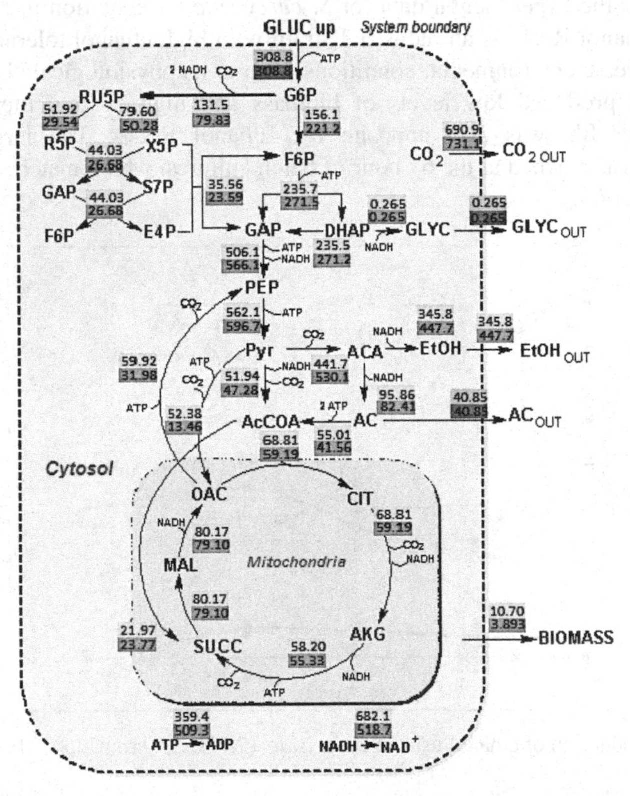

Fig. 2. Metabolic flux distribution in ethanol production: upper values correspond to experimentally constraint FBA; lower values of flux distribution are obtained as a result of reduced expression levels of glucose-6-phosphate dehydrogenase (reaction 2). Glucose, glycerol and acetate fluxes were used as experimental constraints [mmol/h].

3.3 Elementary Mode Analysis for Ethanol Production in *S. cerevisiae*

After running the CASOP tool, a set of importance measures for reactions in the pathway was attained, and are showed in Figure 3. Reactions for glucose uptake and the synthesis of glycolytic intermediate metabolites were essential for both, biomass and product biosynthesis (ω=1). A large set of reactions showed a significant importance for biomass biosynthesis (the regeneration of NADPH in the oxidative PP pathway, the citrate synthase and aconitase activity and the non-oxidative synthesis of 5-carbon sugars in the PP pathway); however, as mass proportion (γ) tends to 1, consistent to merely ethanol biosynthesis, their ω values did drop toward zero. Ethanol biosynthetic reactions (R9 and R10) did display the highest importance, being no needed for biomass production. Interestingly, an important set of reactions showed the lower importance values for ethanol production, and suggested a wide range of possibilities as promissory targets for genetic manipulation, aiming at attaining higher ethanol titers. Among these, it is highlighted the catalytic activity of pyruvate decarboxylase for favoring the synthesis of AcCoA, and/or for contributing to pyruvate accumulation as a direct ethanol precursor; the anaplerotic pyruvate decarboxylation which promotes the metabolic activity throughout the Krebs cycle, by means of the intermediate metabolite oxaloacetate (OXA), can also be considered.

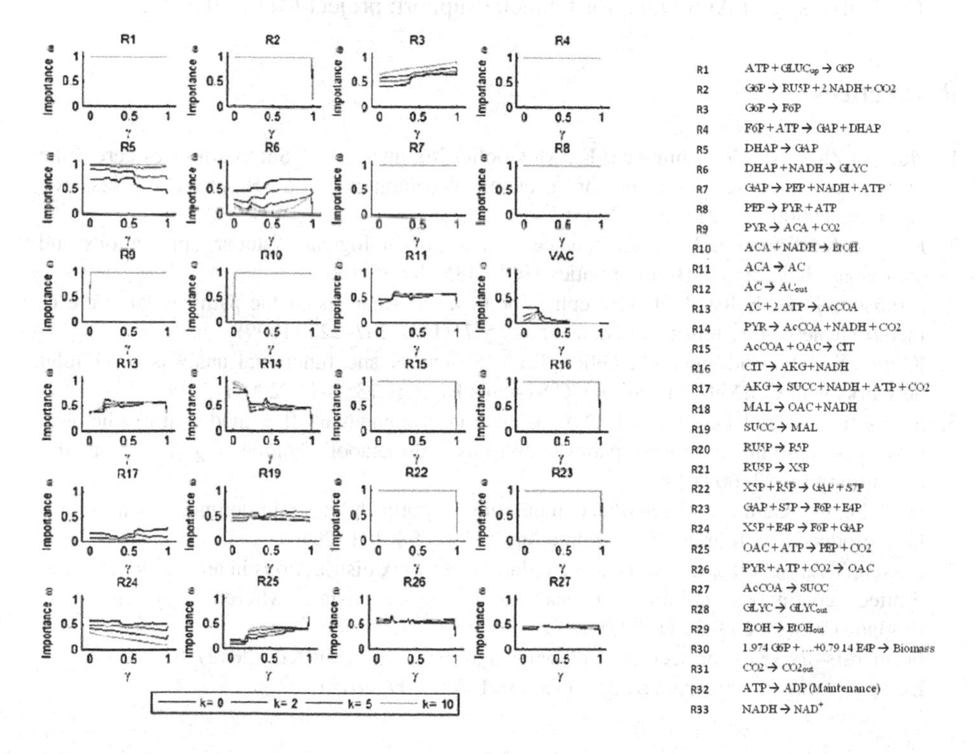

Fig. 3. Reaction Importances in the *S. cerevisiae* metabolic network. K is a parameter adjusting the intensity of quantitative weighting. If k is set to zero each EM is equally weighted. With increasing k, the weight accounts more for yield optimality of EMs.

4 Conclusions

Saccharomyces cerevisiae metabolic capabilities were successfully represented by a stoichiometric metabolic model. The model did estimate a 56% higher ethanol production when the carbon flux is reduced in the oxidative pentose phosphate pathway by means of down regulating the enzyme glucose-6-phosphate dehydrogenase. The combined flux balance and Elementary mode analysis rendered valuable information regarding the contribution of the different metabolic fluxes to the highest ethanol formation, without compromising the synthesis of biomass precursors. It was observed that only the reactions involved in transport and uptake of glucose, as well as the synthesis of glycolytic metabolic intermediates, are essential, and hence are necessary for both, ethanol and biomass production. The major group of reactions that might be considered as metabolic targets are those related to the decarboxylation of pyruvate, so as to eventually attain higher ethanol yields. This combined computational and experimental approach contributed reliable hypothesis prone to be experimentally tested.

Acknowledgment. The authors thank The Committee for Research Development – CODI, University of Antioquia, for financial support; project MDC-10-1-05.

References

1. Huimin Zhanga, S.Y., Shimizua, K.: Metabolic flux analysis of Saccharomyces cerevisiae grown on glucose, glycerol or acetate. Biochemical and Biophysical Research Communications 16, 211–220 (2003)
2. Raman, K., Chandra, N.: Flux balance analysis of biological systems: applications and challenges. Briefings in Bioinformatics 10(4), 435–449 (2009)
3. Simpson, T.W., Follstad, B.D., Stephanopoulos, G.: Analysis of the pathway structure of metabolic networks. Journal of Biotechnology 71(1-3), 207–223 (1999)
4. Klamt, S., Saez-rodriguez, J., Gilles, E.D.: Structural and functional analysis of cellular networks with CellNetAnalyzer. BMC Systems Biology 13, 1–13 (2007)
5. Schuster, S., Dandekar, T., Fell, D.A.: Detection of elementary flux modes in biochemical networks: a promising tool for pathway analysis and metabolic engineering. Trends in Biotechnology 17, 53–60 (1999)
6. Hädicke, O., Klamt, S.: CASOP: a computational approach for strain optimization aiming at high productivity. Journal of Biotechnology 147(2), 88–101 (2010)
7. Nissen, T.L., Schulze, U., Nielsen, J., Villadsen, J.: Flux distributions in anaerobic, glucose-limited continuous cultures of Saccharomyces cerevisiae. Microbiology (Reading, England) 143(1), 203–218 (1997)
8. Fermentis-dry yeast for beer, ethanol, wine and spirits. Ethanol Red. (2012), `http://www.fermentis.com` (accessed: April 09, 2013)

Domain Ontology-Based Query Expansion: Relationships Types-Centered Analysis Using Gene Ontology

Alejandra Segura[1], Christian Vidal-Castro[1], Mateus Ferreira-Satler[2], and Salvador-Sánchez[3]

[1] Universidad del BíoBío, Avda. Collao 1202, Concepción, Chile.
[2] Universidade Federal de Ouro Preto, Rua 36, nº 115, B. Loanda, João Monlevade-MG-Brasil,35.931 – 008.
[3] Universidad de Alcalá, Ctra. Barcelona km. 33.6, 28871 Madrid, España.
{asegura,cvidal}@ubiobio.cl, mateus@decea.ufop.br, salvador.sanchez@uah.es

Abstract. This paper presents a new evaluation approach for aontology-based expansion strategy proposed in previous work. In this work, the evaluation focuses on the search for Learning Object (LO) in specialized repositories. A total of 98 user queries onthe Gene ontology, a knowledge representation largely validated by the scientific community, were used in the test. For each one, new concepts were extracted from the ontology considering different expansion types. The work carried out demonstrates that query expansion delivers better results in terms of novelty and precision, independently of the expansion type. Also, that expansion types resulting from "part-of" relationships present better results in terms of precision and gained novelty, if compared with expansions using "is_a" relationships.

1 Introduction

In Information Retrieval (IR), a well-formulated query has less ambiguity and more likely to get good results. The query expansion is recommended for queries that are not well made [1], queries expressed using a country or domain specific terminology or in short queries in which might be more ambiguity.

Various combinations of techniques such as lexical co-occurrence, clustering, collaborative learning, stemming and knowledge models are used in query expansion [2]. Ontologies and thesaurus are most commonly used knowledge models in IR tasks. Ontologies provide consistent vocabularies and world representations for clear communication in a knowledge domain. In they, knowledge is specified through a formal representation language based on descriptive logic, and this knowledge should represent a consensus view of the domain.

The rest of this paper is organized as follows. Section 2 describes the main lines of work related to knowledge model-based query expansion. Section 3 presents the expansion strategy evaluation, emphasizing the different expansion types used. Evaluation results in terms of novelty, precision and DCG (cumulated gain with

L.F. Castillo et al. (eds.), *Advances in Computational Biology*,
Advances in Intelligent Systems and Computing 232,
DOI: 10.1007/978-3-319-01568-2_27, © Springer International Publishing Switzerland 2014

discount by document rank) are presented in Section 4, and results discussion is placed in Section 5. Finally, in Section 6, main conclusions and interest aspects for future research are highlighted.

2 Related Work

In knowledge models-based query expansion we distinguish two kinds of approachs: those who use knowledge models dependent of a corpus and those who uses independent knowledge models. The first kind raises using models created from a collection of domain documents[3], [4], [5]. Because of this, any change in the collection means that models must be upgraded or rebuilt. In the second kind the knowledge represented in the model is extracted from other sources, e.g., from experience and knowledge of experts. For example, some studies use thesauri [6],[7], [8]and other ontologies[9], [10], [11].

In summary, the proposal of query expansion will differ depending on the following aspects:

- Expansion mechanism type. The expansion can be manual, automatic or interactive. The approaches described above are mostly automatic or interactive.
- Ontology type. Thesaurus, dictionaries and terminologies are used instead of formal domain ontologies.
- Relationships used for expansion. Relationships used to extract new concepts are lexical relationships and in some cases basic ontological relationships, mainly the relation *is_a*.
- Conceptual distance. Most of the approaches do not restricted the conceptual distance. This will depend on the ontology used.
- Number of new concepts. It is common in most of approaches do not restrict the number of new concepts. On the other hand, many strategies left to the user the task of selecting concepts to expand.

3 Expansion Strategy Evaluation

In this paper, a evaluation of the ontology-based query expansion strategy[11] is carried out, in order to analyze the differences in expansion results depending on the connection type used to extract new concepts. The expansion types defined in this strategy are:

- *Father*, i.e. expansion is performedwith the father of the query, *Brother*, i.e. expansion is performed with concepts that share the same father of the query and *Son*, i.e. expansion is performed with concepts whose father is the query,using relationship *is_a*.
- *Whole*, i.e. expansion is performed with concepts that contain the query and *Parts*, i.e. expansion is performed with concepts that are part of the same concept that contains the query, using relationship *part_of*.

- *Exact synonym*, i.e. expansion is performed with conceptsthat are exact synonym and *Other synonym*, i.e. expansion is performed with conceptsthat are another synonym type,using *synonym* lexical relationship.
- Expansion based on domain relations. The GENE ontology models the relationship *"regulate"*; from it, 2 kinds of expansions are defined: *"regulated by"* and *"regulates"*. The first one expands with the concepts that regulate the searched concept. The second one expands the concepts that are regulated by the searched concept.

This paper focuses on the search for digital learning resources in specialized repositories, in particular on the genetics knowledge area using the Gene ontology[12]. We also used the MERLOT repository, since it has a specific collection of over two hundred learning resources labeled in the genetics area.

The new set of test queries was extracted from the content list of 25 academic programs of courses in the genetics area. A total of 459 unique concepts were extracted, of which 34 were in the ontology, 64 had similar concepts and 361 did not exist in the ontology. For each of the 98 concepts, new concepts were extracted from the ontology considering the various types of expansion. The summary of this process is shown in Table 1(a).

New concepts from the ontology were extracted for 86 of the 98 queries. *Brother* and *Son* expansion types produced the highest number of concepts. Both expansion types come from the relationship *is_a*.

For the 86 queries, original queries (without expansion) and expanded queries for each expansion were performed in the repository. The results of this process are summarized in Table 1(b). For 10 queries, no learning objects were retrieved for either, the original query and expanded queries.

Table 1. Average of new concepts extracted (a) and LO extracted from the repository (b) for each query and expansion type

	Original query	*Father*	*Brother*	*Son*	*Whole*	*Parts*	*Regulates*	*Regulated by*	*Exact synonym*	*Others synonym*
(a)	-	1.2	22.1	8.4	1	4.2	1	2.7	2.5	3
(b)	4.78	1.02	6.7	1.25	3.35	4.5	3.33	0.16	0.66	0.63

In total, 748 learning objects were recovered, 463 were unique, once a LO repeated, on average, 1.35 times for a single even in a different expansion type.

For the final relevance evaluation of all LO recovered from each query, the evaluation set was restricted to queries that have results in different types of expansions and involve less effort evaluation. Experts who have experience and/or knowledge in genetics and teaching participated in the experiment. The relevance assessment of a retrieved LO is based on the topic involved in the query and in the content of the LO, i.e. the evaluation type is topically relevance [13]. The evaluation question that the expert must answer in evaluating each LO recovered is: *"Does the learning object satisfies the applied query?"*.

The retrieved LOs for each query, with and without expansion, were evaluated according to 3 ordered levels of response Relevant (R) whose weight is 1.0, Partially

Relevant (PR) whose weight is 0.5 and Not relevant (N) whose weight is 0.0.The weight given to each response level allows to refine the estimate of the precision metric, considering the relevance degree of the results.

Since relevance results perceived by some experts show a low level of correlation, the data used for the analysis represent only the most correlated expert evaluations. The relevance value of a LO is the statistic of central tendency (mode) of relevance values given by the experts. If this is not applicable, we obtained the median of relevance values.

4 Results

The novelty , DCG and precision metrics are calculated using the first 10 retrieved LOs' relevant evaluations, both for the original query and the expanded queries.The novelty metric corresponds to the proportion of new relevant results, that is, results that cannot be retrieved by the original query. The precision metric is the ratio between the relevant retrieved results and the total number of retrieved results, while the DCG metric is an improvement to the traditional precision metric, which reduces the contribution of the accumulated relevance for lower results rankings.

Table 2. Novelty results for each query and expansion type

	T128	T13	T213	T248	T34	T369	T57	$\overline{X}$
Brother		0.600	1.000	1.000		0.667		0.817
Son	0.000						0.333	0.167
Whole	0.800	0.600	1.000		1.000		0.500	0.780
Parts	0.800	0.500	1.000				0.333	0.658
Regulates	0.800				1.000			0.900
Regulated by						0.667		0.667
Others synonym				1.000	1.000			1.000
Exact synonym				1.000				1.000

Following is a summary of the results. Regarding precision, it can be noted that average precision for expanded queries is 0.397, with values varying between 0.13 and 0.75, and a standard deviation of 0.19. The average precision for the original queries is 0.16, with values varying between 0.0 and 0.5, and a standard deviation of 0.181. The DCG metric presents a larger variability, with a standard deviation of 0.463.

The novelty average is 0.748, and its values are between 0.167 and 1, while its standard deviation is 0.269 (see Table 2). The maximum possible value of 1 is returned when the original query does not retrieve any results, so the expanded queries yield all relevant results.

In general, the expansion types that retrieved results in the largest number of queries are the *Brother*, *Sons*, *Parts* and *Whole* expansions, which derive from the *is_a* and *part_of* relationships.The *Exact Synonym* and *Other Synonym* expansions, which derive from the *Synonym* relationshop, only retrieved results for the T34 *cell cycle regulation* and T248 *RNA modification* queries. In these cases, the *Exact synonym* expansion shows higher precision than the *Other synonyms* expansion, but the same novelty metric.The *Regulates*expansion, derived from the *regulate* domain-specific relationship , only retrieves results for two queries. In these cases, it achieves novelty levels similar to other expansion types, but its precision results are slightly better.

5 Discusion

The novelty metric is always greater than zero, indicating that expanded queries help retrieve relevant LO that are different to the results retrieved by means of the original query, regardless of the query expansion type.

Expansion strategies must be careful to avoid reducing precision results. In our case, we obtain good novelty levels, while maintaining or even improving the precision and DCG results. Average precision is 0.283, not taking into account the original queries T34, T213 and T248 which do not retrieve results. This result is less that the average precision of the expanded queries, which is 0.397.

On average, the original queries are formed by two words (not considering stop-words). Results show a high correlation (0.755) between the precision average of the expanded queries and the number of words in the query. This confirms the premise that shorter queries are more ambiguous and, consequently, retrieve less precise results.

6 Conclusion

Evaluation results both complement and ratify the conclusions from the previous strategy expansion evaluation's results. We have demonstrated that our model can provide expanded queries which allow accessing to relevant LO that would be inaccessible without query expansion. , i.e., users obtain new relevant results that would not be retrieved by the original queries. This is specially evident in those cases where the original query does not return any results: in these cases, all relevant results are obtained solely through the expanded queries.

The results of our experiment results show that there are differences in expansion queries depending of the type of relationship used to retrieve the new concepts. The expansion types *Whole*and *Parts* yield good novelty metrics, and in general terms, results for expansion types derived from the *part_of* relationship show good levels of accumulated novelty and precision metrics.The relevance of retrieved LO is affected, among other factors, by the quality of the indexed resources, the repository's ranking algorithm, and the quality of the knowledge model used for the expansion.

Future research should establish the relationship between the query's generality and the expansion types that retrieve the most new and relevant results.

Acknowledgments. This work is supported by DIUBB 115215 3/RS project and DIUBB 125515 3/RS Chile and CIP-ICT-PSP.2010.6.2 project Organic.Lingua, reference: 270999.

References

1. Bhogal, J., Macfarlane, A., Smith, P.: A review of ontology based query expansion. Information Processing and Management: an International Journal 43(4), 866–886 (2007)
2. Mandala, R., Tokunaga, T., Tanaka, H.: Combining Multiple Evidence from Different Types of Thesaurus for Query Expansion. In: SIGIR 1999: Proceedings of the 22nd Annual International ACM SIGIR Conference on Research and Development in Information Retrieval (1999)
3. Pizzato, L.A.S., de Lima, V.L.S.: Evaluation of a thesaurus-based query expansion technique. In: Mamede, N.J., Baptista, J., Trancoso, I., Nunes, M.d.G.V. (eds.) PROPOR 2003. LNCS (LNAI), vol. 2721, pp. 251–258. Springer, Heidelberg (2003)
4. Zazo, Á.F., Figuerola, C.G., Alonso Berrocal, J.L., Emilio, R.: Reformulation of queries using similarity thesauri. Information Processing and Management: an International Journal 41(5), 1163–1173 (2005)
5. Huang, Y.-F., Hsu, C.-H.: PubMed smarter: Query expansion with implicit words based on gene ontology. Knowledge-Based Systems 21(8), 927–933 (2008)
6. Navigli, R., Velardi, P.: An Analysis of Ontology-based Query Expansion Strategies. In: Workshop on Adaptive Text Extraction and Mining (2003)
7. Song, M., Song, I.-Y., Hu, X., Allen, R.: Integration of association rules and ontologies for semantic query expansion. Data & Knowledge Engineering 63(1), 63–75 (2007)
8. Díaz-Galiano, M.C., Martín-Valdivia, M.T., Ureña-López, L.A.: Query expansion with a medical ontology to improve a multimodal information retrieval system. Comput. Biol. Med. 39(4), 396–403 (2009)
9. Zou, G., Zhang, B., Gan, Y., Zhang, J.: An Ontology-Based Methodology for Semantic Expansion Search. In: FSKD 2008: Proceedings of the 2008 Fifth International Conference on Fuzzy Systems and Knowledge Discovery, pp. 453–457 (2008)
10. Lee, M.-C., Tsai, K.H., Wang, T.I.: A practical ontology query expansion algorithm for semantic-aware learning objects retrieval. Computers & Education 50(4), 1240–1257 (2008)
11. Segura, A., Sánchez, N.S., García-Barriocanal, E., Prieto, M.: An empirical analysis of ontology-based query expansion for learning resource searches using MERLOT and the Gene ontology. Knowledge-Based Systems 24(1), 15 (2011)
12. Diehl, A.D., Lee, J.A., Scheuermann, R.H., Blake, J.A.: Ontology development for biological systems: immunology. Bioinformatics 23(7), 913–915 (2007)
13. Borlund, P.: The concept of relevance in information retrieval. Journal of the American Society for Information Science and Technology 54(10), 913–925 (2003)

FS-Tree: Sequential Association Rules and First Applications to Protein Secondary Structure Analysis

Nilson Mossos[1], Diego Fernando Mejia-Carmona[2], and Irene Tischer[1]

[1] School of System Engineering and Computer Science, Universidad del Valle
[2] Graduate School of Biomedical Sciences, Universidad del Valle,
Calle 13 # 100 – 00, Cali, Colombia

Abstract. A protein's structure is determined by its amino acid sequence alone. In order to describe the relation between amino acid and corresponding structural sequences, we use an association rule mining approach. Traditional association rule mining is not appropriate in a sequential context. Therefore, we develop the structure FS-tree to represent subsequences and their frequencies in a sequence database, as well as the underlying construction algorithm.

A FS-tree is a prefix tree, which stores subsequences in a compact way. The sequential context oblige us to introduce a modification of the support concept, the relative support which does not give too much weight to short sequences. A 2-dimensional FS-tree for sequence pairs over different alphabets allows to obtain rules that establish the relation within the pairs.

Mining a 2-dimensional FS-tree of amino acid sequences and corresponding secondary structures, enables us to generate rules for their relation. We analyze hypothetical and observed tree size, inferring that there are short residue sequences acting as determinants of specific secondary structures. The most important rules are related to pure structure sequences, where rules for turn and helices exceed by far the rules for strands, as revealed by a rule composition analysis. By cross validation we verified that residue sequences with high propensity to specific structure sequences apply generally, independant of a specific protein sample. These promising results motivate us to explore FS-tree related analysis in a wider range of applications including the development of rules based prediction algorithms.

Keywords: algorithms, association rules, data mining, prediction, protein, secondary structure, propensity.

1 Introduction

Increasingly, data mining approaches are used to analyze and predict protein structure and function as well as interactions between proteins. Protein secondary structure analysis and prediction were initially studied by Chou and

L.F. Castillo et al. (eds.), *Advances in Computational Biology*, Advances in Intelligent Systems and Computing 232,
DOI: 10.1007/978-3-319-01568-2_28,

Fasman [1,2] and GOR [3,4], who applied mainly statistical methods to determine the propensity of a group of 1-3 amino acids for a specific secondary structure. The propensity approach was extended to longer sequences[5]. Following studies include nearest neighbor algorithms [6,7] and neural networks[8]. The neural network approach was recently modified [9,10], resulting in a significant increase in prediction accuracy. Support vector machines, frequently used in the protein interaction problem, are successfully applied to predict secondary structure [11,12,13,14]. In this paper we explore the potential of association rules in secondary structure analysis and prediction, a data mining technique barely used in this context.

Originally, association rules were determined for a transactional database, where every transaction consists of a transaction ID and an itemset [15,16]. An association rule $P \Rightarrow Q$ is a relation between itemsets P and Q, such that $P \cap Q = \oslash$. The support of a rule $P \Rightarrow Q$ is the frequency of $P \cup Q$, its confidence represents the probability that Q occurs, whenever P occurs. A rule $P \Rightarrow Q$ is defined as interesting, if its support and confidence values exceed a chosen threshold. Rule extraction addresses the frequent itemset mining problem, which is generally resolved by the candidate-generate-and-test approach proposed by Agrawal [16] or by pattern growth introduced by Han[17]. Han employs a prefix tree, the Frequent Pattern Tree (FP-tree), to store itemsets, ordered by item frequency and eliminating not frequent items. The FP-tree is a compact representation of itemsets, because common prefixes share a path of nodes. The rule mining problem for biological sequence analysis has to deal with sequential, not transactional databases. We need to identify frequent subsequences, being most important the order in which symbols occur. Therefore, we create the data structure Frequent Sequence Tree (FS-tree), capable to efficiently store subsequences occurring in a sequential database together with their frequencies. We develop the algorithms for FS-tree construction and frequent subsequences extraction. Applying the FS-tree construction and extraction to paired sequence data (P, Q), where P and Q are sequences of different types, we are able to generate association rules of the form $P \Rightarrow Q$ that establish the relation between P and Q quantified by the corresponding support and confidence values.

Constructing the 2-dimensional FS-tree for data pairs of proteins and their secondary structure sequences we are able to extract association rules in order to determine those amino acid subsequences, that assume a specific structural sequence with high frequency. We explore the bioinformatic potential of this FS-tree, analyzing the obtained tree size and the resulting association rules, and we show that the most interesting rules do not depend on the specific data set used to generate the FS-tree.

The rest of this document is organized as follows: Section 2 contains the informatic development: tree construction and sequence extraction; section 3 applies the FS-tree to protein structure analysis; and finally, section 4 is dedicated to our conclusions.

2 Tree Construction and Sequence Extraction

2.1 FS-Tree Construction

```
Algorithm buildFS-tree(D)
Input:  sequence database D
Output: its frequent sequence
        tree T; level
        frequency array L

Create the root of an FS-tree T,
    and label it as "null"
Initialize L in 0
for all sequences X ∈ D
   S_X = set of suffixes of X
   for all s ∈ S_X
       increasePath(T,L,s)
```

Fig. 1. Algorithm for tree construction

The Frequent Sequence Tree (FS-tree) we propose here, stores efficiently all subsequences from a sequence database D over a finite alphabet Σ, along with their frequencies. The tree is constructed from the suffixes of all sequences in D. Every suffix is represented as path obtained by traversing the tree in depth from its root. Each node N stores a symbol from alphabet $\sum$ and a frequency count. The shortest path between root and N represents a suffix of a subsequence in D, the count establishes its frequency in D. Sequences in D with common suffix s in the database share this suffix in the tree, allowing to refer to a suffix as s, whether it is subsequence in the database or path in the tree. The resulting tree is able to store all subsequences occurring in D. It is compact because common suffixes are shared. The stored suffix frequencies allow easy extraction of frequent subsequences. The algorithm responsible for the tree construction, $\texttt{buildFS}-\texttt{tree}$ is shown in figure 1.

2.2 Frequent Sequence Extraction

In the current situation, the concept of support expressed directly in frequency seems to be not the most appropriate one. The shorter a sequence, the higher is its probability to appear in the database and this high probability could shadow the relative frequent appearance of interesting longer sequences. Therefore, we introduce the concept of relative support of a sequence s as its frequency with respect to the set of all sequences of the same length: A sequence is defined to be frequent, if its relative support exceeds a given support threshold $minsup$. Note that different to the traditional support concept, the relative support does not necessarily decrease with the tree level.

The level frequency L updated during tree construction, is used to extract frequent sequences. The iterative extraction algorithm applies an in-depth search. Due to the inequality

$$\text{relative support}(h) \leq \frac{\text{frequency in } h}{L_{level(h)}} \leq \frac{\text{frequency in } c}{L_{height}}$$

it is possible to prune the search process at a descendant h of a node c whenever $\frac{\text{frequency in c}}{L_{height}} \leq minsup$.

2.3 Time Analysis and Space Requirements

Let D be a database of n sequences over an alphabet Σ of size s and m the maximum length of sequences in D.

Algorithm `buildFP − tree` is analyzed in the worst-case scenario, where all database sequences have length m. For each sequence in D, a set of m suffixes will be generated, with lengths ranging from 1 to m, requiring an execution cost of $\mathcal{O}(m)$. The cost of sending a suffix to `increasePath` is proporctional to its length; therefore, we obtain a total cost of $\mathcal{O}(\Sigma_{j=1}^{m} j) = \mathcal{O}(m^2)$ for constructing the paths for one sequence; considering all n sequences in D, the resulting worst-case cost for `buildFP − tree` is $\mathcal{O}(n \cdot m^2)$. The worst-case for space requirements is $\mathcal{O}(m^s)$, according to the size of a complete $s-$ary tree of depth m.

The frequent subsequence extraction is an in-place algorithm, hence no additional space is required. The execution time, given by the number of visited nodes, is $\mathcal{O}(m^s)$.

2.4 Frequent Sequence Pairs and Rule Extraction

According to our bioinformatics goals, we need to construct a FS-tree for sequence pairs and extract its interesting rules, i.e. given is a database D of sequence pairs (P, Q), where P and Q have same length and are defined over the finite alphabets Σ_1 and Σ_2, respectively. The problem is reduced to the previous case, considering the 2-dimensional alphabet $\Sigma = \Sigma_1 \times \Sigma_2$ and the transposed 2-dimensional sequences $Z = (P, Q)^t$. Subsequence pairs are frequent (section 2.2) if a user defined relative support threshold *minsup* is exceeded. A rule is a frequent subsequence pair (P, Q) for which also a confidence threshold *minconf* holds. The confidence of a frequent pair depends on the frequency in which P occurs as antecedent. In order to obtain the confidence of a frequent pair we have to search the tree in depth first and totalize the frequencies of nodes with antecedent P, extracting rules only if *minconf* is exceeded.

3 First Applications to Protein Structure Analysis

A main problem in protein structure analysis is to determine the tendency of an amino acid sequence P for a secondary structure sequence Q. Our approach applies FS-tree construction and rule extraction, identifying rules $P \Rightarrow Q$, which exceed given thresholds for support and confidence.

3.1 Data Set and Tree Construction

The proposed mining technique is based on data from Protein Data Bank (PDB) which offers high quality and quantity information of protein sequences together with their 3-dimensional atom coordinates[18]. We based our analyses on a subset PDB, referred as PDBsub, that consists of proteins solved by X-ray diffraction, removing sequences with non-standard amino acid residues. PBDsub contains

39.660 protein sequences, composed of 11'039.951 residues, 37,82 % of them in helix, 22,27% in strand and 39.92% in turn. To obtain the secondary structure sequences, we applied the DSSP algorithm [19] an algorithm that assigns the secondary structure from the atomic coordinates of a protein, based on hydrogen bond pattern detection; grouping in one 3 types of helix (H, G, I) labeled as H, 2 strands (B, E) labeled as B and 3 coils (T, S, U) labeled as U. We constructed the FS-tree of pairs, as described in section 2.4, using the alphabet Σ_1 of amino acids (1-letter code) and the alphabet of secondary structures $\Sigma_2 = \{H,B,U\}$.

3.2 Real versus Hypothetical FS-Tree Size for PDBsub

At each tree level k, the hypothetical number of nodes is given by $(20 \times 3)^k = 60^k$, according to the possibility of combining in a sequence of length k, 20 residues with 3 structures. Contrasting the number of hypothetical nodes with the observed ones we obtain table 1.

Table 1. Observed and hypothetical FS-tree size

level k	1	2	3	4	5
subsequences of length k	11'039.951	10'978.935	10'918.836	10'859.045	10'799.491
observed nodes	60	3.538	119.641	1'176.938	2'914.686
hypothetical nodes	60	3.600	216.000	12'960.000	777'600.000

At level 1, all 60 theoretically possible nodes are found in our sample set, which is not surprising, given that our sample contains over 11 million residues in different secondary structures. That means, that every residue occurs in each, helix, strand and turn. The frequency in each structure is shown in figure 2. We observe that Proline and Glycine are the amino acids with the highest frequencies in a specific secondary structure; appearing over three times more in turns than in others. Valine is the only amino acid with the highest frequency for strands. Alanine, Glutamic acid, and Leucine are the amino acids with a dominant frequency for helices. Cysteine, Phenylalanine, Isoleucine, Lysine, Valine and Tyrosine are amino acids whose frequency is not dominant in any secondary structure. Given the frequencies of individual amino acids at the FS-tree level 1, it is easy to calculate the classic Chou and Fasman propensities of amino acids, for any given data set and extent the propensity concept to longer sequences (see section 3.3). This could be an interesting application of the FS-tree structure, especially if we keep in mind that recent publications ([20,21]) state that propensity depends on the selected protein data set. We calculated the propensities of individual amino acids based on the FS-tree and found nearly perfect coincidence between our results and the propensities determined for PDBselect in [21]. At level 2, not all possible nodes are present. Table 2 shows the residue-structure pairs of length 2 which are not observed in our dataset. Each residue pair occurs in any of the considered pure structures (HH, BB, UU) - obviously with very different frequencies. But there are some residue pairs, that are not found in the transitional structures BU, HU, BH or HB. Proline seems to play a special role in pure structures, as many possible rules relating Proline with a transitional structure are not observed.

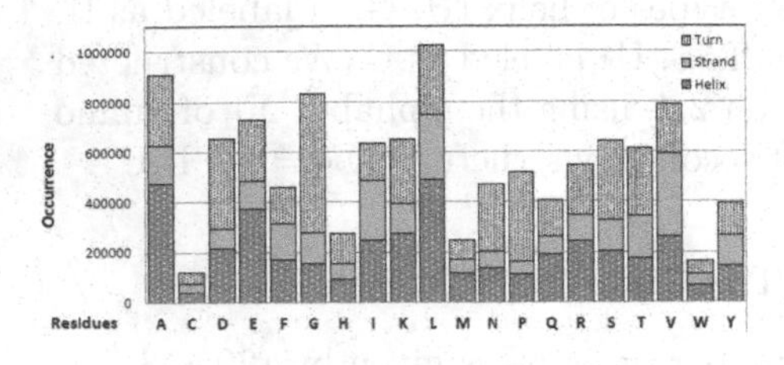

Fig. 2. Analysis of tree level 1: Frequency of residues in secondary structures

Table 2. Analysis of tree level 2: Residue-structures pairs not present in PDBsub

Residue sequence	not combined with	structure sequence
PP.		BU
CP, EP, NP, PH, PI, PR, PV, PY, WP, YP.		HU
MF, PP, QC, WC.		BH
AP, CD, CP, DP, EP, EQ, FP, GP, HP, HW, IP, IW, KP, LP, MP, NP, PA, PC, PD, PE, PF, PG, PH, PI, PK, PL, PM, PN, PP, PQ, PR, PS, PT, PV, PW, PY, QP, RP, SP, TP, VH, VP, WN, WP, WS, WW, YP.		HB

At level 3, only 119.641 out of 216.000 possible nodes occur (55.39%). More than 10 million residue-structure triplets are distributed over the observed nodes, corresponding to an average frequency of 91.26 per residue-structure triplet. At level 4 and higher, the number of observed nodes does not reach the hypothetic quantity, *i.e.* our sample set PDBsub and even PDB (PDBsub is roughly half the size of PDB) are too small to conclude whether all combinations naturally exist. Nevertheless, the node frequency at each level is relatively high (about 3 at levels 5-8), and there are nodes significantly more frequent than others (rules), which suggests that there are short residue sequences acting as determinants of specific secondary structures.

3.3 Analysis of Rule Number and Composition

A rule's support refers to its frequency in the dataset, its confidence indicates the preference of the residue sequence to a specific structural sequence. Together, support and confidence allow to identify relations between residual and structural sequences which are very probable. These interesting rules, easily extracted from the FS-tree, could be considered as an extension of the propensity concept for longer sequences. Different to the classic propensity, this extended concept allows to adjust the desired grade of certainty of selected rules defining adequate support and confidence thresholds. For instance, in the case of sequences of length 1, using a threshold of 10^{-6} for support and 0.41 for confidence, we extracted 16 interesting rules, all of them with a classic propensity greater than 1.1; for thresholds of 10^{-5} and 0.445 respectively, we obtain only 9 rules, all of them with a classic propensity greater than 1.3.

We analized number and composition of interesting rules, in dependence of support and confidence thresholds. As expected, the number of rules decreases as support and confidence increase (see table 3 for some specific values).

Table 3. Rules in dependence of support and confidence

Support	Confidence 50%	60%	70%	80%	90%
10^{-5}	102.934	94.099	84.447	67.647	51.205
$5 \cdot 10^{-5}$	1.264	598	187	57	13
10^{-4}	648	316	67	5	1

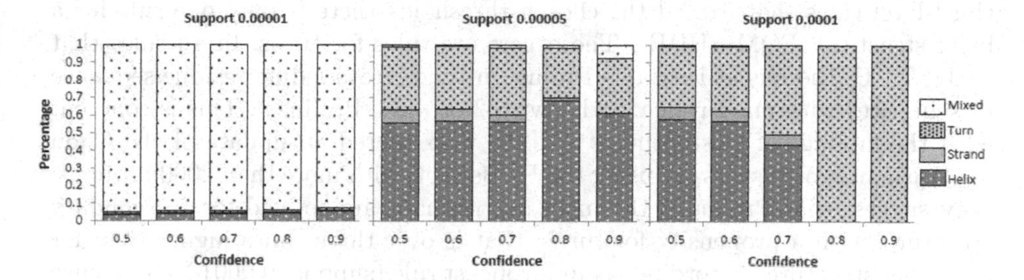

Fig. 3. Rule composition according to secondary structure

Table 4. Rules for thresholds of 0.00005 (support) and 0.8 (confidence)

Residue sequence	structure sequence
PP.	UU
FQM.	HHH
PPP, QPD, PPG, PDG, KPP, GPP, QQP, DPP, NPP, EPP, HPN.	UUU
TVLV.	BBBB
EQVL, LQKV, KLFN, RALA, LAKE, RNLL, ELDK, DEAE, RAKR, DEAA, KSEL, EAEL, EMLR, NAAK, AAKS, DKAI, DLLE, EKLF, EALA, ELAR, VEAA, EEAL, AVRR, DAAV, AAVN, EELL, EALR, LDKA, QKVV, EAAR, EAEK, RAAL, AELL, KALE, AAVR, EAAV, VDAA.	HHHH
RDLK, QPDG, HRDL.	UUUU
VDAAV.	HHHHH
RDLKP.	UUUUH
HRDLK.	UUUUU

In figure 3 we show the rule composition for specific support and confidence values. We observe that the percentage of mixed rules -rules that involve more than one structure type- decreases as support and confidence are increased. Accordingly, the percentage of pure rules -where only one structure is present- increases for higher support and confidence thresholds: the most import rules predict pure structures; *i.e.* certain residue sequences occur frequently in a specific secondary structure type. These sequences could be considered as elementary units for secondary structure determination, or, in other words, sequences with a very high propensity to a specific secondary structure.

The rules obtained for thresholds of 0.00005 (support) and 0.8 (confidence) are shown in figure 4. The most interesting rules represent a " pure", not mixed secondary structure (with only one exception, the rule RDLKP⇒UUUUH), All are short rules, the maximum length is 5. The shortest rule is PP⇒UU; *i.e.* a sequence of 2 Prolines is most likely to be found in a turn. Among the 12 third-level rules that exceed the chosen thresholds there is only one rule for a helix structure (FQM⇒HHH). The others are rules for turns. In all rules that derive UUU, there is at least one Proline in the corresponding residue sequence (4 sequences with one Proline, and 7 with 2 or more Prolines). This is coherent with the findings of Costantini *et al.* [21], who offered an update of the Chou and Fasman propensities, by using the PDBselect, with more than 2000 proteins: they suggest that Proline is the most determinant amino acid for a secondary structure, with a propensity for turns, that is over three times higher than for any other structure. Accordingly, our strongest rule (support 0.00013, confidence 90%) is PPP⇒UUU; *i.e.* PPP is most probably found in turns. This supports the suggestion to use Prolines to increase stability of turns ([22]). At level 4, the dominant structure is helix, present in 37 of 41 rules. Almost all residue sequences for helices contain at least one of the amino acids with the highest propensities for helices (Alanine (A), Glutamic Acid (E) and Leucine (L)); being QKVV⇒HHHH the only exception. At this level we also extract 3 rules for turns, and the only interestig rule for strands TVLV⇒BBBB. At level 5, there are only 3 rules, each one continues a rule of length 4: VDAA**V**⇒HHHH**H**, RDLK**P**⇒UUUU**H** and HRDL**K**⇒UUUU**U**. It is curious that in fifth-level rules, the HHHHH and UUUUU do not contain the most determinant residues for those secondary structures (A, E, L for H and P for U).

3.4 Domain of Interesting Rules

Previously we derived the interesting rules for chosen support and confidence thresholds. In which degree do these rules depend on the specific data set, used to construct the FS-tree? To answer this question we applied a seven-fold cross-validation, dividing PDBsub and its corresponding structure sequences in 7 parts. In each validation process, 6 parts are used for training; the seventh part is reserved to test if the rules apply. We used the training data to construct the FS-tree and to capture the rules. Then we verified in the test set the rule application: for each incident of a rule's residue sequence we verified if it appears with the corresponding structural sequence. The percentage of rules that apply correctly, is reported in table 5.

The table shows that the most important rules (confidence 90%) apply in a very high percentage in the test sets, suggesting that these rules do not dependent on the data set that is used to construct the tree. Therefore we can conclude that the residue sequences with high propensity to specific structure sequences apply generally, not only in a specific context.

Table 5. Percentage of rules generated by the training set, which apply in the test set

	Confidence				
Support	50%	60%	70%	80%	90%
10^{-5}	74.269±0.011	83.142±0.012	89.295±0.011	93.153±0.012	96.835±0.011
$5 \cdot 10^{-5}$	58.527±0.002	66.524±0.002	74.666±0.005	84.671±0.011	93.037±0.010
10^{-4}	58.405±0.002	66.299±0.002	73.431±0.005	82.410±0.009	89.978±0.025

4 Conclusions

We present the FS-tree, a new prefix tree structure, capable to store compactly subsequences of a sequential database. In a FS-tree, a pattern is an ordered sequence of symbols, whereas the traditional FP-tree for transactional databases is based on sets. For sequential databases, the standard concept of support has to be revised. In this context it is important to determine the frequency of a sequence with respect to sequences of equal length, which leads us to the definition of relative support. We define an efficient algorithm for extracting frequent sequences, based on the appropriate strategy to reduce FS-tree search. Applying the FS-tree structure and algorithms to sequence pairs, we are able to generate sequential association rules between sequences of two different alphabets. In order to illustrate the potential of FS-trees in bioinformatics, we apply it to secondary structure analysis of protein sequences.

Our first analysis concerns the number of observed nodes of fixed length. As an overall result we obtain, that neither all hypothetically possible short residue sequences nor all possible structure sequences for a given residue sequence are observed. A possible explanation of these results includes both, the limited number of structurally known proteins as well as the propensity of residue sequences for specific secondary structures. Secondly we analyzed the number and composition of interesting rules, in dependence of the chosen support and confidence thresholds. The most important rules are related to pure structure sequences, where rules for turn and helices exceed by far the rules for strands. The obtained rules provide the residue-structure relations with the highest propensity. By cross validation we showed that the most frequent rules are general, not depending on the specific data set used to generate the tree.

These first analyses show the high potential of the FS-tree structure in bioinformatics and motivate us to explore it in a wider range of applications, which include the development of rules based prediction algorithms.

References

1. Chou, P.Y., Fasman, G.D.: Prediction of protein conformation. Biochemistry 13(2), 222–245 (1974)
2. Chou, P.Y., Fasman, G.D.: Conformational parameters for amino acids in helical, beta-sheet, and random coil regions calculated from proteins. Biochemistry 13(2), 211–222 (1974)

3. Garnier, J., Osguthorpe, D.J., Robson, B.: Analysis of the accuracy and implications of simple methods for predicting the secondary structure of globular proteins. Journal of Molecular Biology 120(1), 97–120 (1978)
4. Garnier, J., Gibrat, J.F., Robson, B.: GOR method for predicting protein secondary structure from amino acid sequence. Methods in Enzymology, vol. 266, pp. 540–553. Academic Press (1996)
5. Strelets, V.B.: New machine learning technique for analysis and prediction of sequence and structure features: Protein secondary structure prediction. Network Science (1995)
6. Salzberg, S., Cost, S.: Predicting protein secondary structure with a nearest-neighbor algorithm. Journal of Molecular Biology 227(2), 371–374 (1992)
7. Salamov, A.A., Solovyev, V.V.: Prediction of protein secondary structure by combining nearest-neighbor algorithms and multiple sequence alignments. Journal of Molecular Biology 247(1), 11–15 (1995)
8. Rost, B.: Phd: predicting one-dimensional protein structure by profile-based neural networks. Methods in Enzymology 266, 525–539 (1996)
9. Pollastri, G., McLysaght, A.: Porter: a new, accurate server for protein secondary structure prediction. Bioinformatics 21(8), 1719–1720 (2005)
10. Zhang, G.Z., Huang, D.S., Zhu, Y.P., Li, Y.X.: Improving protein secondary structure prediction by using the residue conformational classes. Pattern Recognition Letters 26(15), 2346–2352 (2005)
11. Bock, J.R., Gough, D.A.: Predicting protein-protein interactions from primary structure. Bioinformatics 17(5), 455–460 (2001)
12. Bradford, J.R., Westhead, D.R.: Improved prediction of protein-protein binding sites using a support vector machines approach. Bioinformatics 21(8), 1487–1494 (2005)
13. Ward, J.J., McGuffin, L.J., Buxton, B.F., Jones, D.T.: Secondary structure prediction with support vector machines. Bioinformatics 19(13), 1650–1655 (2003)
14. Birzele, F., Kramer, S.: A new representation for protein secondary structure prediction based on frequent patterns. Bioinformatics 22(22), 2628–2634 (2006)
15. Agrawal, R., Imielinski, T., Swami, A.: Mining association rules between sets of items in large databases. In: Proceedings of the 1993 ACM SIGMOD International Conference on Management of Data SIGMOD 1993, vol. 22, pp. 207–216 (May 1993)
16. Agrawal, R., Srikant, R.: Fast Algorithms for Mining Association Rules in Large Databases, vol. 15, pp. 487–499. Morgan Kaufmann Publishers Inc. (1994)
17. Han, J., Pei, J., Yin, Y.: Mining frequent patterns without candidate generation. ACM SIGMOD Record 29(2), 1–12 (2000)
18. Berman, H.M., Westbrook, J., Feng, Z., Gilliland, G., Bhat, T.N., Weissig, H., Shindyalov, I.N., Bourne, P.E.: The protein data bank. Nucleic Acids Research 28(1), 235–242 (2000)
19. Kabsch, W., Sander, C.: Dictionary of protein secondary structure: pattern recognition of hydrogen-bonded and geometrical features. Biopolymers 22(12), 2577–2637 (1983)
20. Fujiwara, K., Toda, H., Ikeguchi, M.: Dependence of alpha-helical and beta-sheet amino acid propensities on the overall protein fold type. BMC Structural Biology 12(1), 18 (2012)
21. Costantini, S., Colonna, G., Facchiano, A.M.: Amino acid propensities for secondary structures are influenced by the protein structural class. Biochemical and Biophysical Research Communications 342(2), 441–451 (2006)
22. Fu, H., Grimsley, G.R., Razvi, A., Scholtz, J.M., Pace, C.N.: Increasing protein stability by improving beta-turns. Proteins 77(3), 491–498 (2009)

Presentation and Evaluation of ABMS (Automatic Blast for Massive Sequencing)*

Nelson Enrique Vera Parra[1], José Nelson Pérez Castillo[2], and Cristian Alejandro Rojas Quintero[3]

[1] GICOGE Research Group - Teacher / Researcher
Distrital University Francisco José de Caldas, Bogotá D.C., Colombia,
neverap@udistrital.edu.co
[2] GICOGE Research Group - Director Center for Scientific Research and Development
Distrital University Francisco José de Caldas, Bogotá D.C., Colombia,
nelsonp@udistrital.edu.co
[3] GICOGE Research Group - Student
Distrital University Francisco José de Caldas, Bogotá D.C., Colombia,
carojasq @correo.udistrital.edu.co

Abstract. In this article is presented and evaluated ABMS (Automatic BLAST for Massive Sequencing) an online bioinformatic free tool designed with the objective to automate and optimize the search process through local alignments for large unknown nucleotide or aminoacid sequence data against local known sequence databases (Swissprot, Uniprot, Refseq, among others). ABMS integrates the following processes: sequence entry management, proteome, genome and transcriptome database management, BLAST execution (blastp, blastx, blastn, tblastn) , and results management; these are presented to the biologist as an unified process, transparent with a friendly web interface. ABMS is built with the following modules: SM (Sequence manager), LBS (Local BLAST Server), SDBA (Sequence database administrator), RM (Results manager).Assessing ABMS with 2 data set (20/500 aminoacid sequences and 20/500 nucleotide) and compare its performance against NBCI's BLAST server showed that: The strength of ABMS for massive sequence analysis and the of NBCI's BLAST (both blastx and blastp) for data sets with more than 20 sequences. The advantages of ABMS against NBCI's BLAST in terms of administration, data set storage, management, download and results feedback.

1 Introduction

Deciphering biological sequences is virtually essential for all branches of Biology research. For several decades the sequencing process was done thanks to the Sanger's method (including human genome project where this method was fundamental). Although its high costs and limitations related to the performance, scalability, speed and

* Work done in collaboration with High Performance Computational Center (**CECAD**) - Distrital University Francisco José de Caldas, Bogotá D.C., Colombia (http://cecad.udistrital.edu.co) and Genetics Institute - National University (**IGUN**), Colombia, (http://www.genetica.unal.edu.co).

L.F. Castillo et al. (eds.), *Advances in Computational Biology*,
Advances in Intelligent Systems and Computing 232,
DOI: 10.1007/978-3-319-01568-2_29, © Springer International Publishing Switzerland 2014

resolution has forced over the past 5 years to take a migration to new procedures called "next generation sequencing" [1-2]. These new sequencing technologies allow a more economic and efficient sequencing which has led to an exponential growth of the sequence data volume. Optimize the sequencing process would be meaningless without the development and optimization of software tools capable of analyze this vast sequenced volume data. One of the main requirements related to genomic and transcriptomic data mining is the sequence comparison using alignments for finding similar sequences in databases of known sequences, this is called annotation (unknown sequence association with known sequences). The most widely used tool for sequence comparison using alignments is BLAST - Basic Local Alignment Search Tool [3-5].

When working with small sets of sequences the annotation process normally can be performed using the BLAST server provided by the NCBI (National Center for Biotechnology Information), however when the biologist need to perform massive annotation processes (large volume of sequences) or when he needs to compare those sequences against different databases offered there, this server is not adequate therefore the biologist is forced to make the annotation process locally , which means to have a high expertise in the installation, configuration and implementation (using command line) of BLAST, importing and updating nucleotide and aminoacid databases and finally the interpretation of results usually in XML format. The tool presented in this paper, called ABMS (Automatic BLAST for Massive Sequencing) presents massive annotation to the biologist as an unified process very easy to use via a web interface. ABMS optimizes the use of BLAST to perform massive annotations and integrates facilities such as database administration and results management.

This document is organized into 4 sections: Initially presents an overview of ABMS and exposes its functionality, then describes its architecture explaining module to module: LBS (Local BLAST Server), SDBA (Sequence database administrator), RM (Results manager), then there is a comparison of the processes and capabilities against NBCI's BLAST, finally conclusions are obtained.

2 General Description

ABMS is a free tool (online access and local installation possibility) that facilitates the process of massive annotation for transcriptomes, genome and proteins. ABMS is optimized to work with large amount of sequences with user interfaces and very intuitive processes for the biologist.

Fig. 1. Welcome screen ABMS

3 Architecture

ABMS is composed by 4 modules which form the workflow. For the execution and integration of this modules its required some additional tools used transversally in all those modules. Below are the 4 modules, the workflow and the transverse tools.

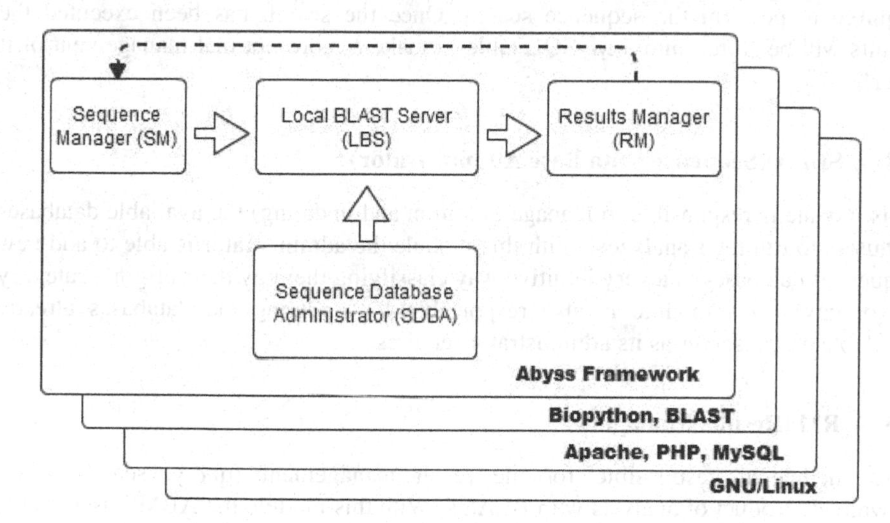

Fig. 2. Structure and workflow for ABMS

3.1 Transverse Software Components

Abbys Framework. A framework for web application development using the PHP programming language and relational databases such as MySQL created by Steven Sierra Forero student from the Distrital University Francisco José de Caldas. It uses a layer based architecture called MVC (Model, View, Controller).

Biopython. Free license project that offers various modules and facilitates to work with bioinformatics data [6].

Apache. A HTTP server with free license.

BLAST(Basic Search Alignment Tool). A tool for finding local regions of similarity using sequence alignments.

MySQL. Management relational database system, multithread and multiusers. Free licence.

GNU/Linux. Free Operating System optimal for servers and execution of bioinformatic tools.

3.2 SM (Sequence Manager)

This module is responsible for the management of FASTA which contains the sequences belonging to the user data set. Though this module the user can upload his sequences, select the ones that he requires to supply to BLAST and delete the ones no longer needed.

3.3 LBS (Local BLAST Server)

This module is responsible for running a search using BLAST against certain databases and store the results into a MySQL table. Through this module the user can choose the sequence files previously uploaded to the server and select the databases (including the most popular as Refseq, Uniprot, Swiss-Prot) against the ones its required to perform the sequence search. Once the search has been executed the results will be stored into a MySQL table to make a more optimal management of it later.

3.4 SDBA (Sequence Data Base Administrator)

This module is responsible to manage (addition and updating) the available databases for users to run their analyzes. With this module the administrator is able to add new sequences databases in a very intuitive way classifying them by their origin a category (taxonomy). This module is also responsible for updating the databases already available on the server as its administrator requires.

3.5 RM (Results Manager)

This module is responsible for the result management (query, search, filter, download) product of analysis with BLAST. With this module the ABMS user is able to filter the results of the analysis, download the sequences in FASTA format, with the possibility to replace the identifiers or even generate a new FASTA file for send it to the Sequence Manager module such that it is possible to re-run an analysis on a specific result. The elements of this module are:

4 Evaluation Methodology

4.1 Data Set

Table 1. Data sets

	Dataset A	Dataset B
Organism	Diploria trigosa	Acropora Digitifera
Sequence Type	Transcriptome	Proteome
Sequence Number	20, 500	20, 500
Nucleotides Number	12671, 371607	7889, 230921
Format	FASTA	FASTA
Database for search execution	Non-Redundant.[7]	Non-Redundant. [7]
Expect Value:	1e-3	1e-3
BLAST type	Blastx	Blastp

4.2 Metrics

Processing time, result number.

4.3 Assessments References

Currently there is not any recognized public software that offers the same functionality as ABMS, however as a reference standard it was taken the NCBI's public server [8] because it is the most used and it is the one with more similar functions.

5 Results

The above table demonstrates the limitation of the public NBCI's server for massive sequence analysis. Both Blastx and Blastp the maximum allowed execution in NCBI was for a sequence number of 20. By contrast, ABMS ran up to 500 sequences analysis without problem. About the number of results showed that the amount od hits are similar to NCBI's Blast to NCBI and ABMS in both cases: Blastp and Blastx. For datasets under and equal to 20 was noted that the average execution time is:

NCBI BLAST: Blastx (2.35 minutes per sequence) / Blastp (1.5 minutes per sequence)

ABMS: Blastx (2.55 minutes per sequence) / Blastp (5.6 minutes per sequence)

Table 2. Dataset comparison results in NBCI's Blast and ABMS

		Dataset A		**Dataset B**	
Number of Sequences		20	500	20	500
Time (Minutes)	NCBI BLAST	47	CPU Limit	30	CPU Limit
	ABMS	51	1427	112	2160
Number of Results	NCBI BLAST	5	CPU Limit	19	CPU Limit
	ABMS	5	247	19	451

6 Conclusions

ABMS is an automatic annotation tools optimized to work with large amounts of sequences and equipped with user interfaces and intuitive processes. Its projection useful for biologists researches is high because the current tools and recognized BLAST public servers have limitation for large datasets (in this case NCBI's Blast was limited to 20 sequences). However for small amounts of sequences NCBI's performance is better than ABMS using Blastp.

ABMS presents facilities to the user both the administration and storage of his dataset as in the interpretation and result downloading. This prevents the biologist to perform processes using the command line. Recognized public BLAST servers do not offer the possibility to add new databases different to those offered in their listings. ABMS allows the administrator to customize the set of databases adding new ones, including unpublished organisms to search with BLAST.

References

1. Metzker, M.: Sequencing technologies – the next generation. Nature Reviews Genetics 11, 31–46 (2010)
2. Martin, J., Wang, Z.: Next-generation transcriptome assembly. Nature Reviews Genetics 12, 671–682 (2011)
3. Altschul, S.F., Gish, W., Miller, W., Myers, E.W., Lipman, D.J.: Basic local alignment search tool. J. Mol. Pub. Med. Biol. 215, 403–410 (1990)
4. Madden, T.L., Tatusov, R.L., Zhang, J.: Applications of network BLAST server. Meth. Enzymol. 266, 131–141 (1996)
5. Camacho, C., Coulouris, G., Avagyan, V., Ma, N., Papadopoulos, J., Bealer, K., Madden, T.L.: BLAST+: architecture and applications. BMC Bioinformatics 10, 421 (2008)
6. Cock, P.J., Antao, T., Chang, J.T., Chapman, B.A., Cox, C.J., Dalke, A., Friedberg, I., Hamelryck, T., Kauff, F., Wilczynski, B., de Hoon, M.J.: Biopython: freely available Python tools for computational molecular biology and bioinformatics. Bioinformatics 25(11), 1422–1423 (2009)
7. Pruitt, K.D., Tatusova, T., Maglott, D.R.: NCBI reference sequences (RefSeq): a curated non-redundant sequence database of genomes, transcripts and proteins. Nucleic Acids Research 35(suppl. 1), D61–D65 (2007)
8. Altschul, Madden, T.L., Schäffer, A.A., Zhang, J., Zhang, Z., Miller, W., Lipman, D.J.: Gapped BLAST and PSI-BLAST: a new generation of protein database search programs. Nucleic Acids Res. 25, 3389–3402 (1997)

In silico Analysis of Iduronate 2 Sulfatase Mutations in Colombian Patients with Hunter Syndrome (MPSII)

Johanna Galvis[1,*], Jannet González[2], Daniel Torrente[2], Harvy Velasco[1], and George Emilio Barreto[2]

[1] Maestría en Genética Humana, Universidad Nacional de Colombia
djgalvisr@unal.edu.co, juana7@gmail.com
[2] Labotarorio de Nutrición y Bioquímica. Pontificia Universidad Javeriana, Colombia
dtorrente@javeriana.edu.co

Abstract. Hunter syndrome or Mucopolysaccharidosis II is an inherited X linked disease, caused by mutations in iduronate 2 sulfatase (IDS), enzyme which catalyzes the initial step reaction of heparan and dermatan sulfate degradation. Allelic heterogeneity in MPSII challenges genotype-phenotype correlation. With the aim of understanding the repercussion of mutations on enzyme structure-function, we performed protein modeling and docking simulations with wild and mutant forms of hIDS. Mutations were obtained from a molecular study conducted in Colombian patients. Point mutations affected substrate-protein interactions. In the case of S71N (attenuated phenotype) further experimentation is required. Novel mutants P160SfsX4, D190Pfs13X and P185GfsX2 have a severely distorted conformation. Detailed analysis of the ligand-protein interaction is also of great significance in designing molecules for treatment. This is the first report of molecular docking performed with wild and mutant forms of iduronate-2-sulfatase as a bioinformatical approach to phenotype genotype correlation in patients with Hunter Syndrome in Colombia.

Keywords: Mucopolysaccharidosis II, Sulfoiduronate sulfatase, Bioinformatic Analysis, Molecular Docking.

1 Introduction

Mucopolysaccharidosis II (MPSII), also known as Hunter syndrome, is a rare, X-linked disorder caused by deficiency of the lysosomal enzyme iduronate-2-sulfatase (IDS), which catalyzes the first step in dermatan (DS) and heparan sulfate (HS) degradation. Specifically, IDS removes O-2 linked sulfate from uronic acid, i.e., GlcA (D-glucuronic acid) in HS, and IdoA (L-iduronic acid) in DS [1]. IDS deficiency caused by mutations in IDS gene leads to abnormal storage of these glycosaminoglicans, leading to pathogenic events which result in multisystemic compromise. MPSII is chronic and progressive, and occurs in people of all ethnicities, with an estimated prevalence of ~1 in 170000 male live births [2]. There are two forms of MPSII:

* Corresponding author.

L.F. Castillo et al. (eds.), *Advances in Computational Biology*,
Advances in Intelligent Systems and Computing 232,
DOI: 10.1007/978-3-319-01568-2_30, © Springer International Publishing Switzerland 2014

neuropathic, with severe intellectual impairment, and non-neuropathic, with minimal or absent cognitive involvement.

The gene encoding IDS is located in Xq28 [3]. IDS enzyme belongs to the highly conserved family of sulfatases [4]. Most frequently gene mutations found in patients with MPSII are point mutations, with substitution of one aminoacid by another having different chemical properties. These changes can affect protein stability, processing, trafficking and even enzyme/substrate interactions. Knowledge about mutations, its structural implications and patient phenotype could be useful in diagnosis, prognosis and treatment of MPSII. However, experimental data about structure and function of IDS mutants are scarce. The crystalized structure of Iduronate 2 sulfatase has not been obtained, and the study of mutations and its implications on protein structure and function has been performed in silico by different groups [5-8].

The present study reports both the modeling of native human IDS and mutations found in a group of Colombian patients with Hunter Syndrome. Finally, molecular docking is performed with each one of these mutants against the natural substrates with the aim of understanding the possible mechanisms of the pathogenesis of this disease.

2 Methodology

2.1 Template Selection, Modeling and Model Assessment

Template search from PDB database was performed by means of Basic Local Alignment Search Tool for proteins (BLASTp). Tridimensional modeling of hIDS was performed using protein threading. Modeling using the Molecular Operating Environment software package [9] indicated that the A chain of human Arylsulfatase A (1AUK) possessed an arrangement of alpha helical structural elements that could be superimposed on hIDS. Furthermore, conserved regions and predicted catalytically active residues were completely covered by 1AUK. Selected force field in MOE was amber99. Energetic and stereochemical evaluation was performed with Structure Assessment tool from Swiss-Model workspace server [10]. Overall geometric and stereochemical qualities were examined by Ramachandran plots generated by RAMPAGE[11]. Measurement of RMSD (Root median square deviation) was performed in MOE, to evaluate accuracy of the model.

2.2 Ligand-Protein Docking

Wild type hIDS docking was performed in Autodock Vina software[12] against its substrates HS and DS. To simulate the posttranslational modification of this residue to FGly, it was performed on the 3D model atom by atom, with builder tool in Pymol™ v 1.3 for educational use[13] . Using Structure Assessment tool, this modified model was evaluated to make sure that no steric or energy alterations were generated. Ligplot+ software package was employed for docking assessment[14], and visualization of its results was done in Pymol™ v 1.3[13].

2.3 IDS Mutants

The source of the different IDS mutants evaluated in this work was the unpublished study Identification of Iduronate 2 Sulfatase Gene Mutations in Colombian Patients with Hunter Syndrome[15]. Direct sequencing of the nine exons [16, 17] and MLPA (Multiplex ligation-dependent probe amplification) [18] allowed the identification of seven mutations suitable of bioinformatic simulations. Protein modeling followed the same steps described above, using wild type IDS model as template. RMSD of mutants was determined with MOE by superimposition to wild hIDS as reference. Subsequently, docking and docking assessment were performed.

3 Results

Human Iduronate 2 sulfatase tridimensional model superposed to its template is shown in figure 1. RMSD was 0.97Å, indicating high accuracy of the model. Geometric evaluation showed that 75.3% of the residues were in the favored region, 16.3% were in the allowed region and 8.3% were in the disfavored region. These results indicate that the φ and ψ backbone dihedral angles in the hIDS model are reasonably accurate.

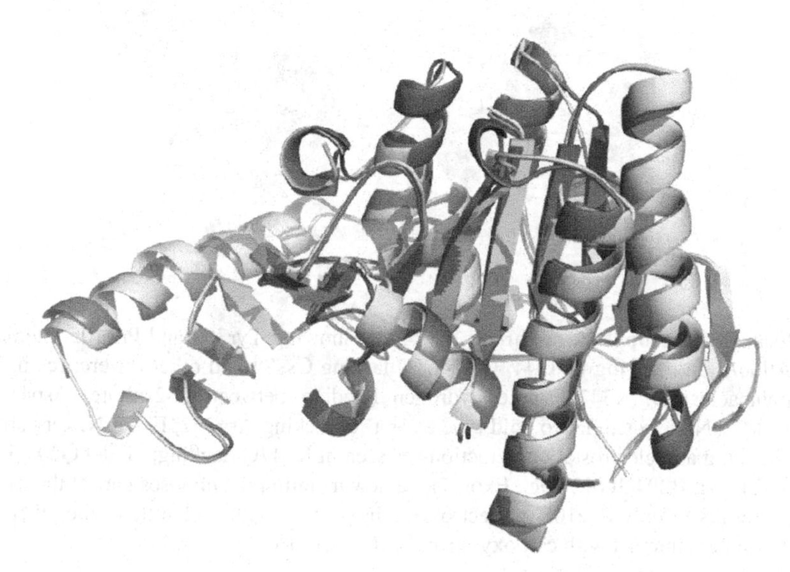

Fig. 1. hIDS model (dark gray) superimposed to its template 1AUK (light gray). Molecular Operating Environment v.2010.

Docking results are shown in Figure 2. Wild type hIDS docking allowed the identification of electrostatic interactions between the substrate and the residues at the enzyme catalytic pocket.

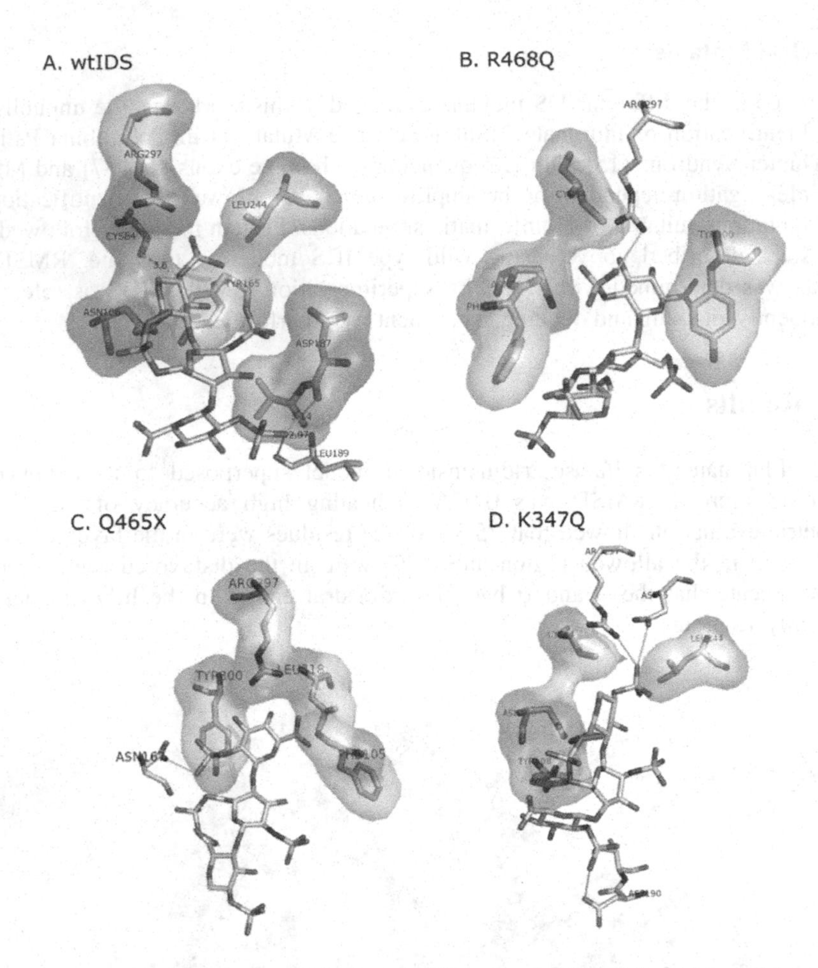

Fig. 2. Docking simulations. A. wt hIDS; B. R468Q showing Tyr300 and Phe105 interacting with GlcA/IdoA, and lacking Leu244; C. Q465X lacking Cys84 and other differences in interacting aminoacids; D. K347Q, note hydrogen bonding between O-2sulfate, Asp45 and Arg247; E. K236N, very similar to wild type except for lacking Arg297; F. S71N, very similar hydrogen bonding and electrostatic interactions as seen in K347Q docking; G. del.Q200_E203, lacking Leu244; H. R294GfsX2 (Del. Exon 7), a new mutation which loses part of the catalytic core and results in only Phe105 in electrostatic interaction with O-2 sulfate (the other aminoacids shown here interact with carboxyl group of GlcA/IdoA.

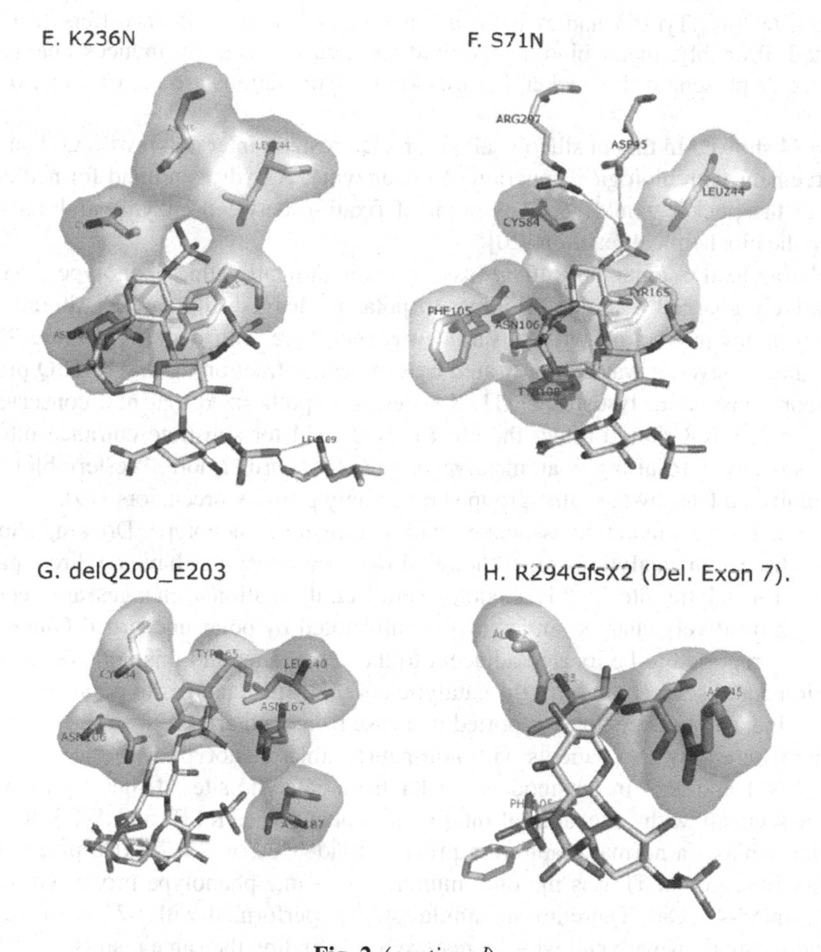

Fig. 2. *(continued)*

In wtIDS significant interactions involved GlcA O-2sulfate with Cys84, Tyr165, and Leu244. Electrostatic interactions were also seen between GlcA carboxyl group and Asn106. At the external portion of the pocket, three hydrogen bonds were observed between different GlcA monomer (in the same HS chain), Leu189 and Asp187. Relating to the different mutants analyzed, results are explained in figure 2.

4 Discussion and Conclusions

The present docking simulation showed that HS/DS bound at the wild hIDS in a conserved catalytic pocket. Studies related to hIDS tridimensional modeling have reported that putative active site residues are Asp45, Asn46, Cys84, Arg88, Lys135, His138, Asp334, His335, and Lys347, but these groups did not performed docking simulations and its arguments were inferred from homology between sulfatases[8],[19] . In contrast, our docking simulation found not only Cys84, but also

Arg297, leu244, Tyr165 and Asn106 as catalytically active, which differs from the reported. Probably, metal binding required for catalytic activity induces changes in aminoacids present at the pocket, but this kind of simulations are out of scope of this study.

Cys84 showed in this in silico analysis an electrostatic interaction with O-2 sulfate, in agreement with biological function of the enzyme. A hydrogen bond formed at the edge of the pocket would serve as point of fixation between enzyme and substrate during the biochemical reaction [20].

Relating to IDS mutants, R468Q (associated with neuropathic phenotype) changes a positively charged residue, Arginine, to polar uncharged Glutamine. Alteration of some residues interacting with substrate was seen here with docking (Figure 2). At experimental level, it was demonstrated by subcellular fractioning that R468Q protein has poor transport to lysosomes[21]. Kato et al. hypothesized that non conservative mutation in R468 should affect the electrostatic field for substrate entrance into the active site cavity resulting in an inactive enzyme [8]. Furthermore, Western blot analysis published later by the same group showed only primary precursors [19].

Q465X is also a mutation associated with neuropathic phenotype. Docking showed Cys84 absence in catalytic core. Although downstream residues have no direct participation in catalytic site in this model, overall conformational changes are seen. In K347Q a positively charged aminoacid is substituted by polar uncharged Glutamine. Lys347 is reported to be located adjacent to the active site, and this non conservative mutation could severely modify the catalytic core. K236N changes to polar uncharged Asparagine, and has been only reported in a case from Bulgaria[22].

There were only two patients with non-neuropathic phenotype with the S71N mutation. S71 and N71 in our models are far from catalytic site. Other reported mutants associated with the normal-intelligence phenotype, R48P and W337R, were founded without a normal maturation process, evidenced by 73–75 kDa precursor in western blot, but A85T was the only mutant from same phenotype processed to mature form(55-45kDa). There are no similar studies performed with S71N, so we performed simulations with and without mature form, finding the same results.

S71N docking resulted in hydrogen bonding between Arg297, Asp45 and O2-sulfate, and electrostatic interactions very similar to wild type IDS. This mutation requires further investigation for better elucidate its relationship with non-neuropathic phenotype. Finally, del.Q200_E203 and R294GfsX2 (del. exon 7) are two novel mutations found in the Colombian patients[15]. Docking analysis showed distorted interactions within the pocket of R294GfsX2, which lacks the last 254 aminoacids, thus losing a great part of catalytic core.

Structural analysis indicated that novel frameshift mutants P160SfsX4, D190Pfs13X and P185GfsX2 lost most of the catalytic domain structure, so it is feasible that these distorted polypeptides will not retain enzymatic activity

Computational analyses are not only useful in contribute to the understanding of disease mechanisms, but also have considerable relevance in designing therapeutic molecules such as chaperones for genetic diseases that result from misfolded/aggregated proteins. Different studies suggest that pharmacological chaperone therapy is an emerging field in lysosomal storage diseases (LSD). In relation to MPSs, Sanfilippo

Syndrome type C and Morquio B have chaperones in different stages of investigation [23-25]. Peripheral IDS mutations such as R468Q, Q465X and S71N could be suitable of pharmacological chaperones therapy, through the rescue from endoplasmic reticulum quality control system and correction of protein processing and trafficking.

This is the first report of molecular docking performed with wild and mutant forms of iduronate 2 sulfatase in Colombia. Even point mutations, not necessarily occurring at core functional region, may affect substrate-protein interactions.

References

1. Neufeld, E.F., Muenzer, J.: The mucopolysaccharidoses. In: Scriver, C.R. (ed.) The Metabolic and Molecular Bases of Inherited Disease. McGraw-Hill, New York (2001)
2. Martin, R., Beck, M., Eng, C., Giugliani, R., Harmatz, P., Munoz, V., Muenzer, J.: Recognition and diagnosis of mucopolysaccharidosis II (Hunter syndrome). Pediatrics 121, e377–e386 (2008)
3. Wraith, J.E., Scarpa, M., Beck, M., Bodamer, O.A., De Meirleir, L., Guffon, N., Meldgaard Lund, A., Malm, G., Van der Ploeg, A.T., Zeman, J.: Mucopolysaccharidosis type II (Hunter syndrome): a clinical review and recommendations for treatment in the era of enzyme replacement therapy. Eur. J. Pediatr. 167(3), 267–277 (2008)
4. Diez-Roux, G., Ballabio, A.: Sulfatases and human disease. Annu. Rev. Genomics Hum. Genet. 6, 355–379 (2005)
5. Saenz, H., Lareo, L., Poutou, R.A., Sosa, A.C., Barrera, L.A.: Computational prediction of the tertiary structure of the human iduronate 2-sulfate sulfatase. Biomedica 27(1), 7–20 (2007)
6. Chkioua, L., Khedhiri, S., Ferchichi, S., Tcheng, R., Chahed, H., Froissart, R., Vianey-Saban, C., Laradi, S., Miled, A.: Molecular analysis of iduronate -2- sulfatase gene in Tunisian patients with mucopolysaccharidosis type II. Diagn. Pathol. 6, 42 (2011)
7. Kim, C.H., Hwang, H.Z., Song, S.M., Paik, K.H., Kwon, E.K., Moon, K.B., Yoon, J.H., Han, C.K., Jin, D.K.: Mutational spectrum of the iduronate 2 sulfatase gene in 25 unrelated Korean Hunter syndrome patients: identification of 13 novel mutations. Hum. Mutat. 21(4), 449–450 (2003)
8. Kato, T., Kato, Z., Kuratsubo, I., Tanaka, N., Ishigami, T., Kajihara, J., Sukegawa-Hayasaka, K., Orii, K., Isogai, K., Fukao, T., et al.: Mutational and structural analysis of Japanese patients with mucopolysaccharidosis type II. J. Hum. Genet. 50(8), 395–402 (2005)
9. Molecular Operating Environment (MOE), 2010.10; Chemical Computing Group Inc., 1010 Sherbooke St. West, Suite #910, Montreal, QC, Canada, H3A 2R7 (2010)
10. Arnold, K., Bordoli, L., Kopp, J., Schwede, T.: The SWISS-MODEL workspace: a web-based environment for protein structure homology modelling. Bioinformatics 22(2), 195–201 (2006)
11. Lovell, S.C., Davis, I.W., Arendall III, W.B., de Bakker, P.I., Word, J.M., Prisant, M.G., Richardson, J.S., Richardson, D.C.: Structure validation by Calpha geometry: phi,psi and Cbeta deviation. Proteins 50(3), 437–450 (2003)
12. Trott, O., Olson, A.J.: AutoDock Vina: improving the speed and accuracy of docking with a new scoring function, efficient optimization, and multithreading. J. Comput. Chem. 31(2), 455–461 (2010)
13. Schrödinger, L.L.C.: The PyMOL molecular graphics system, version 1.3 r1. The PyMOL Molecular Graphics System (2010)

14. Laskowski, R.A., Swindells, M.B.: LigPlot+: multiple ligand-protein interaction diagrams for drug discovery. J. Chem. Inf. Model. 51(10), 2778–2786 (2011)
15. Galvis, J., Contreras, G., Correa, L., Sierra, G., Mansilla, S., Piñeros, L., Prieto, J.C., Corredor, C., Uribe, A., Velasco, H.: Identificación de mutaciones en el gen Iduronato 2 sulfatasa en pacientes colombianos con Síndrome de Hunter. Universidad Nacional de Colombia, Colombia (2012) (unpublished data)
16. Lau, K.C., Lam, C.W.: Molecular investigations of a novel iduronate-2-sulfatase mutant in a Chinese patient. Clin. Chim. Acta 392(1-2), 8–10 (2008)
17. Alves, S., Mangas, M., Prata, M.J., Ribeiro, G., Lopes, L., Ribeiro, H., Pinto-Basto, J., Lima, M.R., Lacerda, L.: Molecular characterization of Portuguese patients with mucopolysaccharidosis type II shows evidence that the IDS gene is prone to splicing mutations. J. Inherit. Metab. Dis. 29(6), 743–754 (2006)
18. Schouten, J.P., McElgunn, C.J., Waaijer, R., Zwijnenburg, D., Diepvens, F., Pals, G.: Relative quantification of 40 nucleic acid sequences by multiplex ligation-dependent probe amplification. Nucleic Acids Res. 30(12), e57 (2002)
19. Sukegawa-Hayasaka, K., Kato, Z., Nakamura, H., Tomatsu, S., Fukao, T., Kuwata, K., Orii, T., Kondo, N.: Effect of Hunter disease (mucopolysaccharidosis type II) mutations on molecular phenotypes of iduronate-2-sulfatase: enzymatic activity, protein processing and structural analysis. J. Inherit. Metab. Dis. 29(6), 755–761 (2006)
20. Negishi, M., Dong, J., Darden, T.A., Pedersen, L.G., Pedersen, L.C.: Glucosaminylglycan biosynthesis: what we can learn from the X-ray crystal structures of glycosyltransferases GlcAT1 and EXTL2. Biochem. Biophys. Res. Commun. 303(2), 393–398 (2003)
21. Villani, G.R., Daniele, A., Balzano, N., Di Natale, P.: Expression of five iduronate-2-sulfatase site-directed mutations. Biochim. Biophys. Acta 1501(2-3), 71–80 (2000)
22. Gucev, Z.S., Tasic, V., Sinigerska, I., Kremensky, I., Tincheva, R., Pop-Jordanova, N., Danilovski, D., Hofer, D., Paschke, E.: Hunter syndrome (Muccopolysaccharridosis Type II) in Macedonia and Bulgaria. Prilozi. 32(2), 187–198 (2011)
23. Feldhammer, M., Durand, S., Pshezhetsky, A.V.: Protein misfolding as an underlying molecular defect in mucopolysaccharidosis III type C. PLoS One 4(10), e7434 (2009)
24. Schitter, G., Scheucher, E., Steiner, A.J., Stutz, A.E., Thonhofer, M., Tarling, C.A., Withers, S.G., Wicki, J., Fantur, K., Paschke, E., et al.: Synthesis of lipophilic 1-deoxygalactonojirimycin derivatives as D-galactosidase inhibitors. Beilstein J. Org. Chem. 6, 21 (2010)
25. Fantur, K., Hofer, D., Schitter, G., Steiner, A.J., Pabst, B.M., Wrodnigg, T.M., Stutz, A.E., Paschke, E.: DLHex-DGJ, a novel derivative of 1-deoxygalactonojirimycin with pharmacological chaperone activity in human G(M1)-gangliosidosis fibroblasts. Mol. Genet. Metab. 100(3), 262–268 (2010)

Phylogenetic Analysis of Four Dung Beetle Species of Neotropical Genus *Oxyternon* (Coleoptera: Scarabaeidae: Scarabaeinae) Based on 28S and COI Partial Regions

Cuadrado-Ríos Sebastián, Chacón-Vargas Katherine,
Londoño-González César, and García-Merchán Víctor Hugo

Group of Ecology, Genetics and Evolution of the University of Quindio. (GEGEUQ).Biology program, Faculty of Basic Sciences and Technologies, Armenia (Quindío-Colombia)
cuadradosebastian@hotmail.com, kchaconv@uqvirtual.edu.co,
cesarlond@gmail.com, victorhgarcia@uniquindio.edu.co

Abstract. *Oxysternon* is a neotropical genus consists of 11 species, distributed between the north of the Tropic of Capricorn and east of the Andes. Were evaluated the monophyly of 4 species of *Oxysternon* using partial sequences of 28S rRNA genes and Cytochrome Oxidase I (COI). Sequences were aligned under the criteria GENEIOUS (Geneious software, 6.1.2 version). Topology was constructed showing phylogenetic relationships using the model of genetic distance of Jukes-Cantor method and the nearest neighbor, Neighbor-Joining to build trees, with 100,000 bootstrap replicates. The aligned sequences were subjected to parsimony analysis using the program TNT 1.1. The most parsimonious topologies showed a pattern related between *O. conspicillatum* with *O. silenus*, relationship sustained with a node bootstrap of 88.2 for the 28S ribosomal marker and 100 for the COI marker. Such groupings confirmed the current taxonomic grouping, shown by internal molecular characteristics, possibly explained by the similar geographic distribution.

1 Introduction

Oxysternon is a neotropical genus, member of the tribe Phanaeini (1), formed by dung beetles characterized by their burrowing habits and robust body. In a taxonomic level are determined by an extension in the form of spina-anteromedial angle metasternum of which extends to the apex of the coxa, besides prolonging pronotal posteromedial angle between the bases of the elytra (2). These are part of the characters defined by Edmonds (3) diagnosed as synapomorphies that define the genus as a monophyletic group.

The genus includes 11 species divided into two subgenera, *Oxysternon* and *Mioxysternon*. This genus is restricted to the Americas and its distribution is between the north of the Tropic of Capricorn and east of the Andes. Were analyzed 4 species of subgenus *Oxysternon*: *O. conspicillatum* and *O. silenius*, located in the southern region of Central America and the Amazon region respectively; *O. festivum* found in the Amazonian north from east of Venezuela to northwestern Brazil, and *O. durantoni* distributed in the northern Amazon (1, 4). The DNA databases provide a wealth of information when trying to assess the phylogeny of certain groups, which can extend the knowledge about the families that have been evaluated only at a taxonomic level.

L.F. Castillo et al. (eds.), *Advances in Computational Biology*,
Advances in Intelligent Systems and Computing 232,
DOI: 10.1007/978-3-319-01568-2_31, © Springer International Publishing Switzerland 2014

The study focuses on analyzing the monophyly relationship of 4 species of *Oxysternon* based on sequences of two partial genes, the 28S Ribosomal subunit and mitochondrial Cytochrome Oxidase I, COI.

2 Methods

In order to perform the phylogenetic analysis of four neotropical species of genus *Oxysternon,* virtual information used was extracted from the genebank of NCBI (http://www.ncbi.nlm.nih.gov) about the following species (recognized by 1, 2 and 3): *O. conspicillatum, O. festivum, O. silenus* and *O. durantoni*. In turn we consulted the information of one external species (outgroup) in relation to the genus *Oxysternon, Phanaeus vindex.*

The information used in the study of the taxa mentioned correspond to molecular markers Cytochrome Oxydase I and 28s Ribosomal, thus we used sequences of 28s rRNA and COI (Table 1). Aligment was perform under the Geneious Alignment of the software GENEIOUS 6.1.2 (Biomatters Ltd) with default values according to the software specifications. Topology of phylogenetic relationships was constructed using the phylogenetic distance model Jukes-Cantor and the methodology for the construction of closest Neighbor-Joining's tree, with a bootstrap of 100.000 replicates, obtaining the taxa cladograms (Fig. 1). The aligned sequences were submitted into a parsimony analysis in the software TNT 1.1 (5), in order to compare the topologies. The consensus tree obtained was resampled using the bootstrap analysis using 10.000 replicates.

Table 1. Consulted references in the genebank (NCBI) about the analyzed species

Species	**Cytochrome oxidase subunit I (COI)**			**Ribomal 28s rRNA**		
	NCBI Reference	**No. of Nucleotides**	**Cited**	**NCBI Reference**	**No. of Nucleotides**	**Cited**
O. conspicillatum	AY131948	746 bp	(6)	AY131792	580 bp	(6)
O. festivum	EU477357	516 bp	(7)	EU432272	569 bp	(7)
O. silenus	EU477362	517 bp	(7)	EU432271	395 bp	(7)
O. durantoni	EU477358	481 bp	(7)	EU432273	609 bp	(7)
P. vindex	EU477348	526 bp	(7)	EU432264	571 bp	(7)

3 Results

The main tree obtained with COI sequences alignment presented a total of 15 gaps with a maximum length of 5pb. In comparison, the number of gaps in the 28S rRNA alignment presented 26 gaps with 6pb as the maximum extent. The phylogenetic relationships obtained are illustrated in the figure 1.

The genus was recovered as monophyletic with inclusion of *Phanaeus vindex* as the outgroup. Both clades (using 28s and COI information) were supported with bootstrap values greater than 50%.

The 28s rRNA tree was founded bootstrap values close to 100, which supported the internal branches. The most parsimonious topologies show a relationship between *O. conspicillatum* with *O. silenus,* related by the values of bootstrap, supported by 88.2 node values (Fig. 1). On the other hand, COI marker tree related these two species with a 100 node value.

As well as, the clade related *O. duratoni* with *O. festivum* was supported by 90 node value for 28s and 84.1 for COI.

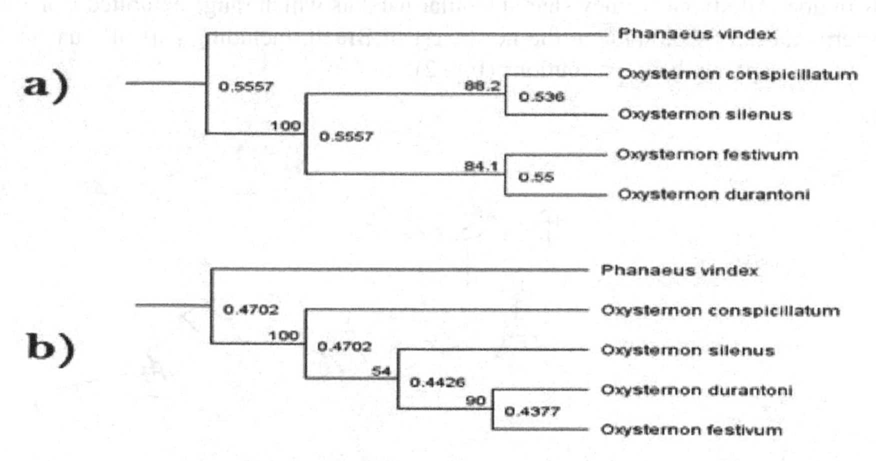

Fig. 1. Phylogenies constructed with the Geneious Tree Builder tool, using the Jukes-Cantor distance model, Neighbor-Joining tree build method. *Phanaeus vindex* as the outgrup. The nodes (left) and branch (right) labels are indicated in the branch roots. Phylogenies built with the COI (a) and 28S rRNA (b) alignments.

4 Discussion

DNA Data

According to the estimates made by Brower (8), the observed divergence on mitochondrial DNA among arthropod species recently separated was usually constant, with a divergence rate of 2.3% per million years. This was consistent with the observed parsimony COI tree.

Multiple Alignments

The multiple sequences alignments performed by Geneious allowed the identification of similar regions among several sequences, resulting phylogenetic relationships among individuals and those regions which have undergone insertion, deletion and substitution processes. Similarly, the ratio is greater than the transition ratio associate to transversions (9, 19). This alignment allowed the use of Jukes-Cantor model (10), resulting a appropriate substitution pattern to molecular data type analyzed.

Phylogenetic Analysis

Currently several studies have been developed by mitochondrial markers, such a Cytochrome b (Cyt-b) and Cytochrome Oxidase I (COI) for analysis of divergence at taxonomic and phylogenetic animal levels (i.e. 11, 12, 13, 14).

Edmonds (1) rejected the hypothesis proposed by Edmonds (3) about the division of the subgenus *Oxysternon* in two groups of species. In order to, our results have more than one synapomorphy linking the three groups, therefore it was conceived as a paraphyletic group. As morphological evidence, our molecular alignments related *Oxysternon festivum* and *Oxysternon duratoni*, regardless Edmonds (1) related them as a paraphyletic group. These results could be associated to the altitudinal similar distribution. Additionally, they shared similar habitats which range exhibited from the Western Andes of Colombia at the northwest of Brazil, including part of Guayana's forests, overlapping their distributions (Fig. 2).

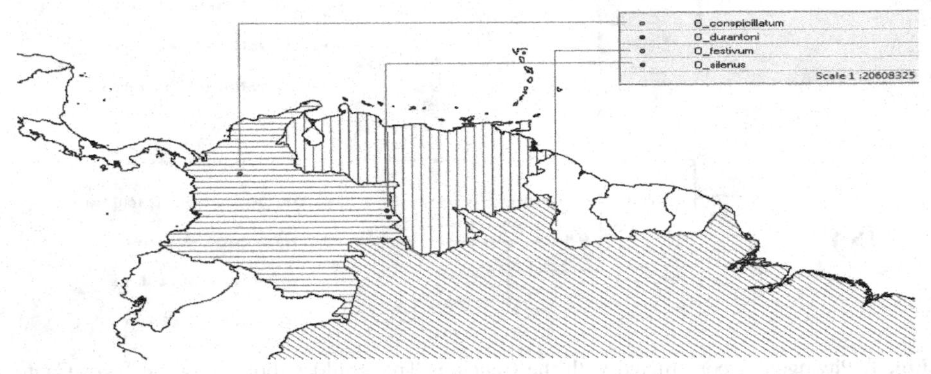

Fig. 2. Distibution of *O. conspicillatum, O. durantoni, O. festivum* and *O. silenus* in South America according data SIB (biodiversity information system http://data.sibcolombia.net/search/oxysternon consulted 19-05-2013)

The grouping proposed by Edmonds (3), which connected *O. conspicillatum* and *O. silenus* as a sympatric clade was supported by synapomorphies clearly defined. This clade was defined by a high bootstrap confidence in the morphological analysis performed.

These molecular relationships were contrasted with morphological characters reported by Edmons (3) and Vítolo (2). *O. conspicillatum* and *O. silenus* share a bipodal cephalic arm, cephalic coniform process in males, size bigger than 15mm and a carenacircumnotal complete, do not erased behind the eye.

O. conspicillatum share more morphological traits with *O. silenus* than with other species included in this paper, result reflected in the molecular analysis (Fig. 1. a) in which the node value was 88.2, showing a high level of reliability. Is important to explain, the characters were not quantitatively analyzed in function of molecular alignments, seeking matches between molecular clusters and taxonomic characters that share the species. Also, the criteria for choosing the characters were based on the traits of the genus and each species from the taxonomic keys of Edmons (1, 3), Vítolo (2, 20) and Medina and collaborators (15).

O. festivum also was close related with the clade between *O. conspicillatum* and *O. silenus.* This is congruent with morphological evidence sharing character carenacircumnotal complete, do not erased behind the eye, bipodal cephalic arm and size bigger than 15mm.

O.duratoni was related with *O. festivum* with a node value of 90 for 28s rRNA and 84.1 node value for COI. This relationship could be measure in function of distribution. Thus, *O. festivum* occurs at north of the Amazon, occupying a broad northern area extending from eastern of Venezuela through the Guianas to Amapa (Brazil) while *O. duratoni* occurs at north and south of Amazonas (1). Similar habitats could implicate characters in common for species which have similar biotic and abiotic factors. Many phenotypic and genotypic traits are given by the interaction with environmental factors, where it is possible that species that are not related share common characters in analogous manner, which serve as a similar function to adapt to the environment. When the species with a phylogeny that share the same habitat, as the case of *O. festivum* and *O. duratoni*, the similarity may be attributable to factors due to common habitat. However, must take into account different evolutionary mechanisms of each species (i.e. speciation), as well as ecological aspects of life history traits, also greatly influence in genetic variation patterns.

For future studies it is important to include more species of the genus to have a broader resolution of the phylogeny. Relict populations have been characterized in the north of Brazil, so it requires a population genetic analysis to determine whether a population is under the influence of reproductive isolation. Similar case can present in *O. striatopunctatum* (16), although this species is part of the subgenus *Myoxisternon.* Similarly, studies using molecular markers can define whether the subspecies described by Olsoufieff (17) and *O. oberthuri* is part of the life stages of *O. conspicillatum* or a subspecies defined by environmental pressures.

A similar molecular marker analysis used in this study can define the phylogenetic relationships within the subgenus *Oxysternon.* In order to, we would expect that *O. spiniferum* confirm your location as most basal species within the South American species because it differs from other species for its cephalic arm monopodal (2, 18). In other scenario, *O. lautum* is the basal through its mild pronotum side roughness, likewise, are expected to *O. palaemon* and *O. smaragdinum* form a clade defined by synapomorphies with *O. silenus* as transverse clypeal process and *O. ebeninum* relates to *O. festivum* thanks to clypeal spiniform process that would support this clade (2, 18). Finally it is necessary to include in these studies the species *O. smaragdinum* and *O. sericeum,* considered by Edmonds (1) and can be synonymous or valid species.

Artificial classifications can be solved by using molecular data that define relationships despite the many morphological similarities in their distribution and show tendencies to associate certain species, so that the division proposed in principle by Edmonds (3) suggest should not be completely ruled out.

5 Conclusions

The sequencing of a partial region of the 28S ribosomal region and mitochondrial COI, gives a new value to the phylogenetic relationships among *O. conspicillatum,*

O. durantoni, O. festivum and *O. silenus* species of genus *Oxysternon*, agreeing with groupings made by Edmonds (1) based on morphological characters; besides, greater relationship between *O. conspicillatum* and *O. silenus* probably by their similar distribution.

References

1. Edmonds, W.D., Zidela, J.: Revision of the Neotropical dung beetle Genus Oxysternon (Scarabaeidae: Scarabaeinae: Phanaeini). Folia Heyrovskyana 11(suppl.), 1–58 (2004)
2. Vitolo, A.: Clave para la identificación de los géneros y especies Phanaaeinas (Coleoptera: scarabaeidae: coprinae: phanaeini) de Colombia. Revista Académica Colombiana de la Ciencia 26(93) (2000)
3. Edmonds, W.D.: Comparative skeletal morphology, systematics and evolution on the Phanaeine dung beetles (Coleoptera: Scarabaeidae). The University of Kansas Science Bulletin 49(11), 731–874 (1972)
4. SIB. Sistema de Información de Biodiversidad, `http://data.sibcolombia.net/species/30210` (consultado May 05, 2013)
5. Goloboff, P., Farris, S., Nixon, K.: TNT (Tree analysis using New Technology) (2000)
6. Monaghan, M., Inward, D., Hunt, T., Vogler, A.: A molecular phylogenetic analysis of the Scarabaeinae (dung beetles). Molecular Phylogenetics and Evolution 45(2), 674–692 (2007)
7. Price, D.L.: Phylogeny and biogeography of the dung beetle genus Phanaeus (Coleoptera: Scarabaeidae). Journal Systematic Entomology 34(1), 137–150 (2009)
8. Brower, A.V.Z.: Rapid morplogical radiation and convergence among races of the butterfly Heliconius erato inferred from patterns of mitochondrial DNA evolution. Proceedings of the National Academy of Sciences of the USA 91(14), 6491–6495 (1994)
9. Hasegawa, M., Kishino, H., Saitou, N.: On the Maximum Likelihood Method in Molecular Phylogenetics. Journal of Molecular Evolution 32, 443–445 (1991)
10. Jukes, T.H., Cantor, C.R.: Evolution of protein molecules. In: Munro, H.N. (ed.) Mammalian Protein Metabolism, vol. 3, pp. 21–132 (1969)
11. Wang, I.A.: Recognizing the temporal distinctions between landscape genetics and phylogeography. Molecular Ecology 19, 2605–2608 (2010)
12. Hellberg, M.E.: No variation and low synonymous substitution rates in coral mtDNA despite high nuclear variation. BMC Evolution Biology 6, 24 (2006)
13. Kartavtsev, Y.P., Lee, J.S.: Analysis of Nucleotide Diversity at the Cytochrome b and Cytochrome Oxidase 1 Genes at the Population, Species, and Genus Levels. Russian Journal of Genetics 42(4), 341–362 (2006)
14. Hebert, P.D.N., Ratnasingham, S.: Barcoding Animal Life: Cytochrome c Oxidase Subunit 1 Divergence samong Closely Related Species. Proc. R. Soc. London, B 270, 1512 (2002)
15. Medina, C.A., Lopera-Toro, A., Vitolo, A., Gill, B.: Escarabajos Coprófagos (Coleoptera: Scarabaeidae: Scarabaeinae) Revista de Colombia. Biota Colombiana 2(2), 131–144 (2001)
16. Hamel-Leigue, A., Herzog, S., Mann, D., Larsen, T.H., Gill, B.D., Edmonds, W.D., Spector, S.: Distribution and natural history of the dung beetle tribe Phanaeini (Coleoptera: Scarabaeidae: scarabaeinae) in Bolivia. Kempffiana 5(2), 43–95 (2009)

17. Olsoufieff, Les Phanaeides, G.: Famille Scarabaeidae - Tr. Coprini. Revue Illustrée d'Entomologie 13, 4–172 (1924)
18. Medina, C.A., Pulido, L.A.: Escarabajos coprófagos (Coleoptera: Scarabaeinae) de la Orinoquia colombiana. Biota Colombiana 10(1&2), 55–62 (2009)
19. Gillespie, J.J., Yoder, J.M., Wharton, R.A.: Predicted Secondary Structure for 28S and 18S rRNA from Ichneumonoidea (Insecta: Hymenoptera: Apocrita): Impact on Sequence Alignment and Phylogeny Estimation. Journal Molecular Evolution 61, 114–137 (2005)
20. Pulido Herrera, L.A., Medina, C.A., Riveros, R.A.: Nuevos registros de escarabajos coprófagos (Scarabaeidae: scarabaeinae) para la región andina de Colombia. Rev. Academia Colombiana Ciencias 31(119), 305–310 (2007)

Limits to Sequencing and *de novo* Assembly: Classic Benchmark Sequences for Optimizing Fungal NGS Designs

José Fernando Muñoz[1,2], Elizabeth Misas[1,2], Juan Esteban Gallo[1,3], Juan Guillermo McEwen[1,4], and Oliver Keatinge Clay[1,5,*]

[1] Cellular and Molecular Biology Unit, Corporación para Investigaciones Biológicas, Medellín, Colombia
[2] Institute of Biology, Universidad de Antioquia, Medellín, Colombia
[3] Doctoral Program in Biomedical Sciences, Universidad del Rosario, Bogotá, Colombia
[4] School of Medicine, Universidad de Antioquia, Medellín, Colombia
[5] School of Medicine and Health Sciences, Universidad del Rosario, Bogotá, Colombia
{josejfmg,Elizabeth.misas,galloucf,oliver.clay}@gmail.com, mcewen@une.net.co

Abstract. Planning of pipelines for next-generation sequencing (NGS) projects could be facilitated by using simple DNA sequence benchmarks, i.e., standard test sequences that could monitor or help to predict ease or difficulty of (a) short-read sequencing and (b) *de novo* assembly of the sequenced reads. We propose that familiar, gene-sized sequences, including but not limited to nuclear protein-coding genes, would provide feasible consensus benchmarks allowing simple visualization. We illustrate our proposal for fungi with candidates from ribosomal DNA (rDNA, used in phylogeny and identification/diagnostics), mitochondrial DNA (mtDNA), and combinatorially constructed conceptual (synthetic) DNA sequences. The exploratory analysis of such familiar candidate loci could be a step toward finding, testing and establishing familiar, biologically interpretable consensus benchmark sequences for fungal and other eukaryotic genomes.

Keywords: Next generation sequencing, Eukaryotic genomes, *De novo* assembly, Benchmarking.

1 Introduction

When one plans pipelines for next-generation sequencing (NGS) projects, it would often be helpful to have available simple reference DNA sequences or benchmarks, i.e., standard or consensus test sequences that could monitor or help to predict ease or difficulty of (a) short-read sequencing and (b) *de novo*

* Corresponding author.

L.F. Castillo et al. (eds.), *Advances in Computational Biology*,
Advances in Intelligent Systems and Computing 232,
DOI: 10.1007/978-3-319-01568-2_32,

assembly of the sequenced reads. Indeed, a genome or genomic region can have patches that are inherently refractory to either endeavour.

We consider separately two main groups of tasks in a genomics pipeline leading to an assembly, which lend themselves to separate benchmarking: tasks that are done before the official read set is obtained, and tasks that are done after one has the reads. The former, *sequencing-related* tasks include choice of sequencing technology and choice of parameters such as read length and insert size; the preparation and possible selection of insert libraries; actual sequencing; and any routine censoring or other processing of raw sequencer output that is then performed in order to arrive at a set of official or presentable read files. The latter, *assembly-related* tasks include choice of assembly program(s) and choice of parameters such as k-mer length (for de Bruijn graph-based programs); actual assembling into contigs and/or scaffolds; and any subsequent censoring or elimination of short or otherwise doubtful contigs or scaffolds from the assembly. The tasks listed in these two groups are the ones we can influence, for which we seek guidance, and for which existing and new benchmarks could be valuable. We propose that familiar sequences, of the length of one or a few known genes, could provide useful and realistic external references that would be helpful in such contexts.

If one looks only at whole-genome assemblies of eukaryotes without considering the structural and functional features that exist along the chromosomes, benchmarks already exist. Well-known examples are N50, NG50 or NG90; state-of-the-art benchmarks of this category are listed and used in the recent Assemblathon 2 report [5]. If one restricts one's attention to protein-coding genes of the nuclear genome, benchmarking can be done by assessing how well one's assembly covers conserved or 'core' genes that one expects to find, e.g., using CEGMA [28,29,5]. We propose, in the case of fungi, to complement such existing benchmarks with benchmarks based on ribosomal DNA (rDNA, used in phylogeny and identification/diagnostics), mitochondrial DNA (mtDNA, of which for example the gene for cytochrome c oxidase subunit 1, *CO1*, is used in barcoding), and combinatorially constructed conceptual or synthetic DNA sequences. We present our proposal using selected examples.

2 Materials and Methods

Reference sequences were selected after performing exploratory studies on DNA sequences of model and pathogenic fungi (*Saccharomyces cerevisiae*, *Aspergillus fumigatus*, and dimorphic fungi from the Onygenales order) that are represented by an intrinsic interest in the scientific community and may therefore already be available prior to a strain's whole-genome sequencing. We retrieved real DNA reference sequences from public databases (GenBank, Broad Institute), and created synthetic DNA reference sequences by generating DNA stretches that have no repeats of length $\geq w$. Reads used in the exploratory studies were taken from Illumina paired-end short-read ($l \geq 100$ bp) files downloaded from the NCBI Sequence Read Archive (SRA), from our own Illumina paired-end reads ($l \geq 100$

bp) or, in the case of ideal reads (see [20]), created from longer public sequences by generating all possible subsequences of a fixed length l ($l \geq 100$ bp). Further details of presented examples are given in Figure legends and descriptions in the main text.

3 Results and Discussion

3.1 Benchmarks from Ribosomal DNA for Optimizing Sequencing-Related Tasks

Although ribosomal DNA (rDNA) of a eukaryotic organism can seriously resist assembly because of its tandemly repeated nature (for example, some of the human genome's known rDNA regions are still missing from the hg19 sequence), we focus here on inherent resistance to sequencing and/or related tasks that lead to the production of an official read set.

Generally, patches along a chromosome that are truly refractory to sequencing (thus causing unsatisfactory read depths) can result from various factors, of which one seems to be extreme or unusual GC (guanine-cystosine) levels [33, and refs. therein]. The rDNA loci in eukaryotes, including fungi and also human [16, Genbank accession U13369], can exhibit some of a genome's most severe GC changes and/or extremely high levels of GC; in the interphase nucleus, rDNA does not reside in typical chromosome territories but in nucleoli [1].

Figure 1 shows an example of how a classic rDNA reference sequence of a fungus can be used as a potential benchmark, to monitor patches along the sequence where reads from a closely related organism (here, another strain of the same species) are thinly spread. If the organism represented by the reads is close enough to the organism represented by the reference sequence, and if one chooses a matching method having appropriate sensitivity and specificity (e.g., general-purpose matching programs such as BLAST or BLAT or more dedicated NGS programs such as BWA [22] or Bowtie 2 [21]), then patches of shallow reads must have been caused during sequencing or sequencing-associated steps.

3.2 Benchmarks from Mitochondrial DNA for Optimizing Assembly-Related Tasks

Mitochondrial genomes of some fungi can be surprisingly difficult to assemble *de novo*, despite their small sizes, which are typically less than 100 kb. For example, two groups that used shotgun methods to sequence strains of yeast (*Saccharomyces cerevisiae*) appear to have encountered problems in obtaining satisfactory assemblies of the mitochondrial genome [12,25].

A short sequence of less than 100 kb might, at first sight, seem like child's play to assemble, but it turns out that some far bigger nuclear genomes are routinely and quite successfully assembled and annotated while the same organisms' mitochondral genome is not. Mitochondrial genomes can contain seriously repetitive and/or low complexity DNA interspersed in a number of noncoding

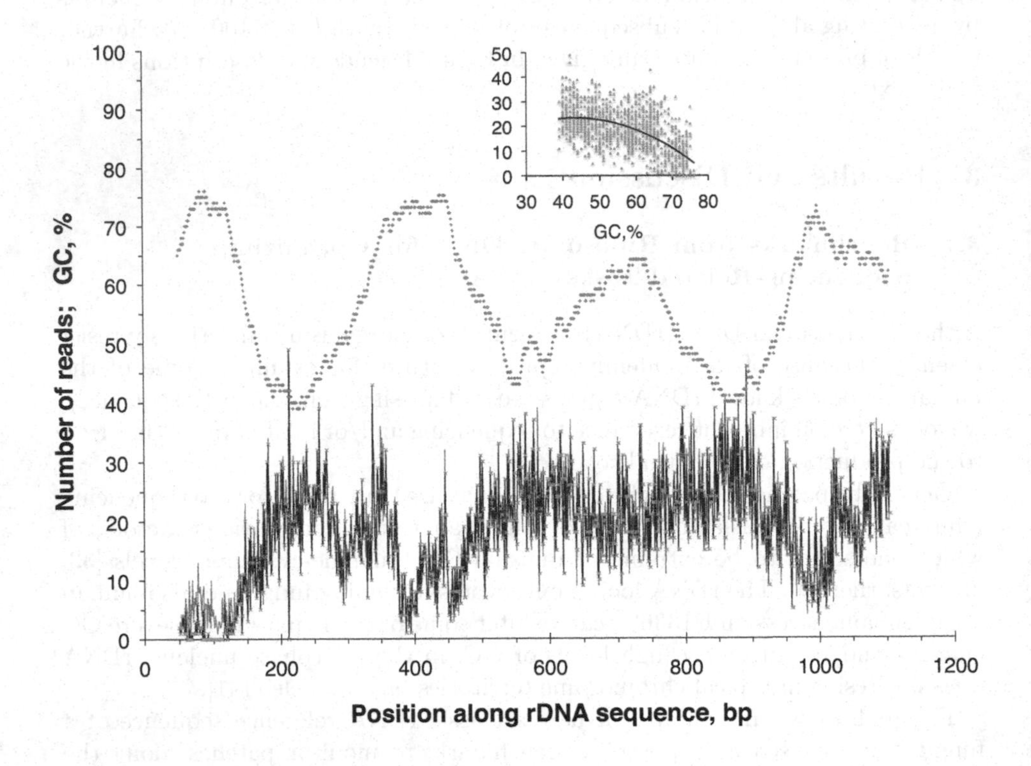

Fig. 1. Positional variation of read counts of *Aspergillus fumigatus* strain A1163 (Illumina GAII paired-end reads of length 101 bp, insert size 300; NCBI SRA accession SRX028559) that match along an rDNA reference sequence. The reference sequence is from *A. fumigatus* strain NRRL 35223 (GenBank accession EF634403, 1154 bp). Matching was done using BLASTN with default settings (black read depth curve). A moving-window GC plot (window 101 bp, step 1 bp) is shown above the match counts, anticorrelating with them. Matching was repeated with the DUST [27] low complexity filter switched off (`-dust no` , light curve superposed on and essentially coinciding with black read depth curve), in order to confirm that the anticorrelation with GC is not an artifact of low-complexity patches. Since the first two valleys of the read coverage plot are found within internal transcribed spacer regions (ITS1: positions 13–195, ITS2: positions 353–521), we also verified that these sequences are highly conserved (close to 100% identity) among *A. fumigatus* strains, i.e., that the valleys are not due to more pronounced ITS sequence divergence as is sometimes seen in higher-level, interspecies alignments [17,18]; we have observed similar fluctuations of read depth when mapping our own Illumina 101 bp reads for an *Emmonsia parva* strain to a reference rDNA sequence for that strain. Positions are midpoint of BLAST match for read match counts and window midpoint for GC. Inset: Scatterplot and quadratic polynomial fit sketching the negative relation between read depth and GC ($R = 0.604$; cf. also a similar general relation observed in [33, Suppl. Figure 3]).

stretches along the sequence, which can confuse assembly programs. In fact, repeats can confuse not only assembly programs, but also (biological) homologous recombination, leading to mitochondrial genome instabilities that can become visible as petite mutant colonies in yeast [3,4,12], so if one is looking for a tough benchmark sequence this may be a good candidate: Figure 2 shows that even an ideal paired-end strategy with no sequencing errors and providing uniform, good coverage by reads cannot bridge the *ori* repeats, and suggests that the resulting (modest) fragmentation may be a fundamental limit [20] that is not dependent on the program used for assembling. This small genome has relatively few genes, of which the *CO1* gene has been proposed for barcoding [31], and it is well-studied and familiar to all biologists. Use of mitochondrial genomes as benchmarks might correspond to benchmarking ideals of a "toughest competitor" [7, p. 10], or to agile practices advocating that one write a test that fails, then be prompted by the failing test to write code that will pass that test ([9, pp. 25–29], [6, Chapters 16–19]).

3.3 Comments on Benchmarks from Nuclear Protein-Coding DNA

Protein-coding genes of the nuclear genome are the most abundant genes of a eukaryotic organism, but they are often less difficult to sequence or assemble than other genomic regions. On the other hand, in some sequencing and assembly projects one may be primarily interested in obtaining the full set of nuclear protein-coding sequences, and much less interested in other regions. A recent reminder that it is important to be clear about one's priorities was given by the results of Assemblathon 2, in which some of the least successful assemblies of vertebrate genomes, as measured by genome-wide metrics, were successful in covering a majority of genes ([5]; see also [20]).

The frequently observed or rediscovered success of simple next-generation sequencing strategies in assembling most protein-coding sequences suggests that the notion of *gene space* is appropriate here [29], a term that was originally used to characterize the dramatic enrichment of genes in a relatively small interval of the GC distribution of genomes such as those of cereal plants [8].

The set of genes that was used to assess coverage by the Assemblathon 2 entries came from a collection of core eukaryotic protein-coding genes (CEGs), proposed by Parra et al. [28,29] as a composite benchmark. Their rigorous filtering protocol yielded 458 genes [28] present in 6 eukaryotic model organisms including human and baker's yeast, of which 248 genes were found to be generally present as single copy genes [29]; the 248 genes were then further divided into 4 groups according to their conservation. Based on mapping of CEGs in 25 previously characterized eukaryotic gen he most conserved group (group 4), consisting of 65 genes, has been proposed as an estimator of the percentage of total protein-coding genes covered by the assembly that should be useful even for highly divergent genomes; a CEGMA mapping package is available for analyzing coverage of arbitrary eukaryotic genomes using CEG genes [29].

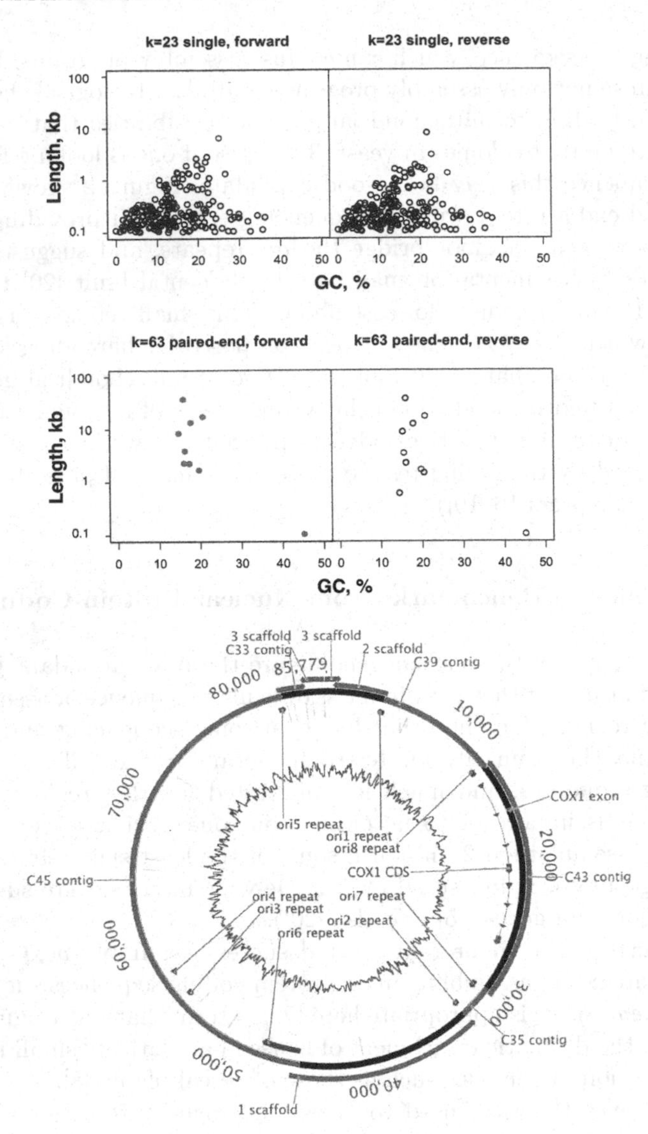

Fig. 2. Simulated NGS strategy designs for the yeast mitochondrial genome. Results of feeding all possible subsequences of length 100 bp of the *Saccharomyces cerevisiae* reference mtDNA genome sequence [14, GenBank acc. AJ011856], as simulated ideal reads at 100× [20], to the SOAPdenovo/GapCloser assembly pipeline [24]. Scaffold/contig properties are shown for oppositely directed sample runs (sense and antisense, for consistency checking) in two contrasting sample scenarios or strategies. The first strategy is single reads at a low k-mer size (23 bp), and is completely unsuccessful; the second strategy is paired-end reads separated by 512 bp at a high k-mer size (63 bp), and is more successful, but still all scaffolds/contigs are prematurely terminated at *ori* repeats, suggesting a fundamental limit of the NGS sequencing design (lower left scatterplot, and corresponding arcs in circular Geneious [19] display with inner ring showing GC). One sample mitochondrial gene, *CO1*, is shown.

3.4 Benchmarks from Artificially Generated DNA Symbol Sequences for Optimizing Assembly Programs

The last class of benchmarks we discuss comes from artificial DNA. Although the use of such artificial or conceptual DNA (DNA *in silico*) is not backed by a tradition of study by molecular biologists (with an few exceptions), it is backed by a classic branch of thinking and proving theorems about repetitiveness in combinatorics, i.e., in mathematics. This tradition usually bears the name of de Bruijn, although it has its roots in the first graph-theoretic result, Euler's 1766 solution of the Königsberg bridge problem [23, p. 40], [2, Chapter 11].

If mitochondrial genomes, such as that of yeast shown in Figure 2, are some of the toughest benchmarks for assembly-related tasks because of their repeats and low-complexity regions, then we can, conversely, aim to find lenient benchmarks that could serve as a minimal requirement for assembly programs by eliminating repeats or, alternatively, by artificially constructing a library of synthetic DNA (syDNA) sequences that are guaranteed to have no repeats longer than a given length on either strand. For example, the first sequence in such a library could be free of repeats greater than some length w, the second sequence in the library could be free of repeats greater than $w+1$, and so on. Different assembly programs could be assessed or 'acceptance-tested' based on their performance profiles when fed such a synthetic sequence library in read-sized fragments (possibly paired, and/or with simulated sequencing errors to assess robustness).

The problem of constructing, or generating, a repeat-free double-stranded DNA sequence for a given word length w was formulated and addressed already early, in an appendix [15] of a 1966 paper addressing the danger of ectopic (illegitimate, out-of-register) homologous recombination [32]. That appendix presented maximal attainable lengths ($\approx 4^w/2$) and theorems pertaining to those lengths.

The generating of *single-stranded* DNA (ss-DNA) sequences without repeats is relatively straightforward, and even implemented in the general-purpose mathematics software Sage (`DeBruijnSequences(4,w).an_element()` and then replace 0,1,2,3 by A,C,G,T; for theory see [2, Chapter 11], and for the elegant generation of a basic example for a given w used by Sage see [30, section 7.3], [13]). However, the study of algorithms for generating *double-stranded* (ds-DNA) repeat-free sequences, in which no words or their reverse complements encounter a match, has not received much attention since 1966. One reason is that the ss-DNA problem is traditionally solved by considering a well-studied class of directed graphs, the de Bruijn graphs, which are used also in several NGS assembly programs. When one considers ds-DNA, however, the underlying structure is less simple: a word needs to be 'glued' to its reverse complement to form a single node or vertex consisting essentially of two node-chambers, connected to other nodes or their chambers by separate directed or doubly-directed edges (bi-directed graph, bi-flow, conjunction product; [26]; cf. also [11]).

Although it would be useful to have an efficient algorithm for directly generating maximal nonrepetitive ds-DNA sequences, e.g., via an analog of the Lyndon word approach used by Sage, one can create nonrepetitive ds-DNA sequences that are not maximal but often sufficiently long for one's purposes, by generating maximal nonrepetitive ss-DNA sequences and then censoring or editing them. In such editing, visualization using the familiar dotplot method is helpful.

We end this subsection with a toy example. The following randomly generated, 1028-nt de Bruijn sequence, viewed as a simple sequence of letters from a 4-letter alphabet, has no repeats of length 5 nt or more:

```
GCCGAGTCATTTTTATATAGGCTGCCTGCTATACCACGTCTCGGCCCAGCCAAAAAGCTGTTTTCGCTACAGTCTGATCACGGAATGCAAGCA
ACAATGACGTTGACACCCACCAGGTTTGCACCTAAGGTGGGCCGGTCGATGCTCTTATCGCATCCAGTGTAGGGCGTGTTGCCATTCCCGCAA
TCTCACCGTTACTTCATGAGACGATCCGTGAACATCTGTACTAGCTCAACTAAACAGGGACATGTCTTGGTATCCCCAATAAAGGCATAAGAC
AGCGATACACTTTAGTTGTTATGGGAGTATTAGCCTATTCAAGATAGCGCACTATGATGGAGAGGCCAGATCTTTCACTCGTATGTGAGTGAC
CATAGACTACCGGGCTTTTGGACTGACTTAGATTGTGTGCGCGGGTCTAATGTATAATCAGTACAACCAACGGCAAAGAACTTGTAAGTCGGA
TATGCCGCTGGGTTGGCCTCTAGGAGCTAGTGCCCTTGAAGCCGTCACACAAGGGTAACGCAGGAAGGACGCTTGCGGCGAAACTGCGTAATA
CTCATATTGATTCTGCAGTTCAGACCGCGTTTCTATCATCAATTCGAGGGGTGCATGGCGGTGATAACACGCCACTGGCAGAAGTGGCTCCAA
GTTAGGTACCTTACCCCCGAATAGTCCCATGCGAGCCCCTACGTACGGGGGATTTGTCAGCAGCTTCGGTAGTAGAAAGTAAATATCTACTGT
CCACAGAGCGGAGGATGAATCGTGCTTAAAATGGTCAAATCCTGTGGATCGGGAAATTAACCCGTAGCACATTACGCGACAAACCGATTATTT
CCATCGAAGAGTTTACATACGAGAATTTAAGCGTCCTTCTTCCGACTCTCCTCCGCCTTTGAGCATTGGGGCACGAACCTCGCGCTCGACCTG
AAAACGACGGTTCCTAGAGATGTTCGTCGCCCGGCTAACTCCCTGGTGTCGTTCTCTGGAACGTGGTTAATTGCTGAGGTCCGGACCCTCAGG
CGCCG
```

However, if we interpret the sequence as one strand of ds-DNA, the situation changes. Indeed, GAAGG should not occur twice and its reverse complement CCTTC should never occur, but the first two underlined regions show that this condition is not met. Palindromes, i.e., sense-antisense 'self-repeats' (third underlined region), should in principle also be absent, as a short read assembly program could get confused about the direction in which it should continue growing a contig. The underlined regions are the only ones with such problems.

We ran this example through the assembly program `succinctAssembly/gossamer`, which implements efficient algorithms for assembly sub-tasks with few additional heuristics [10], on all of the sequence's 16-bp ds-DNA subsequences choosing a k-mer size of 10. This procedure is analogous to the one used for yeast mtDNA illustrated in Figure 2. As expected, the places where the contigs broke off were precisely the 3 underlined regions.

4 Conclusion

Exploratory analyses of familiar reference sequences, such as those we present here as candidate benchmark loci, could take us a step further toward finding, testing and establishing familiar, modestly sized, biologically interpretable consensus benchmark sequences for fungal and other eukaryotic genomes. The results suggest that we could use selected, known genes or genomic regions as guides to monitor, help predict, and thus optimize the success of sequencing and assembly pipelines or designs for whole genomes.

Acknowledgements. We acknowledge funding from Colciencias, Colombia for the project "Comparative genomics and virulence in the pathogenic fungus *Paracoccidioides brasiliensis*", 2213-48925460. We thank Drs. Natalie Fedorova, Ishwar

Chandramouliswaran, and William C. Nierman (J. Craig Venter Institute) for permission to use the *A. fumigatus* A1163 read set; their work was funded in whole or part with federal funds from the National Institute of Allergy and Infectious Diseases, National Institutes of Health, Department of Health and Human Services under contract numbers N01-AI30071 and/or HHSN272200900007C (BioProject IDs 14003, 18733, 46347, 52783, 9521, and 67101).

References

1. Audas, T.E., Jacob, M.D., Lee, S.: Immobilization of proteins in the nucleolus by ribosomal intergenic spacer noncoding RNA. Mol. Cell 45, 147–157 (2012)
2. Berge, C.: Graphs. North Holland, Amsterdam (1989)
3. Bernardi, G.: Lessons from a small, dispensable genome: The mitochondrial genome of yeast. Gene 354, 189–200 (2005)
4. Bernardi, G.: Structural and evolutionary genomics: Natural selection in genome evolution. Elsevier, Amsterdam (2005)
5. Bradnam, K.R., Fass, J.N., Alexandrov, A., Baranay, P., Bechner, M., et al.: Assemblathon 2: evaluating de novo methods of genome assembly in three vertebrate species. Giga Science (submitted, 2013), preprint at `http://arxiv.org/abs/1301.5406`
6. Brooks, F.P.: The Mythical Man-Month: Essays on Software Engineering, with four new chapters, Anniversary edn. Addison-Wesley, Reading (1995)
7. Camp, R.: The Search for Industry Best Practices that Lead to Superior Performance, 1st edn. Productivity Press (2006)
8. Carels, N., Barakat, A., Bernardi, G.: The gene distribution of the maize genome. Proc. Natl. Acad. Sci. USA 92, 11057–11060 (1995)
9. Chromatic: Extreme Programming Pocket Guide. O'Reilly Media, Sebastopol (2003)
10. Conway, T.C., Bromage, A.J.: Succinct data structures for assembling large genomes. Bioinformatics 27, 479–486 (2011)
11. Deng, A., Wu, Y.: De Bruijn digraphs and affine transformations. Eur. J. Comb. 26, 1191–1206 (2005)
12. Dimitrov, L.N., Brem, R.B., Kruglyak, L., Gottschling, D.E.: Polymorphisms in multiple genes contribute to the spontaneous mitochondrial genome instability of Saccharomyces cerevisiae S288C strains. Genetics 183, 365–383 (2009)
13. Duzhin, S., Pasechnik, D.: Automorphisms of necklaces and sandpile groups. Preprint, arXiv:1304.2563v1 (2013)
14. Foury, F., Roganti, T., Lecrenier, N., Purnelle, B.: The complete sequence of the mitochondrial genome of *Saccharomyces cerevisiae*. FEBS Lett. 440, 325–331 (1998)
15. Fraenkel, A.S., Gillis, J.: Proof that sequences of A, C, G, and T can be assembled to produce chains of ultimate length avoiding repetitions everywhere. Prog. Nucleic Acid Res. Mol. Biol. 5, 343–348 (1966)
16. Gonzalez, I.L., Sylvester, J.E.: Complete sequence of the 43-kb human ribosomal DNA repeat: analysis of the intergenic spacer. Genomics 27, 320–328 (1995)
17. Henry, T., Iwen, P.C., Hinrichs, S.H.: Identification of Aspergillus species using internal transcribed spacer regions 1 and 2. J. Clin. Microbiol. 38, 1510–1515 (2000)
18. Hinrikson, H.P., Hurst, S.F., De Aguirre, L., Morrison, C.J.: Molecular methods for the identification of *Aspergillus* species. Med. Mycol. 43 (suppl. 1), S129–S137 (2005)

19. Kearse, M., Moir, R., Wilson, A., Stones-Havas, S., Cheung, M., et al.: Geneious Basic: an integrated and extendable desktop software platform for the organization and analysis of sequence data. Bioinformatics 28, 1647–1649 (2012)
20. Kingsford, C., Schatz, M.C., Pop, M.: Assembly complexity of prokaryotic genomes using short reads. BMC Bioinformatics 11, 21 (2010)
21. Langmead, B., Salzberg, S.L.: Fast gapped-read alignment with Bowtie 2. Nat. Methods 9, 357–359 (2012)
22. Li, H., Durbin, R.: Fast and accurate short read alignment with Burrows-Wheeler transform. Bioinformatics 25, 1754–1760 (2009)
23. Lovasz, L.: Combinatorial Problems and Exercises. North Holland-Elsevier, Amsterdam (1993)
24. Luo, R., Liu, B., Xie, Y., Li, Z., et al.: SOAPdenovo2: an empirically improved memory-efficient short-read *de novo* assembler. GigaScience 1, 18 (2012)
25. Lynch, M., Sung, W., Morris, K., Coffey, N., Landry, C.R., et al.: A genome-wide view of the spectrum of spontaneous mutations in yeast. Proc. Natl. Acad. USA 105, 9272–9277 (2008)
26. Medvedev, P., Brudno, M.: Maximum likelihood genome assembly. J. Comput. Biol. 16, 1101–1116 (2009)
27. Morgulis, A., Gertz, E.M., Schäfer, A.A., Agarwala, R.: A fast and symmetric DUST implementation to mask low-complexity DNA sequences. J. Comp. Biol. 13, 1028–1040 (2006)
28. Parra, G., Bradnam, K., Korf, I.: CEGMA: a pipeline to accurately annotate core genes in eukaryotic genomes. Bioinformatics 23, 1061–1067 (2007)
29. Parra, G., Bradnam, K., Ning, Z., Keane, T., Korf, I.: Assessing the gene space in draft genomes. Nucleic Acids Res. 37, 289–297 (2009)
30. Ruskey, F.: Combinatorial Generation. Working version 1j-CSC 425/520. Available at CiteSeer:10.1.1.93.5967 (2003)
31. Seifert, K.A., Samson, R.A., de Waard, J.R., Houbraken, J., Lévesque, C.A., et al.: Prospects for fungus identification using *CO1* DNA barcodes, with Penicillium as a test case. Proc. Natl. Acad. USA 104, 3901–3906 (2007)
32. Thomas Jr., C.A.: Recombination of DNA molecules. Prog. Nucleic Acid Res. Mol. Biol. 5, 315–337 (1966)
33. Wang, W., Wei, Z., Lam, T.-W., Wang, J.: Next generation sequencing has lower sequence coverage and poorer SNP-detection capability in the regulatory regions. Sci. Rep. 1, 55 (2011)

Optimal Control for a Discrete Time Influenza Model

Paula Andrea Gonzalez Parra[1], Martine Ceberio[2], Sunmi Lee[3], and Carlos Castillo-Chavez[4]

[1] Universidad Autónoma de Occidente, Departamento de Matematicas, Cali, Colombia
pagonzalez@uao.edu.co
[2] The University of Texas at El Paso, Computer Science Department, El Paso, TX 79968-0514, USA
[3] Kyung Hee University, Department of Applied Mathematics, Yongin, 446-701, Republic of Korea
[4] Arizona State University, Mathematical, Computational and Modeling Sciences Center, Tempe, AZ 85287, USA

Abstract. We formulated a discrete time model in order to study optimal control strategies for a single influenza outbreak. In our model, we divided the population into four classes: susceptible, infectious, treated, and recovered individuals. The total population was divided into subgroups according to activity or susceptibility levels. The goal was to determine how treatment doses should be distributed in each group in order to reduce the final epidemic size. The case of limited resources is considered by including an isoperimetric constraint. We found that the use of antiviral treatment resulted in reductions in the cumulative number of infected individuals. We proposed to solve the problem by using the primal-dual interior-point method that enforces epidemiological constraints explicitly.

Keywords: Influenza, Optimal Control, Interior-Point methods, Epidemiology.

1 Introduction

Continuous time models have been used to study influenza outbreaks and the impact of different control policies [4,9,15]. In the case of influenza, the cost of antiviral treatment or the cost of isolation of infectious individuals has also been addressed using continuous time models [11,12]. Recently, the evaluation of influenza public health interventions using discrete epidemiological models has been proposed [2]. We explore the role of heterogeneity via a discrete time epidemiological model involving two interacting groups. An optimal control problem is formulated to evaluate the effect of antiviral treatment in scenarios involving limited or unlimited resources. The optimal control problem is solved using the primal-dual interior-point method, which to the best of our knowledge has not

L.F. Castillo et al. (eds.), *Advances in Computational Biology*,
Advances in Intelligent Systems and Computing 232,
DOI: 10.1007/978-3-319-01568-2_33,

been previously used to solve control problems in epidemiology. This method allows an efficient inclusion of explicit inequality constraints. In this paper, we introduce the epidemiological model and the optimal control problem in Section 2, the basic ideas of interior point methods are introduced in Section 3. The results of selected numerical simulations are presented in Section 4, by considering different scenarios such as different activity or susceptibility levels under limited or unlimited resources.

2 Problem Formulation

The dynamics of many diseases such as measles and influenza are strongly correlated with age [3]. Epidemiological models with age structure have been considered for the continuous case in [3,5,10]. We divide the population into 2 subgroups. Let $N_i(t)$ be the number of individuals in group i at time t, $(i = 1, 2)$ and q_{ij} be the probability that somebody from Group i has contact with somebody from group j. If we assume that both groups are connected $(q_{ij} > 0)$ and we consider proportionate mixing [3], we have $q_{ij} = q_j = \frac{C_j N_j}{\sum_{k=1}^{m} C_k N_k}$, where C_i is the average number of contacts per unit of time. Let $S_i(t)$, $I_i(t)$, $T_i(t)$, and $R_i(t)$ denote the number of susceptible, infectious, treated and recovered individuals in the ith group. We consider a single outbreak and people remain in the same group. We assume that infectious individuals from group i naturally recover with probability σ_i. We consider that the fraction of infected individuals in group i who get treatment each generation is modeled by $\tau_i(t)$. Since treated individuals are still infectious, the fraction of susceptible individuals on group i at time t that get infected at time $t+1$ is modeled by the function:

$$G_i = \rho_i \sum_{j=1}^{2} \left(q_j \left(\frac{I_j(t) + \epsilon_j T_j(t)}{N_j} \right) \right), \tag{1}$$

where ϵ_j represents the effectiveness of treatment for individuals on group j, with $0 < \epsilon_j \leq 1$. We assume that individuals (from any group) who get treatment recover with probability σ. The model with control is given by the following system of difference equations:

$$\begin{aligned} S_i(t+1) &= S_i(t)(1 - G_i(t)) \\ I_i(t+1) &= S_i(t)G_i(t) + (1 - \tau_i(t))(1 - \sigma_i) I_i(t) \\ T_i(t+1) &= (1 - \sigma) T_i(t) + \tau_i(t)(1 - \sigma_i) I_i(t) \\ R_i(t+1) &= R_i(t) + \sigma_i I_i(t) + \sigma T_i(t). \end{aligned} \tag{2}$$

In the absence of control, the model is reduced to an SIR model, the basic reproductive number R_0 is given by [8] $R_0 = \sum_{i=1}^{2} \frac{\rho_i q_i}{1-(1-\sigma_i)}$. Now we introduce the optimal control problem associated with the group-structured model (2). Our goal is to minimize the number of infected individuals in each group over a finite interval $[0, n]$, by using the least amount of treatment. The optimal control problem can be written as:

$$\min \frac{1}{2} \sum_{i=1}^{2} \left(\sum_{t=0}^{n-1} \left(B_{I_i} I_i(t)^2 + B_{\mathcal{T}_i} \tau_i(t)^2 \right) \right), \text{subject to Model (2)}, \tag{3}$$

where n denote the final time. The weight constants B_j, $(j = I_i, \tau_i)$ are a measure of the *relative* cost of interventions over $[0, n]$. In particular, $B_{\mathcal{T}_i}$ denote the relative costs associated with the implementation of antiviral treatment in group i, respectively.

3 Methodology

The problem is solved by using the primal-dual interior-point method [6,8,14]. Interior-Point Methods (IPM) are algorithms used to solve linear and nonlinear optimization problems. Contrary to the simplex method, which finds an optimal solution by testing the adjacent vertices of a feasible set, IPM find optimal solutions by crossing the interior of a feasible region. Computationally, IPM are more efficient than the simplex method because they have polynomial complexity. In addition, the simplex method finds solutions at the corner points only, while IPM may find solutions in the interior as well.

We rewrite Problem (3) as a nonlinear programming problem:

$$\min f(y), \text{ s.t. } E(y) = 0, \ 0 \leq y \leq y_{\max}, \tag{4}$$

where $\mathbf{y} = \begin{bmatrix} y_1 \\ y_2 \end{bmatrix}$, $\mathbf{y}_i = [S_i(1), I_i(1), T_i(1), \tau_i(0), \ldots, S_i(n), I_i(n), T_i(n), \tau_i(n-1)]^T$, for $i = 1, 2$ and the final time n. The objective functional is given by:

$$f(y) = \frac{1}{2} \sum_{i=1}^{2} \left(B_{I_i} \|\tilde{I}_i\|^2 + B_{\mathcal{T}_i} \|\tau_i\|^2 \right),$$

$f : \mathbb{R}^{8 \cdot n} \to \mathbb{R}$, with $\tilde{I}_i = (I_i(0), I_i(1), \ldots, I_i(n-1))^T$ and $\tau_i = (\tau_i(0), \tau_i(1), \ldots, \tau_i(n-1))^T$, for $i = 1, 2$. From Model (2), we get the equality constraint $E : \mathbb{R}^{8 \cdot n} \to \mathbb{R}^{6 \cdot n}$, $E(y) = \begin{pmatrix} E_1(y) \\ E_2(y) \end{pmatrix}$, where $E_i(y)$, for $i = 1, 2$ is defined from (2) [8]. Now we consider a more realistic scenario when treatment supplies are limited. We modify Problem (3) by including the "isoperimetric" constraint [12,13]

$$\sum_{i=1}^{2} \left(\sum_{t=0}^{n-1} \left(\tau_i(t) I_i(t) \right) \right) = k, \tag{5}$$

where k represents the available number of treatment doses and n the final time. Notice that (5) can be written as $\sum_{i=1}^{2} \boldsymbol{\tau}_i^T \tilde{\mathbf{I}}_i - k = 0$. A similar problem has been solved in [12] by considering limited vaccine in a continuous time influenza model. The problem was solved by including a new state variable related to the isoperimetric constraint, which requires boundary conditions at $t = 0$ and $t = n$; The authors of [12] remark that convergence issues have to be addressed. We solve the new problem by using the primal-dual interior-point method, which allows the inclusion of the new constraint more efficiently. The optimal control problem can be written as (4) where the previous equality constraint is modify as $E : \mathbb{R}^{4 \cdot n \cdot m} \to \mathbb{R}^{3 \cdot n \cdot m + 1}$, $E(y) = \left(E_1(y), E_2(y), \boldsymbol{\tau}_1^T \tilde{\mathbf{I}}_1 + \boldsymbol{\tau}_2^T \tilde{\mathbf{I}}_2 - k\right)^T$.

The Lagrangian function associated with Problem (4) is defined by:

$$L(y, w, z_1, z_2) = f(y) + E(y)^T w - y^T z_1 - (y_{\max} - y)^T z_2,$$

where w, z_1, and z_2 are the Lagrange multipliers associated with the equality and inequality constraints, respectively. Therefore the perturbed KKT conditions [8,14] are given by:

$$F_\mu(y, w, z_1, z_2) = [\nabla_y L, E(y), Y Z_1 - \mu e, (Y_{\max} - Y) Z_2 - \mu e]^T = 0, \qquad (6)$$

where $Y = \mathrm{diag}(y)$, $Y_{\max} = \mathrm{diag}(y_{\max})$, $Z_1 = \mathrm{diag}(z_1)$, $Z_2 = \mathrm{diag}(z_2)$, and $e = (1, \ldots, 1)^T \in \mathbb{R}^{8n}$. The primal-dual interior-point algorithm for the nonlinear programming problem (4) is presented in [8]. The results of some numerical simulations both in the case of limited and unlimited supplies for different scenarios are presented in the next section.

4 Numerical Results

In this section, we present some results of selected simulations under various scenarios. For each case, we compare the proportion of infected individuals generated in the absence or in the presence of control. The baseline parameter values are given in [8]. For scenario 1, we consider the case of seasonal influenza. We divide the total population into two groups with different population sizes. Group

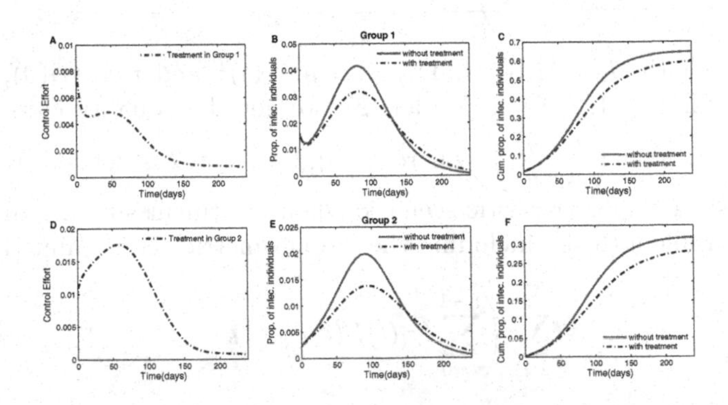

Fig. 1. In scenario 1, Group 1 (12.5 % of the population) is more susceptible but less active than Group 1. Since Group 2 is more active, more effort has to be applied in this group.

1 is given by 12.5% of the population aged 65 or more, and Group 2 is 87.5% of the population aged less than 65 [1]. We assume that $R_0 = 1.27$ and that Group 1 is the high risk population ($\rho_1 > \rho_2$). The final time is 240 days. Figure 1 shows the results for Scenario 1. Since we have Group 2 as the more active one, the optimal control requires more resources for this group than for Group 1; Figure 1D shows that we need to apply twice the treatment for Group 2 than for Group 1 (Figure 1A). The reduction on the final epidemic size is given by 8% and 12% in Groups 1 and 2, respectively.

The case of limited resources is considered in Scenario 2 and 3. We assume that both groups have the same population size. In Scenario 2 we assume same activity level but Group 1 is more susceptible than Group 2, $\rho_1 > \rho_2$. For Scenario 3, we consider same susceptibility but Group 1 is more active than Group 2, $C_1 > C_2$. Figures 2 and 3 show the optimal control function, the proportion of infected individuals, and the cumulative proportion of infected individuals in both groups under each scenario for different values of treatment doses k.

Figure 2 shows the results for Scenario 2. The optimal control solution shows that more resources should be used for Group 1 (Figure 2A and 2C), since this is the high risk group; however the proportion of infected individuals is higher in Group 1 (Figures 2B and 2D). By using different values of k, 3%, 6%, and 13%, the final epidemic size in Group 1 is reduced by 2.4%, 6%, and 16% for each case; for Group 2, it is reduced by 3%, 7%, and 19%. Although the optimal solution allows the use of more resources towards Group 1, the reduction on the final epidemic size is a little higher in Group 2. For small values of k, (3% and 6%), Figures 2A and 2D show that in both groups, the resources should be used at the beginning of the epidemic, 55 and 75 days respectively.

In the case of Scenario 3, Group 1 has a higher activity level than Group 2 but the same susceptibility. Figure 3 shows that the optimal control solution requires the application of more treatment doses in Group 1 (Figures 3A and 3D); however, the proportion of infected individuals is the same in both groups

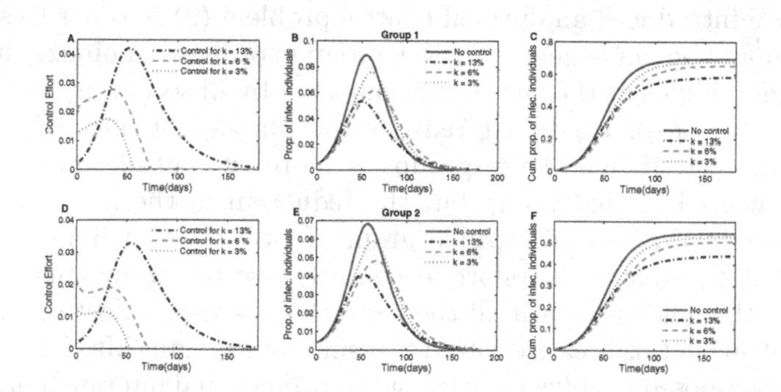

Fig. 2. For Scenario 2, since Group 1 has higher activity level, more resources need to be used towards this group (Figure A and D) however for each value of k the reduction on the final epidemic size is higher in Group 2.

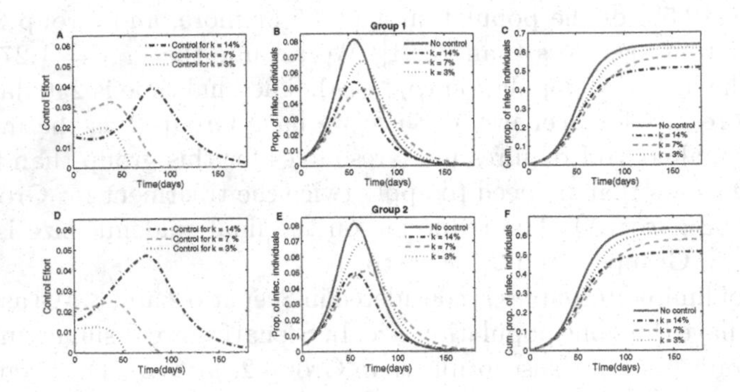

Fig. 3. For Scenario 3, since Group 2 is at higher risk, more resources need to be used for this group. However, since the activity level is the same for both groups, the number of infected individuals is similar for Group 1 and Group 2.

(Figures 3B and 3E). For different values of k (4%, 7%, and 14%), Figures 3C and 3E shows that the final epidemic size is reduced by 5%, 8%, and 15% respectively.

In all scenarios, we find that the use of treatment reduces the number of infected individuals. If one of the groups is more susceptible, more effort has to be implemented in that group, but the reduction in the final epidemic size will be larger in the less susceptible group. In addition, if we consider limited resources, we found that the resources should be used at the beginning of the epidemic until all the resources are used.

5 Conclusions

We formulated a discrete group-structured model under the assumption that people mix more with individuals in the same group and groups are mixing randomly. We introduced an optimal control problem (3) in order to study how treatment should be implemented in each group in order to minimize the number of infected individuals at the end of the epidemic. In all scenarios, we found that the implementation of treatment reduces the number of infected individuals at a minimal cost. If one of the groups is more susceptible, more effort has to be implemented in that group but the reduction in the final epidemic size will be bigger in the less susceptible group. In the case of limited resources, we found that the maximum effort in control have to be implemented at the beginning of the epidemic until all the resources are used. Most of the optimal control problems in this area are solved by using Pontryagins Maximum Principle [7,12,13] We proposed to solve it by using the primal-dual interior-point method. This methodology allows the inclusion of constraints in a simpler way, specially in the case of isoperimetric constraint.

References

1. Brauer, F.: Epidemic models with heterogeneous mixing and treatment. Bull. of Math. Bio. 70, 1869–1885 (2008)
2. Brauer, F., Feng, Z., Castillo-Chavez, C.: Discrete Epidemic Models. Math. Biosc. & Eng. 7, 1–15 (2010)
3. Castillo-Chavez, C., Hethcote, H.W.: Epidemiological models with age structure, proportionate mixing, and cross immunity. J. of Math. Bio. 27, 233–258 (1989)
4. Chowell, G., Ammon, C.E., Hengartner, N.W., Hyman, J.M.: Transmission dynamics of the great influenza pandemic of 1918 in Geneva, Switzerland: Assessing the effects of hypothetical interventions. J. Theor. Biol. 241, 193–204 (2006)
5. Del Valle, S.Y., Hyman, J.M., Hethcote, H.W., Eubank, S.G.: Mixing patterns between age groups in social networks. Social Networks 29, 539–554 (2007)
6. El-Bakry, A.S., Tapia, R.A., Tsuchiya, T., Zhang, Y.: On the formulation and theory of the primal-dual newton interior-point method for nonlinear programming. J. of Optim. Theo. and App. 89(3), 507–541 (1996)
7. González-Parra, P., Lee, S., Velazquez, L., Castillo-Chavez, C.: A note on the use of optimal control on a discrete time model of influenza dynamics. Math. Biosc. & Eng. 8(8), 183–197 (2011)
8. González-Parra, P.: Constraint optimal control for a multi-group discrete time influenza model. PhD. dissertation, The University of Texas at El Paso, El Paso, TX (2012)
9. Herrera-Valdez, M.A., Cruz-Aponte, M., Castillo-Chavez, C.: Multiple outbreaks for the same pandemic: Local transportation and social distancing explain the different "waves" of A-H1N1pdm cases observed in Mxico during 2009. Math. Biosc. & Eng. 8(8), 21–48 (2011)
10. Hethcote, H.W.: An age-structured model for pertussis transmission. Math. Biosc. 145, 89–136 (1997)
11. Lee, S., Chowell, G., Castillo-Chavez, C.: Optimal control for pandemic influenza: the role of limited antiviral treatment and isolation. J. Theor. Biol. 265, 136–150 (2010)
12. Lee, S., Morales, R., Castillo-Chavez, C.: A note on the use of influenza vaccination strategies when supply is limited. Math. Biosc. & Eng. 8(8), 171–182 (2011)
13. Lenhart, S., Workman, J.: Optimal control applied to biological models. Chapman & Hall, CRC Mathematical and Computational Biology series (2007)
14. Nocedal, J., Wright, S.J.: Numerical optimization, 2nd edn. Springer (2006)
15. Rios-Soto, K., Song, B., Castillo-Chavez, C.: Epidemic spread of influenza viruses: The impact of transient populations on disease dynamics. Math. Biosc. & Eng. 8(8), 199–222 (2011)

Transcriptomics of the Immune System of Hydrozoan *Hydractinia symbiolongicarpus* Using High Throughput Sequencing Methods

Alejandra Zárate-Potes and Luis Fernando Cadavid

Institute of Genetics, Department of Biology, Universidad Nacional de Colombia
lfcadavidg@unal.edu.co

Abstract. Cnidarians are ecologically important animals in marine ecosystems like coral reefs, where the incidence of disease has raised in the past years due to numerous environmental changes. In this work, messenger (m)RNA samples from cultured colonies of the hydrozoan *Hydractinia symbiolongicarpus* were sequenced using the Illumina platform. The Trinity program was used to assemble the transcriptome *de novo* and 116.924 contigs were obtained and were annotated by comparisons with public databases. The immunotranscriptome was characterized by a great diversity of transcripts coding for adhesion molecules and peptidases. Through sequencing with the Illumina platform we obtained a transcriptome draft for the organism *H. symbiolongicarpus* from which a large set of immune-related molecules was obtained. ...

Keywords: High Throughput Sequencing, Transcriptomics, Immunology, Cnidaria, *Hydractinia symbiolongicarpus*.

1 Introduction

The Cnidaria as sister group of all bilateria are an important taxon to study the origin and evolution of the immune system in animals [Augustin and Bosch, 2010]. As ecologically important species in marine environments like coral reefs, their structural role has been endangered by the recurrent appearance of disease. For many of these diseases no single etiological agent has been found, suggesting more complex scenarios involving alterations of the structure and diversity of commensal bacteria that inhabit these animals' tissues [Sokolow et al., 2009]. Despite their ecological and evolutionary importance, as well as the high mortality of some taxa due to infectious diseases, still very little is known about their immune responses [Augustin and Bosch, 2010].

The immune system in cnidarians can be divided in three functional modules: First the recognition module, which is composed of intracellular, membrane-bound and secreted receptors that recognize self from non-self and also microbe-associated molecular patterns (MAMPs). Among these, it is worth mentioning LRR domain-containing receptors such as TLRs and NLRs and carbohydrate-binding lectins [Dunn, 2009]. Second the signaling module, which connects the

L.F. Castillo et al. (eds.), *Advances in Computational Biology*,
Advances in Intelligent Systems and Computing 232,
DOI: 10.1007/978-3-319-01568-2_34,

two other modules through signal transduction and includes signaling cascades like the TLR and apoptosis pathways [Miller et al., 2007]. Third the effector module, which consists of molecules that lead either to microbial destruction through inflammatory reactions and the production of antimicrobial peptides or to apoptosis [Dunn, 2009].

To get more insight about the immune system of cnidaria, we sequenced and annotated the transcriptome of *Hydractinia symbiolongicarpus*, a marine, colonial and dioic hydroid living as a surface incrustation of gastropod shells occupied by hermit crabs [Buss and Yund, 1989]. This organism has demonstrated to be an excellent animal model in various biological disciplines. Yet, neither genome draft nor whole transcriptome sequence data is available for it. There is, however, an EST data set of almost 9000 sequences for its sister species *Hydractinia echinata* [Soza-Ried et al., 2010].

With the advent of Next Generation Sequencing technologies the acquisition of sequence data has become faster, cheaper and massive. In this study the transcriptome of model hydrozoan *H. symbiolongicarpus* was sequenced using the Illumina platform. In order to draft its immunotranscriptome, the obtained contigs were compared with public databases to extract sequences already known to participate in the immune system of other animals.

2 Methods

Animal Cultures, RNA Isolation, Library Preparation and Illumina Sequencing

Animals are grown in recirculating Artifical Sea Water (ASW) at a relative density of 1,022 and a temperature of 20±2°C. Colonies were fed 2 to 3 times per week with *Artemia salina* nauplii two days after eclosion. The animals were starved for three days before the experiment to avoid contamination with *A. salina* RNA.Total RNA was isolated using the TRizol™reagent according to manufacturer's recommendations. The sequencing library was constructed by selection of the poly-A mRNA and posterior fragmentation through sonication. The cDNA synthesis was done with random hexamers and 300 bp fragments were purified to add Illumina adaptors, tags and primers. All samples were sequenced in a single HiSeq 2000 Illumina™PET indexed lane.

Data Preprocessing, *de novo* Assembly and Annotation

Raw data quality was evaluated and Illumina adaptors and bases with Phred score under 20 in 3' ends were trimmed. Ribosomal sequence depletion was performed using RiboPicker software [Schmeider, 2012]. Redundant data was removed with the digital normalization tool Diginorm [Brown et al., 2012]. All data left after preprocessing was used for assembly with the k=25 Trinity software [Grabherr et al., 2011]. All contigs obtained in the assembly were compared with the Uniprot/SwissProt and NCBI RefSeq databases using the Blast algorithm with an E-value threshold of 1e-5. Each contig was assigned the name of the best blast hit. Nucleotide BLAST was run using version 2.2.26+ [Altschul et al., 1990].

Whole transcripotme KEGG Orthology annotation was done using the KEGG Automatic Annotation Server (KAAS) [Moriya et al., 2007]. The *H. symbiolongicarpus* proteome was predicted using the Emboss [Rice et al., 2000] translate tool and a Pfam annotation was done. Reciprocal Blasts were performed with the proteomes of the three species of cnidarians for which genomic information is available (*N. vectensis*, *H. magnipapillata* and *A. digitifera*) and for metazoan model organisms *D. melanogaster*, *M. musculus* and *H. sapiens*. To select the transcriptome molecules with a possible implication in immunity a key word list of 128 entries was generated. The key words were searched in the nametags of Uniprot and RefSeq-annotated transcripts.

Table 1. Assembly and Annotation Statistics

CATEGORY	COUNT
Contigs	116.924
Total length of Sequence	114.805.126
N50 length (bp)	1.539
Average length of Contigs	981
Min	201
Median	643
Max	14.407
GC%	36,86
RefSeq	70.666
Uniprot	53.702
Total Unannotated sequences	44.214
Total Annotated sequences	72.710
A. digitifera	7.084
N. vectensis	7.715
H. magnipapillata	10.381
H. sapiens	6.638
M. musculus	6.578
D. melanogaster	5.098
H. echinata	157

3 Results and Discussion

Transcriptome Sequencing and Assembly

After preprocessing, sequence data was used for *de novo* transcriptome assembly with the Trinity software. A total of 67.051.669 reads were obtained and 116.924 contigs were assembled ranging from 201 to 14407 bp in length. A summary of the assembly statistics is shown in Table 1. The coverage of the genome was 22X according to the estimated genome size, suggesting that the sequencing depth was enough for eukaryotic transcriptome reconstruction as was recently assessed in a published study, where the authors concluded that using more than 60 million reads for a transcriptome assembly raises the accumulation of sequencing errors in the assembly and has a low rate of new gene discovery [Francis et al., 2013].

Transcriptome General Annotation

All assembled contigs were compared via BLAST to the Uniprot/Swissprot database, the NCBI RefSeq database and the already annotated transcriptomes of *N. vectensis* and *H. magnipapillata*. The total number of contigs annotated for each database is shown in Table 1. 72.710 contigs had significant BLAST hits, while 44.214 had no hits against the databases. Putative orthologous genes in other species with available genomic sequence data were evaluated by reciprocal blast. Comparisons were made between the transcriptome of *H. symbiolongicarpus* and other species including predicted proteomes of *N. vectensis*, *H. magnipapillata*, *A. digitifera*, *H. sapiens*, *M. musculus*, *D. melanogaster* and the *H. echinata transcriptome*, the only other sequence data available for this genus (Table 1). 25.180 transcripts were found to be putative orthologs from other cnidarian species.The species with which *H. symbiolongicarpus* shares more putative orthologous sequences (10.381) is *H. magnipapillata*, also its closest related species for which there is a genome available.

To assess transcriptome completeness, all assembled contigs were evaluated for KEGG orthologies using the KAAS. Presence of genes from basic metabolic pathways was evaluated (Table 2). This annotation step is also evidence that the experiment captured a representative portion of the organism's transcriptome.

Once the general annotation of the transcriptome was completed we used a key word list including characteristic terms of the immune system that were searched in the annotation names of contigs assigned with Uniprot and RefSeq. A summary of this is shown in Table 3. Although canonical TLR and NLR receptors and signaling pathway members have been found in anthozoans [Miller et al., 2007] we couldn't find any TLR or NLR with a canonical architecture in the *H. symbiolongicarpus* transcriptome draft. Many of the pathway participating molecules were also not found. However, many transcripts encoding only proteins containing LRR domains were found. 50 transcripts encoding only one NACHT domain were also found.

Table 2. Comparative KEGG orthology counts for selected pathways in different species

PATHWAY	*H. symbiolongicarpus*	*N. vectensis*	*A. digitifera*	*H. magnipapillata*	*H. sapiens*	*M. musculus*	*D. melanogaster*
Glycolysis/Gluconeogenesis	29	29	25	30	38	37	29
Citrate Cylce (TCA Cycle)	22	28	22	29	23	23	22
Pentose Phosphate	17	17	19	19	18	18	16
Amino Sugar and Nucleotide Sugar Metabolism	30	30	30	35	36	35	28
Fatty Acid Biosynthesis	6	4	4	7	6	5	4
Purine Metabolism	104	95	94	105	120	121	97
Valine, Leucine and Isoleucine Degradation	38	38	24	39	38	38	27
Basal Transcription Factors	31	30	22	31	34	34	32

The C-type lectins are carbohydrate-binding proteins that participate in cell-cell adhesion, recognition and innate immunity in mammals. In the *H. symbiolongicarpus* transcriptome we found 34 transcripts coding for putative C-type lectins. As it has been observed in *N. vectensis* [Wood-Charlson and Weis, 2009], cnidarians may have large copy numbers of C-type lectins that are probably involved in the immune system. An exclusively cnidarian lectin bound to nematocysts that was found in our study was Nematogalectin [Steele et al., 2011].

The transcriptome was particularly enriched in probable cell adhesion molecules containing multiple cadherin domains, integrin-like or selectin-like architectures.

Astacins are metalloproteinases widely distributed among the metazoa, which participate in tissue differentiation, matrix assembly, digestion and development, among others. Most of its members are secreted and are expected to have an action in the intercellular space. The cnidarian phylum has shown the exclusive presence of some types of these proteins, observed in hydrozoan species [Sterchi et al., 2008]. In *H. symbiolongicarpus* 22 putatively astacin-encoding transcripts were identified.

The Bactericidal Permeability Increasing (BPI) protein is an antimicrobial peptide with activity against Gram-negative bacteria. Among its functions are LPS endotoxin neutralization and clearance. The pathway that activates the overexpression of this protein seems to be triggered by microbial exposure and in presence of proinflammatory cytokines [Schultz and Weiss, 2007]. It has been found in cnidarians *N. vectensis* and *H. magnipapillata* [Miller et al., 2007]. Twenty transcripts with high sequence similarity to BPI were found in the *H. symbiolongicarpus* transcriptome. This molecule remains a candidate of great interest for functional tests in this cnidarian model.

Table 3. Genes Relevant for the Immune System. (CAN) refers to the canonical domain content according to Uniprot. (*) refers to domains present in different architectures of various molecules

RECOGNITION MODULE	(Can) Domain Content	Domain Content	Transcripts	Function
1. NACHT domain containing proteins	Pyrin/CARD+NACHT +LRR	Death+NACHT NACHT	2 50	Innate immune response Defense response to bacterium
2. C-type lectin domain family	Lectin C	Lectin C	34	immune system process
3. Nematogalectin	Collagen+Collagen +Gal Lectin	Collagen+Collagen +Gal Lectin	5	Possible Cell Adhesion
4. Leucine rich repeat containing protein	NA	LRR4+LRR5 +LRR8+PDZ +Miro+Ras +zf-C3HC4-3(*)	135	LRR domain-mediated complex assembly
5. Cadherin domain containing proteins	Cadherin	Cadherin (1-35)	192	Possible Cell Adhesion
6. a)Integrin alpha-like	FG-GAP +Integrin-alpha2 +VWA	FG-GAP +Integrin-alpha2 Integrin-alpha2 Integrin-alpha2+VCBS	8 5 2	Integrin complex
b)Integrin beta-like	EGF2+Integrin-b-cyt +Integrin-B-tail +Integrin-beta	Integrin-b-cyt Integrin +b-cytIntegrin-beta Integrin-B-tail +Integrin-b-cyt +Integrin-beta Integrin-beta	4 4 2 40	Integrin complex
EFFECTOR MODULE	**(Can) Domain Content**	**Domain Content**	**Transcripts**	**Function**
1. a) Astacin-like metalloendopeptidase	Astacin	Astacin	55	Metallopeptidase activity
b)Similar to Astacin-like metalloendopeptidase	N/A	Astacin-ShK Astacin-MAM Astacin-Gal Lectin(2)	19 3 1	Possible Metallopeptidase activity
2. Astacin-like metalloprotease toxin	Astacin	Astacin	22	Metallopeptidase activity
3. a) Bactericidal permeability increasing protein	LBP BPI CETP +LBP BPI CETP C	LBP BPI CETP +LBP BPI CETP C	18	Lipopolysaccharide binding
Bactericidal permeability increasing protein-like	N/A	LBP BPI CETP C LBP BPI CETP	21 7	Possible Lipopolysaccharide binding

Lectins and other cell adhesion molecules seem to play a central role in the recognition module of the immune system of *H. symbiolongicarpus*. In the mutlicellular ancestor of metazoans cell adhesion molecules could have played an important role in the distinction between self and non-self. This may be the reason why the transcriptome of this basal animal is enriched with this kind of molecules, which could have immunological functions. Apoptosis, MAPK, Notch, Wnt and PI3K are signaling pathways of the immune system found to be complete in this organism and they could be playing a role in the signaling of immune stimuli. The effector module is composed of proteinases such as Astacin and other proteases that together with BPI may regulate bacterial population sizes.

4 Conclusion

We present here a draft of the transcriptome and a list of candidate molecules to participate in the immune system of *H. symbiolongicarpus* that will later be experimentally validated. We have generated 116.924 new sequences from this organism previously underrepresented in public databases. This is a first approach to unravel the integrity of this organisms immune system with a snapshot of all the expressed molecules at a given time, which gives us more information than what was available before and is very useful to generate and prove new hypothesis about this biological problem.

Acknowledgments. The authors would like to thank the High Performance Computing Center (CECAD Centro de Computación de Alto Desempeño) of the Universidad Distrital Francisco José de Caldas, Bogota D.C. (Colombia), the Faculty of Enginieering and the GICOGE research group leaded by Professor Nelson Pérez PhD for the collaboration in the implementation of the software and the computational resources used in this work. This work was done by the Evolutionary Immunology and Immunogenetics Group of the Universidad Nacional de Colombia and the authors thank all its members for the feedback along the research process.

References

[Altschul et al., 1990] Altschul, S.F., Gish, W., Miller, W., Meyers, E.W., Lipman, D.J.: Basic local alignment search tool. Journal of Molecular Biology 215(3), 403–410 (1990)

[Augustin and Bosch, 2010] Augustin, R., Bosch, T.C.G.: Cnidarian immunity: A tale of two barriers. In: Söderhäll, K. (ed.) Invertebrate Immunity, Landes Bioscience and Springer Science+Business Media (2010)

[Brown et al., 2012] Brown, T., Howe, A., Zhang, Q., Pyrkosz, A.B., Brom, T.H.: A reference-free algorithm for computational normalization of shotgun sequencing data (2012)

[Buss and Yund, 1989] Buss, L.W., Yund, P.: A sibling species group of hydractinia in the north-estern united states. J. Biol. Ass. UK 69, 857–874 (1989)

[Dunn, 2009] Dunn, S.R.: Immunorecognition and immunoreceptors in cnidaria. ISJ 6, 7–14 (2009)

[Francis et al., 2013] Francis, W.R., Christianson, L.M., Kiko, R., Powers, M.L., Shaner, N.C., Haddock, S.H.D.: A comparison across non-model animals suggests an optimal sequencing depth for de novo transcriptome assembly. BMC Genomics 14(1), 167 (2013)

[Grabherr et al., 2011] Grabherr, M.G., Haas, B.J., Yassour, M., Levin, J.Z., Thompson, D.A., Amit, I., Adiconis, X., Fan, L., Raychowdhury, R., Zeng, Q., Chen, Z., Mauceli, E., Hacohen, N., Gnirke, A., Rhind, N., di Palma, F., Birren, B.W., Nusbaum, C., Lindblad-Toh, K., Friedman, N., Regev, A.: Full-length transcriptome assembly from rna-seq data without a reference genome. Nature Biotechnology 29(7), 644–652 (2011)

[Miller et al., 2007] Miller, D.J., Hemmrich, G., Ball, E.E., Hayward, D.C., Khalturin, K., Funayama, N., Agata, K., Bosch, T.C.G.: The innate immune repertoire in cnidaria - ancestral complexity and stochastic gene loss. Genome Biology 8(4), R59 (2007)

[Moriya et al., 2007] Moriya, Y., Itoh, M., Okuda, S., Yoshizawa, A.C., Kanehisa, M.: Kaas: an automatic genome annotation and pathway reconstruction server. Nucleic Acids Research 35 (Web Server), W182–W185 (2007)

[Rice et al., 2000] Rice, P., Longden, I., Bleasby, A.: Emboss: The european molecular biology open software suite. Trends in Genetics 16(6), 276–277 (2000)

[Schmeider, 2012] Schmeider, R., et al.: Identification and removal of ribosomal rna sequences from metatranscriptomes. Bioinformatics 28, 433–435 (2012)

[Schultz and Weiss, 2007] Schultz, H., Weiss, J.P.: The bactericidal/permeability-increasing protein (bpi) in infection and inflammatory disease. Clinica Chimica Acta 384(1-2), 12–23 (2007)

[Sokolow et al., 2009] Sokolow, S.H., Foley, P., Foley, J.E., Hastings, A., Richardson, L.L.: Editor's choice: Disease dynamics in marine metapopulations: modelling infectious diseases on coral reefs. Journal of Applied Ecology 46(3), 621–631 (2009)

[Soza-Ried et al., 2010] Soza-Ried, J., Hotz-Wagenblatt, A., Glatting, K.-H., del Val, C., Fellenberg, K., Bode, H.R., Frank, U., Hoheisel, J.D., Frohme, M.: The transcriptome of the colonial marine hydroid hydractinia echinata. FEBS Journal 277(1), 197–209 (2010)

[Steele et al., 2011] Steele, R.E., David, C.N., Technau, U.: A genomic view of 500 million years of cnidarian evolution. Trends in Genetics 27(1), 7–13 (2011)

[Sterchi et al., 2008] Sterchi, E., Stocker, W., Bond, J.: Meprins, membrane-bound and secreted astacin metalloproteinases. Molecular Aspects of Medicine 29(5), 309–328 (2008)

[Wood-Charlson and Weis, 2009] Wood-Charlson, E.M., Weis, V.M.: The diversity of c-type lectins in the genome of a basal metazoan, nematostella vectensis. Developmental and Comparative Immunology 33(8), 881–889 (2009)

Fuzzy Model Proposal for the Coffee Berry Borer Expansion at Colombian Coffee Fields

Nychol Bazurto Gómez, Carlos Alberto Martínez Morales, and Helbert Espitia Cuchango

Computer Science Investigation Group, District University, Bogotá, Colombia

Abstract. This paper propose a fuzzy logic model about coffee borer propagation behavior at a colombian context, increasing the information about coffee borer propagation, beyond its growth (topic that has considerable information), looking forward to generate an impact taking into account the existing harm level generated by this plague on this important national product.

1 Introduction

The country's economy is supported by various export products, some are more important than other. At the economy index GDP (Gross Domestic Product) is impacted by the behavior of the coffee production [1], [2], looking to improve the production techniques, ensuring crops with a reasonable amount of healthy fruits. Among the various pests affecting the coffee, the most important is the coffee borer (Hypothenemus hampei). This insect has specific characteristics to breed [12] and propagate at the coffee, in which outstand climatic conditions as a relevant factor[3].

Around the coffee pests issues a lot of different analyzes have been carried out, supported by agronomic institutions, describing the behavior of the coffee borer, specifically the factors leading to the expansion of this crop by accelerating their migration between plants [4],[5],[6],[11]. However, despite the existing information about the models related to the coffee borer (which are some [6],[7],[8]), most of them model the population growth, but no the dynamics of migration movement in the coffee plantation.

2 Studies and Information about the Coffee Berry Borer

As mentioned above the coffee borer is a plague that attacks the coffee fruits (in fact is the most important thing that affects it), the control of coffee borer can achieve a saving of U.S. 120 million [2]. The life cycle that of this insect is estimated in 28 days approximately. According to studies conduced by Cenicafé in the coffee borer are typically generated a greater number of females than males (10:1), which facilitates their proliferation, taking into accound that females are the only ones that can fly, as well depending on the current climate which they live, these fly or not.

L.F. Castillo et al. (eds.), *Advances in Computational Biology*,
Advances in Intelligent Systems and Computing 232,
DOI: 10.1007/978-3-319-01568-2_35,

3 Fuzzy Model Proposal

Then there will be a clarification and justification for the choice of variables and model definition that way.

3.1 Variables:

Regarding the analysis of documents provided by various colombian entities in the field,they were highlighted the following variables:

- ⋆ Climate: It is clear to farmers that dry climates favor the development of this pest and strengthens its drilling in the berry, so Cenicafé certain levels determined this action whit respect to certain temperatures taking[9]:
- ⋆ Altitude: The altitude encourages the increase or not the impact of these phenomena, but consideration of the factors, climate and altitude is ultimately needed to have it as a single variable [10].
- ⋆ Crop age: The crop can´t produce berry initially only after the first year, which does not allow propagation of the CBB.
- ⋆ Collection quality: The grain amount remaining in each plant has qualified for the MIB [4] having a good, fair or poor quality.

4 Model

Knowing the used variables, the model description is brief. The factors (variables) mentioned above, these are the systems inputs, taking a set of rules (144) governing the possible combinations (the most important) that affect the output variable significantly. The expected output is the infestation risk level, as seen in Figure 1:

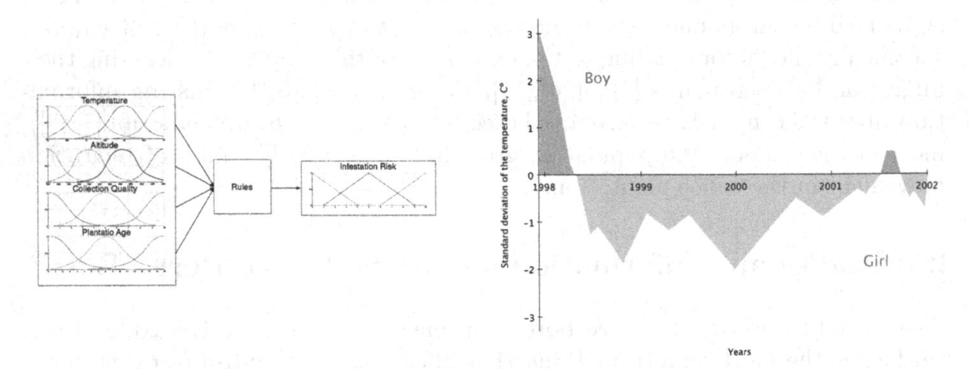

Fig. 1. a) Proposed Fuzzy Model b) Climate variability associated with the event of the girl and the boy [11]

It handles a range to 10, considering as 100 percent of infestation, therefore, each unit is ten percent of this risk.

5 Comparison Data

To make a proper fit to the proposed model took into account the data presented in the document MIB(Manejo Integrado de la Broca, ICA [4]) which presents sample data from the years of 1998 – 2002 in different regions of country. In order to make a better fit to the proposed model was determined to start with the data for the department of Antioquia. with the information in Table 1. The phase

Table 1. Phased data for the department of Antioquia [4]

Phase	Crop age	Recollection Q	Infestation.L.
Phase I	5.97	2.66	4.16
Phase II	5.52	2.82	2.02
Phase III	4.93	2.81	2.29
Phase IV	4.29	2.56	3.58
Phase V	4.12	2.9	3.36

data represent the years in which the sample was taken Phase I is equivalent to the year of 1998, Phase II corresponds to the year 1999, etc. In preliminary analysis the objective was fit the model to only these data, assuming a single temperature for each of the years in the same locality. Despite entering the same data collection and age than the actual sample is almost static behavior in the system response, so we tried to identify the impact of the variables in each of the phases, as our variable temperature (it is known that the altitude does not change, crop age study has as collection, so that the temperature is not the only factor that is expressed in the study). In this case we identify what were the weather conditions at each stage (climatic variations in each year, which could be affected and therefore studies the model with a constant temperature for each phase does not apply), ie climate data were sought each year compared to seasonal impact of boy's phenomenon, as well as when was the girl's phenomenon.

On the basis of analysis document climate variability by the Girl's Phenomenon or the Boy's Phenomenon [11] could identify periods in which it has had no impact of these two phenomena. This way you could set some temperature (in Celsius grades) conditions for each phase: Phase I-19.6320, Phase II-18.9520, Phase III-18.8520,Phase IV-18.9320 and Phase V-19.1320.

6 Optimization Process

In addition to obtaining a fuzzy inference model was revised to optimize the possibility, given the size of the set of rules with which one set (144 rules). Additionally, it is sought to adjust the ranges of the membership functions, such as data generated by the model is accurate.

6.1 Genetic Algorithm

This optimization method to explore the field in a comprehensive solution, generating a higher accuracy (higher accuracy), but their convergence can be slowed down by the same reason. It doesn't requires a starting point, but if you must put a limit to the area exploration, for the case used in this article, the limits were [0 1], corresponding to the range that defines the membership functions that were seen in figure 1.

A hundred iterations were performed on the fuzzy model, choosing the best result iteration (which emphasized the smallest difference between the simulated and real data). Given this set, the rules are considered in it, would be favorable and some of them would be dominant over the other. To this was used as a criterion, discard those that had an impact on the model below 0.3, reducing the set finally.

6.2 Gradient Method

This method requires a starting point, this allows an agile convergence (loss of accuracy), at a minimum, which isn't always global minimum (depending on the given initial point may fall into a local minimum, losing their effectiveness). The set of initial values in the fuzzy model in place were the real data as shown in Table 1.

7 Results and Analysis

Entering the inputs and outputs expected, specifying the set of rules, shall be assessed by cases the behavior model, with respect to the data used in the MIB [4]. In view of that this handles tables per department, each of these will be a test case for the model, as illustrated in Table 1, the MIB handles values for two of the variables and a infestation percentage , if you enter each line of these conditions in the fuzzy system,it is expected to receive an approximate value, taking into account the weather conditions must be intuited, to adjust this. Comparing actual data versus simulated data (Table 2) achieves see how the behavior of the results generated by the model are close to the actual data provided in the study. Denoting the fuzzy model suitable for the biological system of the CBB and the proliferation of coffee crops.

With the results of the preliminary analysis and the subsequent adjustments that led to the search for temperature information for each phase to these periods, it can be said that the temperature at which the coffee is exposed, become a key role in the infestation risk, this corroborates the literature stating that the CBB (Hypothenemus hampei) reduces its life cycle under conditions of higher temperature and therefore tends to reproduce much faster, this can mix with in impact on flowering the coffee the amount of rainfall it receives as well as the amount of sun. In higher rainfall conditions is reduced the number of flowers and therefore coffee berries, which also makes the life cycle of the CBB be normal or prolonged.

To subsequent years remains the effect of the girl but with less intensity to reach a neutral state by the end of 2002. For this reason a priori determined the temperatures for these periods: In 1998 a temperature of 19,6320; 1999-18,9560; 2000-18,8520; 2001-18,9320 and 2002-19,1320. With this information the model has presented the following results for each phase:

Table 2. Comparing table actual data vs results

Year	Real Infestation L.	Simulated Infestation L.	Difference.
1998	4.16	4.2478	0.0878
1999	2.02	1.9915	0.2850
2000	2.29	2.2876	0.0024
2001	3.58	3.4062	0.1738
2002	3.36	2.9179	0.4421

Table 3. Comparative table initial and optimized ranges for climate variable

Variable	Range	Initial center	Optimized center
Climate	Null situation	18	18
	Low situation	19,010	19,300
	Moderate situation	20,000	19,620
	Critical situation	21,010	21,150

With respect to optimization, by applying the genetic algorithm to the set of rules as described above, there was obtained a reduction of 25 percent in this, decreased to 108 rules. Furthermore, the application of the gradient method to the membership functions,it allowed an adjustment in the membership functions. Table 3 shows the shift of the centers of the Gaussian functions with respect to the original model. These values allowed the approximating model and generate identical to the real data.

8 Conclusion

- ⋆ In proposing the model showed the importance of identifying the effect of recurrent cycles of the Boy's Phenomenon and Girl's Phenomenon, to take preventive measures that reduce the impact of higher degree of spread of the CBB in drought conditions.
- ⋆ The behavior of the CBB, in periods preceded by the Girl's phenomenon, increases due to the wet conditions, but without the precipitation that involves the death of the insect. Thus transitions Girl's Phenomenon to Boy's Phenomenon will be helping the proliferation of the CBB.

* Given the current conditions of temperature increase due to global warming can envisage a scenario where the CBB can expand more easily in long periods, it is possible that the increase in altitude coffee plantations (planting in regions with cold temperatures) can reduce the impact of global warming on the coffee plantation and CBB.
* Fuzzy system optimization through genetic algorithms was effective, allowing equivalent performance, through a reduced rule set.

References

1. Corporación colombiana de investigación agropecuaria Corpoica, Investigación sobre los efectos del cambio climático en la distribución altitudinal insectos plagas del café y sus enemigos naturales en la zona cafetera de Colombia. Informe, Centro de investigación EE Caribia, Versión 0, pp. 5-9 (Octubre 2011)
2. Bustillo Pardey, A.: Una revisión sobre la broca del café, Hypothenemus hampei (Coleoptera: Curculionidae: Scolytinae), en Colombia. Revista Colombiana de Entomología 32(2) (Julio-Diciembre 2006)
3. Jaramillo, J., Chabi-Olaye, A., Kamonjo, C., Jaramillo, A., Vega, F., Poehling, H., Borgemeister, C.: Thermal tolerance of the Coffee Berry Borer Hypothenemus hampei: Predictions of Climate Change Impact on a Tropical Insect Pest. PLoS ONE. Sean Rands 4 (Agosto 2009)
4. Matheus, H., Gaviria, M., Jurado, O.: Avances en el manejo integrado de la broca del café Hypothenemus hampei Ferr., en Colombia. Instituto Colombiano Agropecuario, ICA, Primera Edición, Grupo Transferencia de Tecnología, ICA, 7 (Marzo 2004)
5. Ruiz, R., Baker, P.: Life table of Hypothenemus hampei (Ferrari) in relation to coffee berry phenology under Colombian field conditions. Sci. Agric. Piracicaba, Brazil 67, 658–668 (2010)
6. Barrera, J.F., Gómez, J., Lopez, E., Herrera, J.: Muestreo adaptativo para La Broca del café (Hypothenemus hampei). Entomología Mexicana 3, 535–539 (2004)
7. Cure, J., Rodríguez, D., Cotes, J., Gutierrez, A., Cantor, F.: A coffee agroecosystem model: I. Growth and development of the coffee plant. Ecological Modelling 3, 3626–3629 (2011)
8. Cure, J., Rodríguez, D., Cotes, J., Gutierrez, A., Cantor, F.: A coffee agroecosystem model: II. Dynamics of coffee berry borer. Ecological Modelling 248, 203–214 (2013)
9. Benavides, P.: Vuelos de la broca del café durante la cosecha principal Cartilla,Centro Nacional de investigaciones de Café
10. Constantino, M.: La broca del café.. un insecto que se desarrolla de acuerdo con la temepratura y la altitud Cartilla, Centro Nacional de investigaciones de Café
11. Jaramillo-Robledo, A., Arcila, J.: Variabilidad climática en la zona cafetera colombiana asociada al evento de el niño y su efecto en la caficultura. Fondo Nacional del Café, Avances Técnicos Cenicafé (2009)
12. Arcila Moreno, A.: Periodo crítico del ataque de la broca del café Cartilla Cenicafé, Entomología

Analysis of Structure and Hemolytic Activity Relationships of Antimicrobial Peptides (AMPs)

Jennifer Ruiz[1,*], Jhon Calderon[1], Paola Rondón-Villarreal[2], and Rodrigo Torres[3]

[1] School of Bacteriology and Clinical Laboratory, Universidad Industrial de Santander (UIS)
[2] School of Electrical, Electronics and Telecommunications Engineering, UIS
[3] School of Chemistry, Grupo de Investigación en Bioquímica y Microbiología (GIBIM), UIS
jennifer.ruiz@correo.uis.edu.co

Abstract. Antimicrobial peptides (AMPs) have become in a potential source of last generation antibiotics, constituting a diverse group of molecules that participate in the innate immunity of multiple organisms. These molecules share some biochemical characteristics that can be used for identification and prediction of design of new AMPs by computational biology techniques. In spite of promising potential as antibiotics of AMPs, they are often cytotoxic for eukaryotic cells, being a limitation for their use as pharmaceuticals. Hemolytic concentration 50 (HC 50) constitutes one of the most used indicators of toxicity. In the present study, a relationship between HC 50 and physicochemical properties of peptides was analyzed. For this aim, we use a set of descriptors of 18 peptides, which were computed through computational biology tools and analyzed in order to determine relationship and behavior of these descriptors to predict cytotoxicity of AMPs.

Keywords: Antimicrobial peptides, Hemolytic, HC50.

1 Introduction

Antimicrobial peptides (AMPs) are essential components of innate immunity from several biological organisms, from insects to human being, acting as an effective and unspecific defense line against pathogen such as bacteria, fungi, parasites and viruses [1–3]. These biological molecules are a heterogeneous group of compounds with length between 12 and 48 residues of amino acid. In general, these peptides are cationic, relatively hydrophobic and with tendency to form amphipathic -helix in solvents as fluoro-ethanol that mimics cell membrane [1,4].

In the last years, AMPs have become an alternative to conventional antibiotic in the treatment of infections caused by multi-resistant microorganisms because their interaction is mainly mediated by interaction with lipids from cell membrane, which make difficult acquisition of microbial resistance [3]. Nevertheless,

* Corresponding author.

L.F. Castillo et al. (eds.), *Advances in Computational Biology*, Advances in Intelligent Systems and Computing 232,
DOI: 10.1007/978-3-319-01568-2_36,

several of these peptides are toxic at concentrations necessary for treatment of severe infections, which evidences importance for knowing and determining both action mechanisms and structural characteristics that have influence in toxicity [1, 5]. Data mining allows collecting data for the analysis of antimicrobial potential and its selectivity with respect to host cells [6]. Although action mechanism of AMPs on the cell have no been elucidated, it is believed that these interact with polar lipid heads from bacterial membranes causing different effects, such as: 1) pore formation, 2) membrane lysis, 3) formation of lipid-protein domains, 4) induction of non-laminar phases, and 5) disintegration among negatively charged lipids from zwitterionic ones [7, 8].

Cell membrane provides cell protection and shape to cells, and is a selective barrier for cell exchange with the extracellular environment. This is formed by different lipids of which biophysical nature is significant for its structural and functional conformation, and even for providing a better interaction with AMPs [9,10]. For instance, red cells are mainly composed by lipids such as sphingomyelin (SM) and phosphatidylcholine (PC) and non-charge, while bacterial membranes possess higher quantities of phosphatidylglycerol (PG), cardiolipin (CL) and phosphatidylserine (PS) and with negative net-charge at physiological pH [4].

2 Methodology

2.1 Data Used in This Study

Initially, we carried out an extensive review in order to obtain data of HC50, defined as hemolytic concentration of peptides to obtain 50% of lysis of erythrocytes (HC50) under physiological conditions. From this review, we obtained 18 peptides with their corresponding HC50. AMPs were divided in 3 groups: little toxics $< 20\mu M$ (LTP), moderately toxics $> 20\mu M$ and $< 100\mu M$ (MTP), and highly toxics $> 100\mu M$ (HTP).

2.2 Computation of Descriptors

We computed the following descriptors from peptides to predict their toxicity: net charge, isoelectric point, hydrophobicity, molecular weight, stability, length, tendency to aggregate as beta-sheet (AGG), amyloid aggregation (AMYLO), beta-turn aggregation (BETA), and tendency to form beta-turns (TURN) and alpha-helixes (HELIX). For the computation of physico-chemical descriptors we used algorithms developed in our research group, and the software TANGO [11] for predictors for aggregation parameters at physiological conditions (pH 7.4, ionic strength= 0.116 and 38°C).

2.3 Correlation between Descriptors and HC50

Results obtained for every peptide were correlated with HC50 values for each group of peptides, processed and graphed with the software SPSS 13.0.

3 Results and Discussion

In this study, we analyzed different peptides with their respective HC50 values (see Table 1). These were analyzed in different groups according their toxicity (See Methods). All the peptides showed a similar tendency among the different evaluated groups.

Table 1. Antimicrobial peptides used in this study and their corresponding HC50

Peptide	Amino acid sequence	HC50 (μM)	Ref.
Mellitin	GIGAVLKVLTTGLPALISWIKRKRQQ	1.7	[12]
MG-H1	GIKKFLHIIWKFIKAFVGEIMNS	2.9	[13]
Pis-1 (native)	FFHHIFRGIVHVGKTIHRLVTG	11	[14]
Cupiennin1d*	GFGSLFKFLAKKVAKTVAKQAAKQGAKYVANKHMQ	14.5	[12]
MG-H2	IIKKFLHSIWKFGKAFVGEIMNI	16	[14]
Tritrpticin	VRRFPWWWPFLRR	19.98	[15]
Pr-3	VSRRRRRRGGRRRR	37.17	[16]
PDD-B	INWLKLGKKILGAL	45	[17]
PMM	INWKKIASIGKEVLKAL	80	[17]
Ranatuerin-2Ara	GLMDTVKNAAKNLAGQLLDTIKCKMTGC	100	[18]
MP	INWLKLGKKMMSAL	100	[17]
Brevinin-2Tsa	GIMSLFKGVLKTAGKHVAGSLVDQLKCKITGGC	100	[19]
Esculentin-1Arb	GLFPKFNKKKVKTGIFDIIKTVGKEAGMDVLR TGIDVIGCKIKGEC	120	[18]
Ranatuerin-1	SMLSVLKNLGKVGLGFVACKINKQC	140	[17]
Ranatuerin-2ARb	GILDTIKNAAKTVAVGLLEKIKCKMTGC	150	[18]
Ranatuerina-2CSa	GILSSFKGVAKGVAKDLAGKLLETLKCKITGC	160	[20]
Esculentina-1ARa	GIFSKINKKKAKTGLFNIIKTVGKEAGMDVIRA GIDTISCKIKGEC	180	[18]
Palustrina-3AR	GIFPKIIGKGIVNGIKSLAKGVGMKVFKAGLNN IGNTGCNNRDEC	200	[18]

Peptides were ordered in increasing order according to their HC50 values.

The most important descriptors used for estimation of relationships between AMP and toxicity have been net charge and hydrophobicity of peptides. In this work, we prove peptides highly charged were more toxic with high values of hemolytic activity. Some studies have found that even although the relationship between net charge and biological activity is non-linear, it has been demonstrated some correlation between net charge and hemolytic activity of peptides, showing HTP and LTP high positive charge. In general, an increase in cationicity can result in high hemolytic capacity, which diminishes of biological activity of the AMP [4,9].

On the other hand, hydrophobicity data were positively correlated with hemolytic activity. Nevertheless, dispersion analyses were not significant, because data were very disperses. This is a very important property of peptides, because determines interaction of peptides with membranes and its action mechanism [21,22]. Peptides with high hydrophobicity showed a tendency to be grouped in the LTP. Hung-Ta Choua et al. (2008) observed that high hydrophobicity was correlated with hemolytic activity.

Moreover, we found that uncommon descriptors used for prediction of toxicity, such as isoelectric point, length and molecular weight, stability and tendency to form alpha helix and beta-turns can be correlated with hemolytic activity. For instance, peptides with high molecular weight show low toxicity, probably because of their large sizes interfere in the interaction with the target cell membranes [23]. In general, peptides highly toxics were more stable and with tendency to form alpha- helices and beta-turns. Additionally, descriptors of beta-sheet aggregation, tendency to amyloid-aggregation and aggregation of beta-turns showed an small correlation with hemolytic activity of antimicrobial peptides.

4 Conclusions

The analysis of relationship between peptide structure and its hemolytic activity confirmed the effect of charge and hydrophobicity on the biological activity studied. Other physicochemical descriptors used in this study, such as length or isoelectric point or tendency to form some secondary structures (e.g. alpha-helix or beta-turn aggregation) did not show good correlations. Finally, this work opens a new field of study on the development of new descriptors and computational tools useful for prediction not only of antimicrobial activity of peptides if not its toxicity (e.g. hemolytic activity)

References

1. Bolintineanu, D.S., Kaznessis, Y.N.: Computational studies of protegrin antimicrobial peptides: a review. Peptides 32(1) (2011)
2. López-García, B., Ubhayasekera, W., Gallo, R.L., Marcos, J.F.: Parallel evaluation of antimicrobial peptides derived from the synthetic PAF26 and the human LL37. Biochem. Biophys. Res. Commun. 356(1) (2007)
3. Tsai, C.W., Hsu, N.Y., Wang, C.H., Lu, C.Y., Chang, Y., Tsai, H.H.G., Ruaan, R.C.: Coupling molecular dynamics simulations with experiments for the rational design of indolicidin-analogous antimicrobial peptides. J. Mol. Biol. 392(3) (2009)
4. Teixeira, V., Feio, M.J., Bastos, M.: Role of lipids in the interaction of antimicrobial peptides with membranes. Prog. Lipid Res. 51(2) (2012)
5. Frecer, V.: QSAR analysis of antimicrobial and haemolytic effects of cyclic cationic antimicrobial peptides derived from protegrin-1. Bioorg. Med. Chem. 14(17) (2006)

6. Juretić, D., Vukičević, D., Petrov, D., Novković, M., Bojović, V., Lučić, B., Ilić, N., Tossi, A.: Knowledge-based computational methods for identifying or designing novel, non-homologous antimicrobial peptides. Euro. Biophys. J. 40(4), 371–385 (2011)
7. Arouri, A., Kiessling, V., Tamm, L., Dathe, M., Blume, A.: Morphological changes induced by the action of antimicrobial peptides on supported lipid bilayers. J. Phys. Chem. 115(1) (2011)
8. La Rocca, P., Biggin, P.C., Tieleman, D.P., Sansom, M.S.: Simulation studies of the interaction of antimicrobial peptides and lipid bilayers. BBA 1462(1-2) (1999)
9. Epand, R.F., Schmitt, M.A., Gellman, S.H., Epand, R.M.: Role of membrane lipids in the mechanism of bacterial species selective toxicity by two alpha/beta-antimicrobial peptides. BBA 1758(9) (2006)
10. Bahnsen, J.S.B., Franzyk, H., Sandberg-Schaal, A., Nielsen, H.M.R.: Antimicrobial and cell-penetrating properties of penetratin analogs: effect of sequence and secondary structure. BBA 1828(2) (2013)
11. Fernandez-Escamilla, A.M., Rousseau, F., Schymkowitz, J., Serrano, L.: Prediction of sequence-dependent and mutational effects on the aggregation of peptides and proteins. Nat. Biotechnol. 22(10) (2004)
12. Kuhn-Nentwig, L., Muller, J., Schaller, J., Walz, A., Dathe, M., Nentwig, W.: Cupiennin 1, a new family of highly basic antimicrobial peptides in the venom of the spider Cupiennius salei (Ctenidae). J. Biol. Chem. 277(13) (2002)
13. Tachi, T., Epand, R.F., Epand, R.M., Matsuzaki, K.: Position-dependent hydrophobicity of the antimicrobial magainin peptide affects the mode of peptide-lipid interactions and selective toxicity. Biochemistry 41(34) (2002)
14. Lee, S.A., Kim, Y.K., Lim, S.S., Zhu, W.L., Ko, H., Shin, S.Y., Hahm, K.S., Kim, Y.: Solution structure and cell selectivity of piscidin 1 and its analogues. Biochemistry 46(12) (2007)
15. Schibli, D.J., Nguyen, L.T., Kernaghan, S.D., Rekdal, O.Y., Vogel, H.J.: Structure-function analysis of tritrpticin analogs: potential relationships between antimicrobial activities, model membrane interactions, and their micelle-bound NMR structures. Biophys. J. 91(12) (2006)
16. Pérez-Cordero, J.J., Lozano, J.M., Cortés, J., Delgado, G.: Leishmanicidal activity of synthetic antimicrobial peptides in an infection model with human dendritic cells. Peptides 32(4) (2011)
17. Cerovský, V., Slaninová, J., Fucík, V., Hulacová, H., Borovicková, L., Jezek, R., Bednárová, L.: New potent antimicrobial peptides from the venom of Polistinae wasps and their analogs. Peptides 29(6) (2008)
18. Ali, M.F., Lips, K.R., Knoop, F.C., Fritzsch, B., Miller, C., Conlon, J.M.: Antimicrobial peptides and protease inhibitors in the skin secretions of the crawfish frog, Rana areolata. BBA 1601(1) (2002)
19. Conlon, J.M., Al-Ghaferi, N., Abraham, B., Sonnevend, A., Coquet, L., Leprince, J., Jouenne, T., Vaudry, H., Iwamuro, S.: Antimicrobial peptides from the skin of the Tsushima brown frog Rana tsushimensis. Comp. Biochem. Physiol. Toxicol. Pharmacol. 143(1) (2006)
20. Subasinghage, A.P., Conlon, J.M., Hewage, C.M.: Conformational analysis of the broad-spectrum antibacterial peptide, ranatuerin-2CSa: identification of a full length helix-turn-helix motif. BBA 1784(6) (2008)
21. Lehrer, R., Barton, A., Daher, K.A., Harwig, S.S.L., Ganz, T., Selsted, M.E.: Interaction of Human Defensins with Escherichia coni Mechanism of Bactericidal Activity. J. Clinic. Invest. 84 (August 1989)

22. Vermeer, L.S., Lan, Y., Abbate, V., Ruh, E., Bui, T.T., Wilkinson, L.J., Kanno, T., Jumagulova, E., Kozlowska, J., Patel, J., McIntyre, C.A., Yam, W.C., Siu, G., Atkinson, R.A., Lam, J.K.W., Bansal, S.S., Drake, A.F., Mitchell, G.H., Mason, A.J.: Conformational flexibility determines selectivity and antibacterial, antiplasmodial, and anticancer potency of cationic α-helical peptides. J. Biol. Chem. 287(41) (2012)
23. Polyansky, A.A., Vassilevski, A.A., Volynsky, P.E., Vorontsova, O.V., Samsonova, O.V., Egorova, N.S., Krylov, N.A., Feofanov, A.V., Arseniev, A.S., Grishin, E.V., Efremov, R.G.: N-terminal amphipathic helix as a trigger of hemolytic activity in antimicrobial peptides: a case study in latarcins. FEBS Letters 583(14) (2009)

Candidates for New Molecules Controlling Allorecognition in *Hydractinia symbiolongicarpus*

Henry J. Rodríguez[1,2] and Luis Fernando Cadavid[1,2]

[1] Department of Biology, Universidad Nacional de Colombia
[2] Institute of Genetics, Universidad Nacional de Colombia,
Bogotá, Colombia
{hjrodriguezv,lfcadavidg}@unal.edu.co

Abstract. The sessile and colonial invertebrates have the capacity of distinguished between self and non-self tissues within of the same species. These allorecognition phenomena have been amply studied in the cnidarian *Hydractinia symbiolongicarpus*, where encounters between colonies result either in fusion or rejection. Allorecognition in endogamic lines of *H. symbiolongicarpus* is governed by two linked and polymorphic loci, *alr1* and *alr2*, where colonies sharing at least one allele fuse whereas those sharing none reject. However, this model doesn't fully predict the outcomes of encounters between wild-type animals, suggesting the existence of additional molecules controlling allorecognition in this species. In order to identify these molecules, we constructed three histocompatibility groups and used a comparative proteomics approach to identify proteins differentially expressed. We identified 48 proteins differentially expressed among the groups and 3 of them had structural features that make them candidates to participate in the control the allorecognition in *H. symbiolongicarpus*.

Keywords: Allorecognition, *Hydractinia*, proteomics, Fibrinogen, EGF-like.

1 Introduction

Sessile and colonial invertebrates have the capacity to discriminate between self and allogeneic tissues [1]. Perhaps the best studied model for invertebrate allorecognition is *Hydractinia symbiolongicarpus* (Cnidaria; Hydrozoa), a colonial hydroid that grows over the shells inhabited by pagurid hermit crabs, and is distributed in the east coast of North America. *H. symbiolongicarpus* is constituted for three types of tissues: the polyps, the stolonal mat and the stolons. The polyps are structures responsible of the feeding and reproduction, the stolonal mat communicate the polyps through of a system of canals called the gastrovascular system and the stolons are extensions of gastrovascular system [2]. Encounters between colonies growing on the shells are frequent, and result in either fusion or rejection, depending on the genetic relationships between them. The fusion is characterized for the establishment of a common gastrovascular system and the formation of an stable chimera, while the rejection is characterized for the absence of continuity between the colonies

L.F. Castillo et al. (eds.), *Advances in Computational Biology*,
Advances in Intelligent Systems and Computing 232,
DOI: 10.1007/978-3-319-01568-2_37, © Springer International Publishing Switzerland 2014

gastrovascular systems and the discharge of the nematocytes causing damage in the opposite colony [2]. The genetic control of allorecognition in inbreed lines of *H. symbiolongicarpus* lies in two linked, co-dominant and polymorphic loci, *alr1* and *alr2* [3–5]. Colonies sharing at least an allele in *alr1* and *alr2* will fuse, while colonies sharing no alleles will reject [3]. However, these fusibility rules not always explain the phenotypes in encounters between wild-type animals [5]. It is likely that inbreeding and selection has masked the effects of other allorecognition molecules which would be expressed otherwise. Thus, we have searched for alternative allorecognition molecules in *H. symbiolongicarpus* by comparing the proteome profile between compatible and incompatible colonies derived from a backcross population.

2 Materials and Methods

2.1 Generation of a Backcross Population and Construction of the Fusibility Groups

The animals were maintained according to the conditions specified in [3]. Two wild-type colonies of *H. symbiolongicarpus* were crossed to generate a F1 generation, and a female individual from this generation was crossed with its male parent to obtain a backcross population of 41 individuals. Backcross individuals were tested for fusibility against each other by the colony fusibility assay [3]. Two fusibility groups were established such that colonies within a group fuse to each other, but reject the individuals from the other group. A third fusibility group was composed of colonies fusing individuals from the first two groups.

2.2 Two-Dimensional Gel Electrophoresis (2D-DIGE)

Total protein was extracted from the pools by homogenization with lysis buffer (30mM Tris-HCl pH 8.8, 7M urea, 2M thiourea, 4% (w/v) CHAPS and 1X protease inhibitor cocktail (cOmplete ULTRA Tablets, Mini EDTA-free, ROCHE)) and centrifugation to 13.000 rpm for 15 min to 4°C. Crude extracts were purified and precipitated with phenol and acetone/methanol, respectively [6], and labeled with a CyDye DIGE fluorchrome and run in a 2D-DIGE for triplicate through of proteomic service provided by *Applied Biomics* (Hayward, USA).

2.3 Identification of the Proteins Differentially Expressed between Pools

Thirty-six protein spots from 2D-DIGE showing a differential expression were picked from the gels with the Spot Picker (GE Healthcare) and digested with trypsin [7]. The resulting peptides were subjected to MALDI-TOF/TOF and identified by database search and *de novo*. For the former, the software X! Tandem [8] was used with a custom database obtained from the *H. symbiolongicarpus* transcriptome [unpublished data]. For the latter, the software PepNovo [9] and BLAST comparisons were used [10].

3 Results

We constructed three fusibility groups, A, B and C. The fusibility groups A and B were constituted by five individuals that fused colonies from the same group but rejected colonies from the other group. The fusibility group C was constituted by a single individual that fused individuals from both group A and B, with the exception of two encounters where showed an atypical phenotype (Table 1).

The proteome of the three fusibility groups were compared through 2D-DIGE, and 96 spots differentially expressed were found. Thirty-six spots having higher values of overexpression or underexpression between groups A and B were chosen for identification with mass spectrometry. Identification by database search and *de novo* yielded 48 proteins that were classified into different categories, finding that the proteins more abundance were the categories Other proteins (29%) and Other enzymes (25%), among which are voltage-dependent anion-selective channel, pathogen-related protein-like, phosphoenolpyruvate carboxykinase (PEPC) and glutathione S-transferase, etc. (Fig. 1).

Three of the identified proteins contained recognition domains and were considered as candidates for allorecognition molecules. The first had Fibrinogen β/γ, C-terminal globular domain similar to Ficolin-2 [*Crassostrea gigas*] and Tenascin-R-like [*Amphimedon queenslandica*]. The second had human growth factor-like EGF domain and an EGF-like domain and was similar to Tenascin-X-like [*Hydra magnipapillata*] and Teneurin transmembrane protein 2 [*Xenopus tropicalis*]. The third protein had three Thrombospondin type 1 (TSP-1) domains, a von Willebrand factor type A (vWF) domain and a FlgD Tudor-like domain and was similar to Rhamnospondin-1 and 2 [*H. symbiolongicarpus*] and Hemicentin-1 [*C. gigas*].

Table 1. Fusibility assays matrix. The individuals of fusibility group A are histocompatible with the individual 15, while the individuals of fusibility group B are histocompatible with the individual 3. The fusibility group C is compound for the individual 43. The results of fusibility assays are showed as fusion (F), rejection (R) and inflammatory fusion (Fi).

ID	3	9	15	22	23	36	40	43	50	53	70
3	-	-	-	-	-	-	-	-	-	-	-
9	F	-	-	-	-	-	-	-	-	-	-
15	R	R	-	-	-	-	-	-	-	-	-
22	F	F	R	-	-	-	-	-	-	-	-
23	R	R	F	R	-	-	-	-	-	-	-
36	R	R	F	R	F	-	-	-	-	-	-
40	R	R	F	R	F	F	-	-	-	-	-
43	F	F	F	F	F	F	F	-	-	-	-
50	F	F	R	F	R	R	R	Fi	-	-	-
53	F	F	R	F	R	R	R	Fi	F	-	-
70	R	R	F	R	F	F	F	F	R	R	-

The protein with the Fibrinogen domain was similar to other fibrinogen-containing proteins only in the C-terminal end. The protein that presents two EGF-like domains might be highly variable as there were around 15 different sequences in the *H. symbiolongicarpus* transcriptome. The differences between these sequences are concentrated in the extremes N-terminal and C-terminal. Finally, the protein containing three TSP-1 domains, a vWF domain and a FlgD Tudor-like domain have two important characteristics: First, this protein has a novel domain architecture [11], and second, the FlgD Tudor-like domain is a hybrid structure between the Tudor domain and Fibronectin type III (Fn-III) domain. The structure of Fn-III is constituted by two β-sheets which forms a sandwich-like structure similar to that found in Immunoglobulin domains [12].

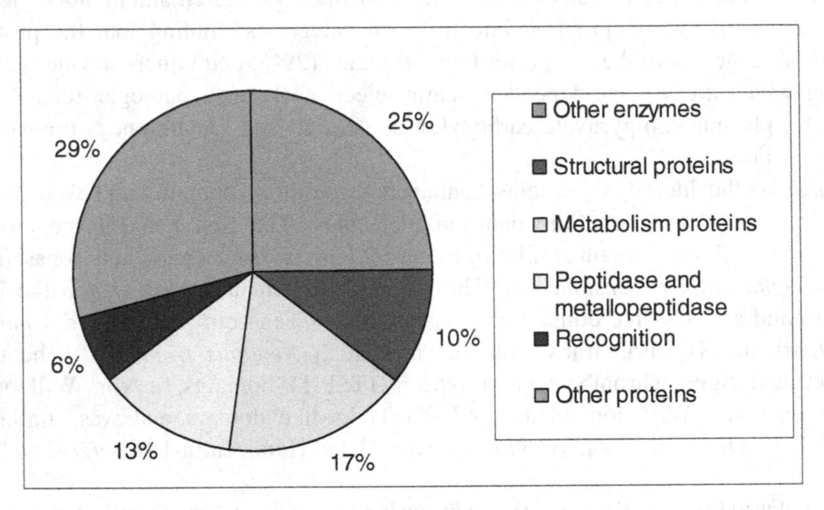

Fig. 1. Distribution of identified proteins for database search and *de novo* approximations. N=48 proteins. The identified proteins were distributed in enzymes (25%), structural proteins (10%), metabolism proteins (17%), peptidases and metallopeptidases (13%), recognition proteins (6%) and other proteins (29%).

4 Discussion

Comparative analysis of the proteome of fusibility groups resulted in three candidate proteins that might function as allorecognition determinants. Two of them, the one having a Fibrinogen domain and the other with two EGF-like domains might be relevant in the search of alternative allorecognition proteins as these domains have been observed in other allorecognition systems. Specifically, the Fibrinogen domain has been found in the proteins v-Themis-A and –B that participate in the self-incompatibility system of *Ciona intestinalis.* These proteins are localized in the vitelline coat of the egg and with their counterparts in the sperm surface, s-Themis-A and –B, determinate the success of fertilization [13]. Likewise, the EGF-like domain has been found as part of FuHC receptor of *Botryllus schlosseri*, a molecule that control the allorecognition in this species [1]. Further, its high variability makes this protein a strong candidate to control allorecognition in *H. symbiolongicarpus*.

References

1. Rosengarten, R.D., Nicotra, M.L.: Model systems of invertebrate allorecognition. Current Biology 21, R82–R92 (2011)
2. Cadavid, L.F.: Self-discrimination in colonial invertebrates: genetic control of allorecognition in the hydroid Hydractinia. Developmental and Comparative Immunology 28, 871–879 (2004)
3. Cadavid, L.F., Powell, A.E., Nicotra, M.L., Moreno, M., Buss, L.W.: An invertebrate histocompatibility complex. Genetics 167, 357–365 (2004)
4. Nicotra, M.L., Powell, A.E., Rosengarten, R.D., Moreno, M., Grimwood, J., Lakkis, F.G., Dellaporta, S.L., Buss, L.W.: A hypervariable invertebrate allodeterminant. Current Biology 19, 583–589 (2009)
5. Rosa, S.F.P., Powell, A.E., Rosengarten, R.D., Nicotra, M.L., Moreno, M.A., Grimwood, J., Lakkis, F.G., Dellaporta, S.L., Buss, L.W.: Hydractinia allodeterminant alr1 resides in an immunoglobulin superfamily-like gene complex. Current Biology 20, 1122–1127 (2010)
6. Arevalo-Ferro, C., Hentzer, M., Reil, G., Görg, A., Kjelleberg, S., Givskov, M., Riedel, K., Eberl, L.: Identification of quorum-sensing regulated proteins in the opportunistic pathogen Pseudomonas aeruginosa by proteomics. Environmental Microbiology 5, 1350–1369 (2003)
7. Shevchenko, A., Tomas, H., Havlis, J., Olsen, J.V., Mann, M.: In-gel digestion for mass spectrometric characterization of proteins and proteomes. Nature Protocols 1, 2856–2860 (2006)
8. Craig, R., Beavis, R.C.: TANDEM: matching proteins with tandem mass spectra. Bioinformatics (Oxford, England) 20, 1466–1467 (2004)
9. Frank, A., Pevzner, P.: PepNovo: de novo peptide sequencing via probabilistic network modeling. Analytical Chemistry 77, 964–973 (2005)
10. Camacho, C., Coulouris, G., Avagyan, V., Ma, N., Papadopoulos, J., Bealer, K., Madden, T.L.: BLAST+: architecture and applications. BMC Bioinformatics 10, 421 (2009)
11. Geer, L.Y., Domrachev, M., Lipman, D.J., Bryant, S.H.: CDART: protein homology by domain architecture. Genome Research 12, 1619–1623 (2002)
12. Kuo, W.-T., Chin, K.-H., Lo, W.-T., Wang, A.H.-J., Chou, S.-H.: Crystal structure of the C-terminal domain of a flagellar hook-capping protein from Xanthomonas campestris. Journal of Molecular Biology 381, 189–199 (2008)
13. Harada, Y., Takagaki, Y., Sunagawa, M., Saito, T., Yamada, L., Taniguchi, H., Shoguchi, E., Sawada, H.: Mechanism of self-sterility in a hermaphroditic chordate. Science (New York, N.Y.) 320, 548–550 (2008)

Escherichia coli´s OmpA as Biosurfactant for Cosmetic Industry: Stability Analysis and Experimental Validation Based on Molecular Simulations

Sonia Milena Aguilera Segura[1], Angie Paola Macías[1], Diana Carrero Pinto[1], Watson Lawrence Vargas[1], Martha Josefina Vives-Florez[2], Harold Enrique Castro Barrera[3], Oscar Alberto Álvarez[1], and Andrés Fernando González Barrios[1]

[1] Grupo de Diseño de Productos y Procesos (GDPP), Departamento de Ingeniería Química, Universidad de los Andes, Bogotá, Colombia
{sm.aguilera37,a-macias,dm.carrero41,wvargas,oalvarez, andgonza}@uniandes.edu.co

[2] Centro de Investigaciones Microbiológicas (CIMIC), Departamento de Ciencias Biológicas, Universidad de los Andes, Bogotá, Colombia
mvives@uniandes.edu.co

[3] Grupo de Comunicaciones y Tecnología de la Información, Departamento de Ingeniería de Sistemas, Universidad de los Andes, Bogotá, Colombia
hcastro@uniandes.edu.co

Abstract. The development of biosurfactants has increased due to their biodegradability, low toxicity and specificity. The aim of this work is to analyze the behavior of the outer membrane protein A (OmpA) of *Escherichia coli* by Molecular Dynamics simulations and to perform experimental validation when used as stabilizer for dodecane/water emulsions. Trajectories were analyzed with the analysis tools provided by the GROMACS package. OmpA was purified from *E. coli* K-12 W3110/pCA24N strain in medium with isopropylthio-β-galactoside. Oil in water emulsions with different concentrations of OmpA were prepared in batch processes. MD trajectories with OmpA reached stability after 1 ns with an average RMSD value of 5.6 nm and they showed that OmpA remains stable in emulsion. An inverse effect related to protein concentration was found on the stability of the emulsion. OmpA displayed a significant role as a stabilizer for dodecane/water emulsions as the presence of OmpA increased their stability up to 7.5 h.

Keywords: Molecular Dynamics, OmpA, *E. coli*, biosurfactant, emulsions.

1 Introduction

Nowadays, all efforts coming from industry and scientific community have focused on finding technologies and products that are friendlier with the environment. Surfactants are widely used for stabilizing systems like emulsions in products on food, personal care products, and cosmetics, among others. However, most of commercial

L.F. Castillo et al. (eds.), *Advances in Computational Biology*,
Advances in Intelligent Systems and Computing 232,
DOI: 10.1007/978-3-319-01568-2_38, © Springer International Publishing Switzerland 2014

surfactants are derived from oil industry; some of them are toxic for the environment and do not degrade easily. As opposite, biosurfactants are surface active compounds synthetized by microorganisms and they have a potential advantage over traditional surfactants in many fields spanning environmental, food, biomedical, and other industrial applications [1].

Engineered peptides have turned into an interesting target as they allow a variety of functionalities based on the diversity of the amino acid sequence [2]. The fact that they can be built from twenty different naturally occurring amino acids provides an infinite amount of possibilities in terms of functionality, three-dimensional structure and response to different physical-chemical conditions. Several authors [3, 4] have developed peptides with a variety of properties that allow them to form films, foams, and nanovesicles, among others.

Some transmembrane proteins can also be tested as biosurfactants due to its amphiphilic nature: they must interact with the periplasm and membrane medium, which are clearly dissimilar in kind. *Escherichia coli's (E. coli)* outer protein membrane A (OmpA) contains hydrophobic and hydrophilic chains and has been found to play a big role during biofilm formation in *E. coli* [5]. The classic folding model of OmpA contains a 170-residue N-terminal domain consisting of eight anti-parallel β-strands, as well as a 155-residue periplasmic C-terminal domain [6]. OmpA homolog (85%), AlnA of Acinetobacter radioresistens, have demonstrated the ability to stabilize hydrocarbons in water emulsion [7]. Therefore, the aim of this work is to analyze the behavior of OmpA of *E. coli* by Molecular Dynamics simulations (MD) and to perform experimental validation when used as stabilizer for dodecane/water emulsions.

2 Materials and Methods

2.1 MD Simulations

MD simulations were carried out using the GROMACS package version 4.6.1 with united atom GROMOS96 53a6 force field [8]. The GROMOS 53a6 force field describes appropriately proteins and DNA, and has been parameterized on the basis of free enthalpies of hydration and apolar solvation which play an important role in protein folding [8]. Currently, a complete 3D model of *E. coli's* OmpA is unavailable. Only the transmembrane domain has been studied both by crystallography and NMR. Therefore, a complete model of the protein was obtained with the I-TASSER server (Figure 1). OmpA model was placed in a rectangular box with the box-edges at least 1 nm apart from the protein surface. The system was solvated with 13729 Simple Point Charge (SPC216) waters and 195 dodecane molecules. Solvent layers of water and dodecane were placed in the protein box mimicking the hydrophobic-hydrophilic regions of the protein. Electrostatic interactions were treated by the Particle Mesh Ewald (PME) method, with a Coulomb cut-off of 1.4 nm, a sixth order interpolation and Fourier spacing of 0.12 nm. The van der Waals interactions were treated using the Lennard-Jones potential combined with a switching function, with a cut-off distance of 0.8 nm and a switching distance of 0.7 nm.

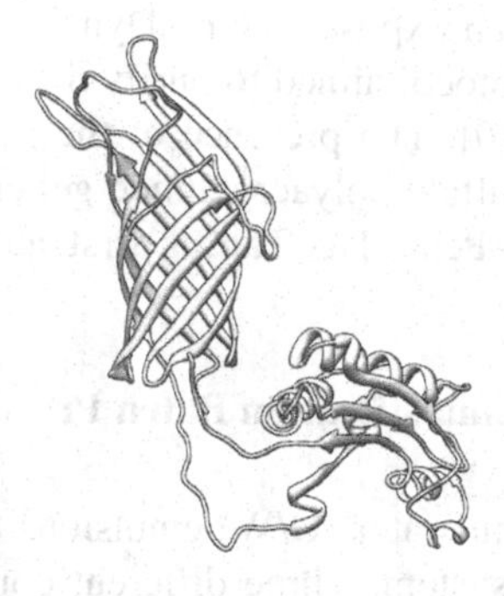

Fig. 1. Complete OmpA model provided by the I-TASSER server. The assembly of both domains reproduces adequately the topology of OmpA in the outer membrane and the periplasm. The hydrophobic β-barrel domain and the hydrophilic periplasmic domain mimic the overall amphiphilic structure of a surfactant.

Energy minimization was performed using the steepest descent algorithm until convergence and when the maximum force was smaller than 100 kJ mol-1 nm-1. After minimization, restrained simulations to stabilize and distribute the solvent molecules around the protein were performed for 200 ps at 300 K with a 2-fs time step. Initial velocities were generated from a Maxwell distribution at 300 K. Finally, a 3-ns MD simulation was performed with an integration time step of 2 fs. A LINCS algorithm was used to constrain the bonds, being 12 the highest order in the expansion of the constraint coupling matrix. Temperature and pressure coupling were handled using the leap-frog stochastic dynamics (sd) integrator and the Parrinello-Rahman method, respectively. Obtained trajectories were analyzed with the analysis tools provided by the GROMACS package. Overall stability of the n-dodecane/water/ OmpA system was measured by estimating the root mean square displacement (RMSD), the radius of gyration (RGYR) of the protein and the solvent accessible surface area (SASA).

2.2 Production of O/W Emulsions

Protein Purification

E. coli K-12 W3110/pCA24N *OmpA*$^+$ [9] was grown overnight at 37°C in LB agar-plates (5 g/L yeast extract, 10 g/L bacto tryptone, 10 g/L NaCl) containing 50 μg/mL chloramphenicol. A colony was inoculated in 50 mL of LB medium with chloramphenicol and incubated at 37°C and 16 hours at 250 rpm. 19.5 ml of fresh LB medium were inoculated with 500 μl of the previous culture at the same conditions. When reaching $OD_{600}nm=0.7$, 2-mM isopropylthio-β-galactoside (IPTG) was added in order to induce the expression of the protein. Cells were exposed to IPTG for three hours and the sample was sonicated afterwards at 38% amplitude for 40 cycles (20sx40s) on ice. Obtained samples were centrifuged and the supernatant was recovered. Supernatant was then exposed to the Dynabeads® TALON® kit (Invitrogen) following the reported protocol aimed to purify the protein as OmpA was cloned with an added histidine tail [10]. The presence of the purified protein was corroborated with a sodium dodecyl sulfate polyacrylamide gel electrophoresis (SDS-PAGE) and a Western blot with anti-Penta-His-Tag as first antibody and anti-mouse IgG peroxidase as second antibody.

Emulsion Formulation and Emulsification Batch Process

For the preparation of the oil-in-water (O/W) emulsions the OmpA was dissolved in water by stirring in a vortex system. Three different concentrations of OmpA were tested: 0.034%, 0.045%, and 0.054% w/v. After dissolving the OmpA, the dispersed phase was added to complete a volume of 2 mL (5% disperse phase) and then it was stirred for 10 min until a homogeneous emulsion was obtained. This homogenization was performed six times for each emulsion in a sonifier (Sonics & Materials Inc, VCX-750), using sonication pulses with 39% of amplitude for 50 s followed by a pause of 40 s. The dispersed phase consisted of dodecane, since it constitutes a model system in industrial emulsions. An emulsion without OmpA was also prepared as a negative control. The average droplet size (hydrodynamic diameter) was measured between 10 and 500 minutes after emulsion preparation to study the evolution of this parameter in time and to evaluate coalescence phenomenon. Hydrodynamic diameter and particle size distribution were measured with a particle size analyzer (Zetasizer Nano ZS), which uses dynamic light scattering (DLS) technique (λ=633 nm, temperature = 25 °C, angle = 173°). This equipment calculates a hydrodynamic diameter as a diameter of a sphere equivalent in terms of its diffusion due to Brownian movement, analyzing the intensity fluctuations of scattered light in time.

3 Results

3.1 Molecular Dynamics Simulations

MD simulations with OmpA reached enough stability after 1 ns with an average system's RMSD value of 5.6 ± 0.1 nm for the last 2 ns (Figure 2). RMSD contributions are mostly because of diffusion of dodecane and water molecules (RMSD values of 3.61 ± 0.19 and 6.16 ± 0.01, respectively), whereas OmpA remains practically invariant (RMSD of 0.44 ± 0.05). Dodecane molecules move to form a single agglomerate around the hydrophobic domain of the protein while water molecules diffuse to fill the released spaces by the dodecane molecules and remain stable around the hydrophilic loops and C-terminal domain of the protein. MD trajectories showed that OmpA has the capacity to remain stable between the dodecano-water interfaces. Furthermore, it has the ability to contribute to the overall stability of the system by reducing the hydrophobic-hydrophilic interactions between dodecane and water molecules, as does a surfactant. We also wanted to evaluate the stability of the protein removing the potential bias generated by RMSD analysis caused by a translation of the protein when calculating it. RGYR for the system displays low variation with an average value of 2.68 ± 0.03 nm, indicating not only that OmpA protein remains stable and folded in emulsion but also that the agglomeration of solvent around contributes to the protein stability. The solvent accessible surface area (SASA) of the protein remained stable over the time, with average values of 121.5 ± 2.2 and 90.7 ± 1.7 nm for the hydrophobic and hydrophilic surfaces, respectively.

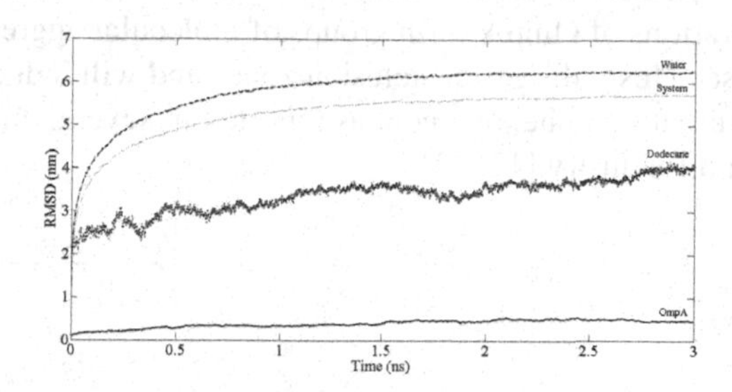

Fig. 2. RMSD values for the system, OmpA, dodecane and water molecules along the trajectories. Labels are on top of each series.

3.2 Purified OmpA as Stabilizer of Dodecano-Water Emulsions

Both SDS-PAGE and Western blot experiments showed a single 31 kDa band that corresponds to the molecular weight for OmpA as reported by le Coutre [11]. Therefore, the purified protein was used in the formulation of the proposed emulsions. The time that takes the emulsion to destabilize by forming two phases constitute an easy yet eloquent approach to evaluate the protein. We measured such time finding that the presence of OmpA augment the stability of the emulsion as in all cases the time obtained was bigger than the negative control. Moreover, we found a negative effect of the concentration on the stability of the emulsion (Table 1). In order to find the underpinnings behind this trend it is necessary to determine the droplet size distribution.

Table 1. Two phases formation time at different concentrations of OmpA for dodecane/water emulsions

(% w/v)	*Time (h)*
0.034	7.5
0.045	5
0.054	4
0.000	0.5

3.3 Average Droplet Size

As shown in Figure 3, protein concentrations of 0.034% and 0.045% w/v achieve to form a monodisperse droplet distribution which contributes to a higher stability of emulsions. As opposite, the protein concentration of 0.054% w/v forms a polydisperse distribution that has a negative effect on the stability. The average droplet size evaluation is not considerable as a function of time, only after long exposure periods. This means that the surfactant achieves to stabilize the droplets avoiding a destabilization process such as coalescence. However, as shown in Table 1, is it clear that the emulsions present a creaming phenomenon which is associated with a difference in the densities between phases. Furthermore, lower protein concentrations achieve longer periods before the creaming phenomenon is observed. It is possible that higher concentrations of OmpA form groups of molecular aggregates of OmpA in the continuous phase. This will create exclusion zones and will induce the grouping of droplets before the creaming phenomenon, as reported in several emulsions stabilized by surfactants from oil industry [12, 13].

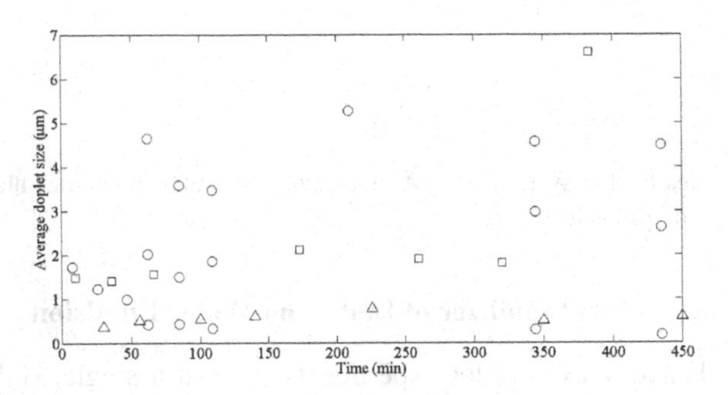

Fig. 3. Average droplet size distribution measurements at the concentrations of OmpA tested. Triangles, squares, and circles markers correspond to 0.034%, 0.045%, and 0.054% w/v concentrations, respectively.

4 Discussion

MD simulations have proven to be a useful tool to predict the potential ability of OmpA to increase the stability of dodecane-water emulsions. However, they have limitations on the study of larger scale systems (μm) and in the determination of dynamic properties of such systems like average droplet size or polydispersion. There is a need of a thermodynamic rather than a dynamic criterion to predict the ability of molecules to stabilize systems and to form stable emulsions over the time. Surfactants have the role of change the Gibbs free energy on the oil-water interfaces [14]. Therefore, a good criterion would be to measure how the presence of the biosurfactant will affect the Gibbs free energy of the system.

OmpA has shown the potential as a biosurfactant. Further work can be directed towards improving the ability of OmpA of stabilizing O/W emulsions. The hydrophilic-lipophilic balance (HLB) has been widely used as a criterion for emulsion formulation and the use of surfactants [15]. Several modifications can be performed on the protein oriented to improve its HLB to stabilize O/W emulsions. SASA can also be used as a formulation criterion since it actually takes into account the groups that interact with the solvent, as opposite of the HLB that takes into account the whole molecule.

We have demonstrated the significant role of *E. coli's* outer membrane protein A as a potential stabilizer for dodecane/water emulsions as the presence of OmpA increased the stability of dodecane/water emulsions up to 7.5 h. Also, we showed the pertinence of using MD simulations as a rational tool for prediction aiming to save experimental work. RMSD, RGYR, SASA, and MD trajectories were able to describe the potential emulsifier properties of the protein and to predict the capacity to stabilize dodecane/water emulsions.

References

1. Banat, I., et al.: Microbial biosurfactants production, applications and future potential. Applied Microbiology and Biotechnology 87(2), 427–444 (2010)
2. Zhao, X.: Design of self-assembling surfactant-like peptides and their applications. Current Opinion in Colloid & Interface Science 14(5), 340–348 (2009)
3. Jones, D.B., Middelberg, A.P.J.: Mechanical Properties of Interfacially Adsorbed Peptide Networks. Langmuir 18(26), 10357–10362 (2002)
4. Dexter, A.F., Malcolm, A.S., Middelberg, A.P.J.: Reversible active switching of the mechanical properties of a peptide film at a fluid–fluid interface. Nature Materials 5, 502–506 (2006)
5. Barrios, A.F.G., et al.: Hha, YbaJ, and OmpA regulate Escherichia coli K12 biofilm formation and conjugation plasmids abolish motility. Biotechnology and Bioengineering 93(1), 188–200 (2006)
6. Sugawara, E., et al.: Secondary structure of the outer membrane proteins OmpA of Escherichia coli and OprF of Pseudomonas aeruginosa. Journal of Bacteriology 178(20), 6067–6069 (1996)
7. Toren, A., et al.: The Active Component of the Bioemulsifier Alasan from Acinetobacter radioresistens KA53 Is an OmpA-Like Protein. Journal of Bacteriology 184(1), 165–170 (2002)
8. Oostenbrink, C., et al.: Validation of the 53A6 GROMOS force field. European Biophysics Journal 34(4), 273–284 (2005)
9. Kitagawa, M., et al.: Complete set of ORF clones of Escherichia coli ASKA library (A Complete Set of E. coli K-12 ORF Archive): Unique Resources for Biological Research. DNA Research 12(5), 291–299 (2006)
10. Waterborg, J.: The Lowry Method for Protein Quantitation. In: Walker, J. (ed.) The Protein Protocols Handbook, pp. 7–9. Humana Press (2002)
11. le Coutre, J., et al.: Proteomics on Full-Length Membrane Proteins Using Mass Spectrometry†. Biochemistry 39(15), 4237–4242 (2000)
12. Aronson, M.P.: The role of free surfactant in destabilizing oil-in-water emulsions. Langmuir 5(2), 494–501 (1989)
13. Izquierdo, P., et al.: Formation and Stability of Nano-Emulsions Prepared Using the Phase Inversion Temperature Method. Langmuir 18(1), 26–30 (2001)
14. Salager, J.: Cuaderno FIRP 300 A. Surfactantes. Tipos y usos. Laboratorio Firp. Universidad de los Andes, Venezuela (2002)
15. Griffin, W.C.: Classification of surface-active agents by "HLB". Journal of Cosmetic Science 1, 311–326 (1949)

Molecular Cloning, Modelling and Docking with Curcumin of the Dengue Virus 2 NS5 Polymerase Domain

Leidy Lorena García Ariza[1], Germán Alberto Téllez Ramirez[1], Héctor Fabio Cortes Hernández[2], Leonardo Padilla Sanabria[1], and Jhon Carlos Castaño Osorio[1]

[1] Molecular Immunology Group GYMOL, Biomedical Research Center, Quindio´s University
gymol@uniquindio.edu.co
[2] Chemical Group of Environmental Research and Development QUIDEA, Quindio´s University

Abstract. The NS5 protein form Dengue virus 2 (DENV2) has an ARN dependant ARN polymerase activity (RdRp) and it is an important target to develop new treatments against dengue. We had amplified by PCR and cloned in the plasmid pGEX-5X-1 the genetic sequence of the polymerase domain from NS5 of DENV2. This domain was expressed as a fusion protein with Glutathion-S transferase in *E. coli* BL21, and was detected by western blot. A structural model of the cloned the polymerase domain was built by homology modelling and it was refined by KOBAMIN, 3Drefine, FG-MD and ModRefiner; an in silico docking was done with curcumin using Autodock Vina, an interaction between the oxygen of the hydroxyl group of the curcumin and the Lys 92 located in the cavity B of the NS5 polymerase domain was seen. This interaction could explain the inhibitory effect of the curcumin in the Dengue 2 virus replication.

Keywords: NS5 protein, Dengue virus type 2, Polymerase Chain Reaction, recombinant proteins, fusion proteins, cloning, expression, RNA polymerase, structural model, molecular docking simulation, curcumin.

1 Introduction

The NS5 protein is the biggest and more conserved protein of the Flavivirus family, it has two domains with independent activities, the C-terminal has an RdRp activity and the N-terminal a S-adenosylmethionine transferase activity [1]; both activities are essential for viral cycle and replication [2].

The RdRp protein family had three conserved sub-domains, palm, fingers and thumb; the palm domain contain three motives A, B and C; the A and C are related with the Mg^{2+} and/or Mn^{2+} and the motive B with the selectivity by ribonucleotides triphosphates. The ARN polymerase catalytic domain from Dengue 3 might be a target for the design of new antiviral compounds against dengue [3].

L.F. Castillo et al. (eds.), *Advances in Computational Biology*,
Advances in Intelligent Systems and Computing 232,
DOI: 10.1007/978-3-319-01568-2_39, © Springer International Publishing Switzerland 2014

The curcumin is a hydrophobic polyphenol derived from the rhizome of the *Curcuma longa,* the curcumin produced an *in vitro* inhibition in the Plaque forming units in DENV2 [4]. We had cloned and expressed the gen for the RdRp domain of the NS5 protein from DENV2; a structural model was generated by homology modelling and an *in silico* docking with curcumin was done.

2 Materials and Methods

2.1 Genetic Amplification

The RdRp domain from the 8298 and the 10235 nucleotides according to the reported DENV2 genome (GenBank: ACN42713.1) was amplified by PCR using the plasmid pBAC-DENV-FL-GFP-T7-RO8 that contains the NS5 genetic sequence. The amplification was done using the forward primer (5´- GGATCCGACACAAGAAAGCCAC TTAYGAGCC-3´) and reverse primer (5´-GTCGACTGCTTTYTACCACAGGA CTCCTG-3´). The amplification program was 1 cycle 94°C, 5 minutes; 40 cycles of 95°C for 45seconds, 70°C for 3 minutes; and a 72°C 10 minutes final extension.

2.2 Cloning of the RdRp Domain of the NS5 Protein from DENV2

The amplified genetic sequence was clones in the expression plasmid pGEX-5X-1 GST Expression Vector cat No. 28-9545-53 GE Healthcare, using digestion with BamHI and SalI; and directional ligation in the vector. The plasmid was transfected in *E.coli* DH5α y *E. coli* BL21 and sequenced using the service from Macrogen USA.

2.3 Expression of the RdRp Domain of the NS5 Protein from DENV2

The expression of the RdRp was evaluated in *E.coli* BL21 under different IPTG concentrations (0.1; 0.3; 0.5; 1 and 2 μM), time intervals from the induction time (0.5; 1; 2; 3; 6; 7; 8; 9; 12; 13; 16; 18; 20; 21; 22 and 24 hours), 37°C temperature and 170 rpm agitation. A 10% poliacrylamide gel electrophoresis (SDS-PAGE) and Western blot (WB) for the lysates was done. For the WB an anti Dengue virus SC-70959 mouse monoclonal IgG2a antibody (Santa Cruz biotechnology) was used as a primary antibody, as a secondary antibody was used a goat Anti-Mouse IgG tag with alkaline phosphatase (Sigma Aldrich A3562).

2.4 Geometry Optimization of Curcumin

Curcumin was designed and optimized in Hyperchem, using molecular mechanics and semiempirical methods by the algorithms of Fletcher-Reeves and Polak-Ribiere. The theoretical values of the IR spectrum bands of curcumin were compared with the experimental data previously obtained and the mean percentages of error were calculated.

2.5 Homology Model Building and Structural Evaluation of the RdRp Domain of the NS5 Protein from DENV2

The model for the DENV2 RdRp domain was build using as template the polymerase domain NS5 protein from DENV3 (PDB: 2J7U) [5], using the program Swiss-Model workspace [6]. The structure the RdRp domain was refined using KOBAMIN [7], 3Drefine [8], FG-MD [9] and ModRefiner [10].The model quality was evaluated by QMEAN and Ramachandran plot. The molecular docking between curcumin and the RdRp domain of DENV2 was performing using AutoDock Vina [11] and the result was analyzed using Autodock tools.

3 Results and Discussion

3.1 Amplification and Cloning of the Gene of the RdRp Domain from DENV2

A Gene fragment of 1985 bp was amplified corresponding to the RdRp domain with a 99% identity compared with the genomic sequence from the GenBank: AAC59274 (Fig. 1A) and the amplified product was cloned into the pGEX-5X-1 plasmid (Fig. 1B).

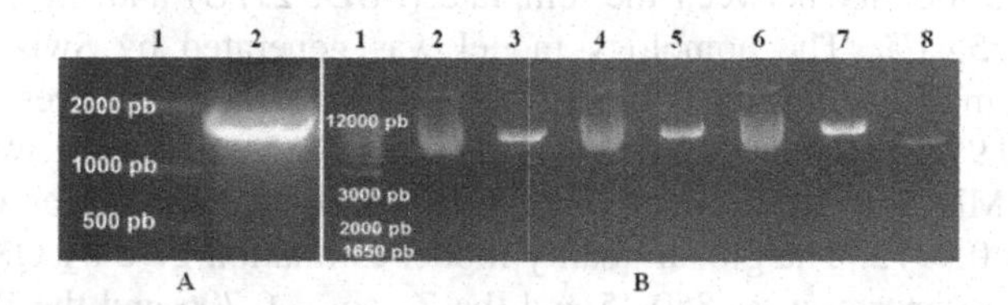

Fig. 1. Amplification and cloning of RdRp domain of the NS5 protein from DENV2. 1% agarose DNA gel electrophoresis. **A**: 1: molecular weight marker (MWM), 2: PCR amplification product of the RdRp domain (1985 bp). **B**: 1: MPM; 2, 4, 6: pGEX-5X-DENV2-NS5-D; 3, 5, 7: pGEX-5X-DENV2-NS5-D SalI digested (6885 bp); 8: pGEX-5X-1 BamHI and SalI digested (4900 bp).

3.2 Expression of the RdRp Domain of DENV2

The RdRp domain was express as a fusion protein with Glutathion S-transferase (GST) in *E. coli* BL21 the expression was seen from the 12 hours after the induction with 0,1mM IPTG (isopropyl β-D-1-thiogalactopyranoside) as 103KD with an expression peak at 21 Hours (Fig. 2A). The RdRp domain expression was confirmed by WB (Fig. 2B).

3.3 *In Silico* Docking for the Curcumin and the Structural Model of the RdRp Domain

3.3.1 Curcumin Geometry Optimization

The determined mean percentage error indicates that the method with the lowest error was the RM1 (Fletcher-R 0.75% y Polak-R 0.61%) (Fig. 3B). Therefore, yields the most stable conformation of curcumin in which the structure is located with the minimum potential energy, that is, on balance; this energy reduction is due to the adjustment of the atomic coordinates of the structure.

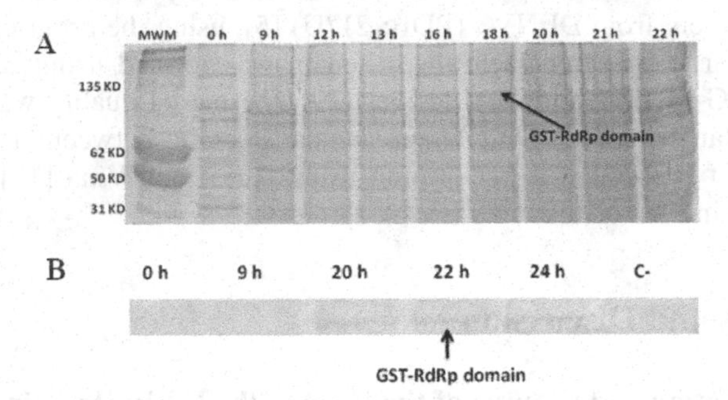

Fig. 2. GST-RdRp domain expression in the lysate bacteria after 0.1mM IPTG induction. **A:** 10% SDS-PAGE. MWM (Molecular Weight Marker) and hours after IPTG induction. **B:** WB hours after IPTG induction.

3.3.2 Structural Modelling of the RdRp Domain of DENV2

The percentage identity between the template (PBD: 2J7U) and the DENV2 RdRp Domain was 75.41%. The homology model was generated by Swiss model work space and this model was used as input for KOBAMIN the result present a reference GDT-TS of 0.991 and GDT-HA 0.917. The KOBAMIN output was refined by 3Drefine, FG-MD and ModRefiner (Fig. 3A). The final RMSD was 0.463 and the TM-score was 0.9972. The global quality model estimation give by QMEAN score6 was 0.699, Dfire energy was -850.45 and the Z-score -0.706 and the Ramachandran plot generated had a value of 99% of the residues in permitted regions.; then the refine structure the QMEANscore6 was 0.658, Dfire energy was -881.79, Z-score -1.164 and the Ramachandran plot generated had a value of 99.2% of the residues in permitted regions, which indicates that it has a more energetically stable conformation [12].

3.3.3 Molecular Docking between Curcumin and RdRp Domain of DENV2

The docking results show nine posible conformations of the curcumin within the cavity B of the RdRp domain (Fig. 3C), the most favorable interaction had -6.2Kcal/mol and occurred between the oxygen of the hydroxyl group of the curcumin and the Lys92 located in the cavity B (Fig. 3D), other interactions are related with residues close to this cavity. The Lys92 residue is essential for the viral replication process, in other works the mutation of this residue in the RdRp of DENV3 abolish viral replication by a decreased in the NS3/NS5 interaction but did not affect the RdRp activity [13].

A possible inhibition of the curcumin to the RdRp domain of DENV2 might be explain by this interaction, and give insights about the posible effects of the curcumin over the inhibition in the production of plaque forming units in DENV2 seen in previous works [4].

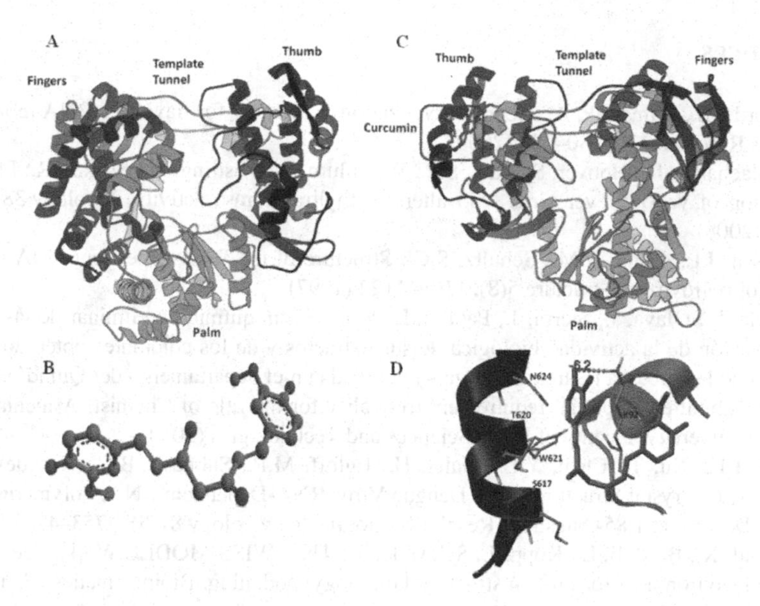

Fig. 3. Molecular modelling and docking between curcumin and DENV2 RdRp domain. **A**: Structural model of the RdRp domain; Thumb subdomain, Palm subdomain, Fingers subdomain, NLS (Nuclear Localization Sequence). **B**: Optimized structure of the curcumin by RM1. **C**: Molecular docking of the curcumin in the cavity B of RdRp domain. **D**: molecular interaction between the oxygen of the hydroxyl group of the curcumin and hydrogen of the amino group of the aminoacid Lys92.

4 Conclusions

The genetic sequence of the RdRp domain of DENV2 was cloned in the expression plasmid pGEX-5X-1, the RdRp domain was express as a fusion protein with GST at 21 hours after 0.1mM IPTG induction.

A structural model of the RdRp domain of DENV2 was built by homology modelling the model show good quality values assessed by ModRefiner, QMEAN6 and Ramachandran plot, with a 0.463 RMSD and a 0.9972 TM-score. The molecular docking between the curcumin and the cavity B show a posible interaction with the residue Lys92 and the curcumin. This interaction may explain the inhibitory effects of the curcumin over DENV2.

Acknowledgements. This work was supported to the scholarship-internship program INNOVATIVE YOUNG RESEARCHERS COLCIENCIAS call 525-2011 (proposal p-2011-0273). We thank the Neuroscience Group from the University of Antioquia for the donation of the pBAC-DENV-FL-GFP-T7-RO8 plasmid.

References

1. Villordo, S., Gamarnik, A.: Genome cyclization as strategy for flavivirus RNA replication. Virus Research 139, 230–239 (2009)
2. Bhattacharya, D., Hoover, S., Falk, S.P., Weisblum, B., Vestling, M., Striker, R.: Phosphorylation of yellow fever virus NS5 alters methyltransferase activity. Virology 380, 276–284 (2008)
3. Hansen, J.L., Long, A.M., Schultz, S.C.: Structure of the RNA-dependent RNA polymerase of poliovirus. Structure 5(8), 1109–1122 (1997)
4. García, L., Olaya, J., Sierra, J., Padilla, L.: Análisis fitoquímico preliminar de las hojas y evaluación de la actividad biológica de sus extractos y de los colorantes obtenidos del rizoma de la cúrcuma (Curcuma longa l.) cultivada en el departamento del Quindío. Degree work submitted as partial requirement to qualify for the title of Chemist. Armenia, Quindio´s University. Faculty of Basic Sciences and Technologies (2010)
5. Yap, T.L., Xu, T., Chen, Y.L., Malet, H., Egloff, M.P., Subhash, B.C., Vasudevan, G., Lescar, J.: Crystal Structure of the Dengue Virus RNA-Dependent RN A Polymerase Catalytic Domain at 1.85-Angstrom Resolution. Journal of Virology 81(9), 4753–4765 (2007)
6. Arnold, K., Bordoli, L., Kopp, J., Schwede, T.: The SWISS-MODEL Workspace: A web-based environment for protein structure homology modelling. Bioinformatics 22, 195–201 (2006)
7. Chopra, G., Kalisman, N., Levitt, M.: Consistent refinement of submitted models at CASP using a knowledge-based potential. Proteins 78(12), 2668–2678 (2010)
8. Bhattacharya, D., Cheng, J.: 3Drefine: Consistent Protein Structure Refinement by Optimizing Hydrogen-Bonding Network and Atomic-Level Energy Minimization. Proteins: Structure, Function and Bioinformatics 81(1), 119–131 (2013)
9. Zhang, J., Liang, Y., Zhang, Y.: Atomic-Level Protein Structure Refinement Using Fragment-Guided Molecular Dynamics Conformation Sampling. Structure 19(12), 1784–1795 (2011)
10. Xu, D., Zhang, Y.: Improving the Physical Realism and Structural Accuracy of Protein Models by a Two-step Atomic-level Energy Minimization. Biophysical Journal 101, 2525–2534 (2011)
11. Morris, G., Huey, R., Lindstrom, W., Sanner, M., Belew, R., Goodsell, D., Olson, A.: Autodock4 and AutoDockTools4: automated docking with selective receptor flexiblity. J. Computational Chemistry 16, 2785–2791 (2009)
12. Benkert, P., Biasini, M., Schwede, T.: Toward the estimation of the absolute quality of individual protein structure models. Bioinformatics 27(3), 343–350 (2011)
13. Zou, G., Chen, Y.L., Dong, H., Lim, C.C., Yap, L.J., Yau, Y.H., Sochat, S.G., Lescar, J., Shi, P.Y.: Functional Analysis of Two Cavities in Flavivirus NS5 Polymerase. Journal of Biological Chemistry 286, 14362–14372 (2011)

In Silico Analysis for Biomass Synthesis under Different CO_2 Levels for *Chlamydomonas reinhardtii* Utilizing a Flux Balance Analysis Approach

David Orlando Páez Melo[1], Rossmary Jay-Pang Moncada[1], Flavia Vischi Winck[1,2], and Andrés Fernando González Barrios[1]

[1] Grupo de Diseño de Productos y Procesos (GDPP), Departamento de Ingeniería Química, Universidad de los Andes Bogotá, Colombia
{do.paez647,r-jay-pa,andgonza}@uniandes.edu.co
[2] Laboratório Nacional de Biociências (LNBio), Centro Nacional de Pesquisa em Energia e Materiais, Brazil
flavia.winck@lnbio.cnpem.br

Abstract. Flux Balance Analysis (FBA) is an approach used to study biochemical networks and predict the flow of metabolites through a metabolic network model. Since the microalgae *Chlamydomonas reinhardtii* has been targeted as a model organism to understand the capabilities of microalgae and to evaluate the feasibility to derive products, this project was aimed to simulate via FBA the effects of different CO_2 inputs (at 2.5%, 5%, 8% and 10% in air) on the growth rate and compare the results with experimental data. For our analysis, the iRC1080 model given by Chang *et al* [9], consisting of 2180 reactions and 1068 metabolites, was validated and fitted to experimental conditions and finally a sensitivity analysis was performed to identify the most sensitive reactions (and genes) to the modifications of CO_2 levels. Our experimental results showed that the growth rate was saturated at the all working conditions, biomass increased significantly to high CO_2 concentration, being the 10%-condition the highest biomass production value (0.463 +/-0.202 g/L), nearly 334% greater compared to cells cultured under reference-CO_2 concentration (air), and it was also noticed that cells increased their production capacity at high CO_2 levels. FBA simulations were unable to reproduce the saturation trend of the experimental data due to the model is composed of linear equations that only increase the objective function; nevertheless the magnitudes of the values were consistent (about 0.03 h^{-1}), indicating that nonlinear approaches are necessary to describe better models. Finally 87 reactions with associated genes were identified as sensitive to the CO_2 input opening the possibility of improvement studies.

Keywords: Flux Balance Analysis, Chlamydomonas, Biomass, Biotechnology.

1 Introduction

Microalgae have the potential to be a great candidate for the challenges of biotechnology in a number of areas including nutrition, aquaculture, pharmaceuticals, and

L.F. Castillo et al. (eds.), *Advances in Computational Biology*,
Advances in Intelligent Systems and Computing 232,
DOI: 10.1007/978-3-319-01568-2_40, © Springer International Publishing Switzerland 2014

biofuels thanks to the advances in engineering and molecular biology. Now, genetic manipulation is possible and a best understanding of metabolic routes is being developed in order to achieve their full processing capabilities [1]. The question is now, why to work with microalgae instead of bacteria or other microorganism? Although the answers may have different perspectives, in here microalgae is shown as an important unicellular eukaryotic organism, able to synthesize macromolecules through photosynthesis using the sunlight and CO_2 to produce mostly proteins, lipids, carbohydrates and pigments, constituting what is known as biomass [2].

Perhaps the current greatest interest is to produce biofuels from algal components and day to day is gaining considerable attention due to the higher efficiency regarding biofuels from crops plants, such as soya and canola [3]. Biomass can be transformed into fuels by different chemical processes [4]. The microalgae *Chlamydomonas reinhardtii* has been targeted as a model organism to comprehend gene expression regulation and to identify gene function [2], although this microalga is not considered as a good lipid-storing, it was found that under N-starvation this capability is favored [5]. Therefore, in order to increase biomass production, currently, several reports are aimed to maximize the content of biofuel precursors by changing the cultivation conditions, identifying chemical inducers of metabolic intermediates, implementing multistage growth systems and making mutagenesis in specific parts of the genome [1]. One proposed way to evaluate and simulate metabolic changes occurring in microalgae cells consists of Flux Balance Analysis (FBA), presented as a mathematical tool to understand metabolic pathways. It considers all known metabolic reactions in an organism and the genes that encode each enzyme; thereby it makes possible to predict the growth rate of an organism or the rate of production of a metabolite with biotechnological importance [6]. FBA is used to predict fluxes by using linear programming with the knowledge of reaction stoichiometry, biomass composition and additional constraints, such as limits on uptake/excretion rates and thermodynamic constraints [7]. Many attempts have been published trying to describe the metabolic pathways for Chlamydomonas, however only primary metabolism has been well documented [7-8]. Recently, Chang *et al* have reconstructed a genome-scale metabolic network for *C. reinhardtii* based on the organism's metabolism and genome annotation providing 2180 reaction, 1068 unique metabolites and encompasses 83 subsystems distributed across 10 compartments [9].

2 Methods and Materials

2.1 Cell Culture and Harvest

The *Chlamydomonas reinhardtii* strain was cultivated in HSM medium without acetic acid addition at 27° C. Experiments comparing cells cultivated under different CO_2 concentration at 0.04%, 2.5%, 5%, 8% and high CO_2 (10 % in air) were performed in a bioreactor R'ALF Plus solo 6.7 L (Bioengineering, Inc., USA) in photoautotrophic conditions under constant illumination with cool-white fluorescent light (average 400 μE/m2s), with continuous shaking at 60 rpm, in a batch mode with starting culture volume of 4 L in opened system and no pH correction. Inoculation was carried out with a 10 mL pre-inoculum taken from a sample at steady state. Cells were harvested at the late exponential phase of cell growth O.D.750 nm ~0.9 for biomass growth and dry weight determination. Cell growth monitoring was performed daily with measurements of O.D. at 750 nm and cell counting using Neubauer counting chamber. For quantification of dry weight, cell pellet from 150 mL culture was washed three times with 2 mL de-mineralized water. Petri dish containing the biomass were kept overnight at 90° C and the dry weight was calculated as the difference in the weight of plates with and without biomass.

2.2 Growth Simulations at Steady State

Based on the iRC1080 model given by Chang *et al* [9], the metabolic network was reconstructed in the stoichiometric matrix and adjusted; thus, the boundaries of the fluxes vector were fixed according experimental condition growth. Irreversible reactions were treated as independent reactions, acetate source inputs were eliminated, light condition was fixed at cool-white fluorescent (57.54 mE/gDW.h, equivalent to 400 μE/m2s), the rate of starch degradation was settled to represent aerobic respiration in light [9] and the CO_2 fluxes were changed to evaluate each condition. Biomass function was optimized by simulation procedures consisted of FBA [6] and was implemented in Xpress IVE® by setting constraints as follow:

$$\max_Z c^T v \quad (1)$$
$$\textit{subject to } \boldsymbol{Sv} = \boldsymbol{0}$$
$$LB \leq \boldsymbol{v} \leq UB$$
$$c^T \in R^n \mid c^T = [0\,0\,0 \ldots 1 \ldots 0\,0\,0] \mid pos(1) = \textit{Biomass reaction}$$
$$\boldsymbol{v} \in R^n$$
$$\boldsymbol{S} \in R^{m \times n}$$
$$LB \in R^n$$
$$UB \in R^n$$
$$n: \textit{number of reactions} = 2893$$

$$m: \textit{number of metabolites} = 1706$$
$$\boldsymbol{v} = \{v_{R1}, v_{R2}, v_{R3}, \ldots, v_{R2893}\}$$
$$R_i = \sum_{j=1}^{m} a_{ij} * M_j[k]$$
$$M_j[k] = j - \textit{Metabolite in the } k - \textit{compartment}$$

Where S is the stoichiometric matrix, v is the flux vector, LB: lower bound, UB: Upper bound, C is a vector of zeros that sets the objective function.

For our simulations, biomass experimental data were used to calculate CO_2 fluxes and based on the resistance to mass transfer phenomenon that occurs from the source of CO_2 into the cell, this approach was used:

$$\textit{Flux } CO_2 = K_L a(C^* - C) \quad (2)$$

Where $K_L a$ is the volumetric transfer coefficient that was calculated by Carvalho *et al* [10]. For their study, the calculated value for $K_L a$ was 7.0×10^{-3} min^{-1} and K_L 1.11 x 11^{-4} m/s, when CO_2 was bubbled into the culture. On the other hand, Badger *et al* reported concentrations inside cell [CO_2 - HCO_3-] of 0.95mM for low-CO_2 and 0.5 mM for high-CO_2 [11]. Also it was necessary to modify the reaction coefficients associated with light flux in order to increase the percentage of light metabolically useful by increasing the photon requirement. The experimental growth rates were calculated as previously described [12] based on O.D. and cell count.

2.3 CO_2-Sensitive Genes

In order to define potential genes that could play a big role in CO_2 level response, it was defined a variation flux coefficient, ρ, that identifies what reactions are being significantly affected under different CO_2 fluxes via FBA. This coefficient was set to be significant when $\rho \geq 0.01$. For this analysis, the whole metabolic network was examined and classified based on the proposed criterion at 0%, 2%, 8%, 5% and 10% CO_2 concentration. Fluxes for two conditions were subtracted ($\Delta \gamma_{Ri}$) (0% and 2.5%, 2.5% and 5%, so forth) and normalized by the Δflux associated with the CO_2 input to the system ($\Delta \gamma_{CO2\ input}$)

$$\rho = \frac{\Delta \gamma_{Reaction\ i}}{\Delta \gamma_{CO2\ input}} \tag{3}$$

For the identified reactions, protein-coding genes were associated by using KeGG database [13].

3 Results

3.1 FBA Approach Yields Linear Results to Increased CO_2

Our experimental results revealed that cells cultivated under different CO_2 condition showed no significant differences in cell growth rates when assessed by cell counting and absorbance measurements. However, it is shown that at high concentrations of CO_2 the biomass production increases considerably, achieving its highest values at 10% CO_2 (dry weight 0.46333 +/-0.2025 g/L). There is a biomass increment of approximately 334% compared to cells cultured under reference-CO_2 concentration (air) as shown in Figure 1. The FBA model was validated according to the results given by Chang *et al*, when growth rate was predicted and compared with experimental data at different photon flux [9]. FBA results show that there is a linear tendency at CO_2 input increments. At higher CO_2 fluxes, the objective function magnitude increases proportionally and shows that the growth rate (h^{-1}) at 10% CO_2 is an order of magnitude greater when compared with the condition of 2.5% as shown in Figure 2.

3.2 Fluxes Associated to Reactions from Mitochondrial Transport and Glycolysis/Gluconeogenesis are the Most Affected in Number at Changing CO_2 Conditions

We found 87 reactions through the sensibility that showed a significant flux variation when changing the CO_2. It was shown that the CO_2 input affects different metabolic pathways, including the synthesis of amino acids, glycerolipids, glycolysis and carbohydrate degradation, fixation of carbon, etc. In table 1, it is presented the number of reactions associated with metabolic pathways; likewise some of the identified genes (see Supplemental Support 1).

Table 1. Protein-coding genes affected by changes in CO_2

Metabolic route	No.	Associated genes	Metabolic route	No.	Associated genes
Transport in mitochondria	12	AOC6, AOT7, MITC14, MPC1, MITC14	Purine metabolism	2	
Transport in chloroplast	6	AOT7, MIP1, MIP2, MOT20	Pentose phosphate pathway	2	RPI1, RPE1
Transport in glioxysome	1		Oxidative phosphorylation	1	ATP2, ATP3, ATP4, ATP5, ATP6, ATP12B, ATP15
Transport in *eyespot*	3		Glyoxylate metabolism	1	GLYK
Extracellular transport	4	MIP1, MIP2, PAT3, PAT4, PAT1, TB5	Glycolysis/gluconeogenesis	10	PYK1, PYK5, PGK1, PGH1, PGM2, PGM5
Extracellular exchange	5		Glycine/serine/threonine metabolism	8	
Demand reactions	2		Glycerolipid metabolism	3	AGA1, DGD1, LCI28
Tricarboxylic acid cycle	3	ACH1, MDH3, CIS1	Galactose metabolism	1	SNE3
Starch/sucrose metabolism	4	STA1, STA6, UGP1, GAD1, UGD1	Glutamate metabolism	1	GLN1
Retinol metabolism	2		Fructose and mannose metabolism	1	FBP1
Pyruvate metabolism	7	ADH1, PAT1, PAT2, AST3, AST1	Carbon fixation	2	PRK1, RBCS1, RBCS2A, RBCS2B
Pyridine metabolism	3	FAP67, FAP103, RSP23	Metabolism of amino sugars and nucleotides	1	
			Butanoate metabolism	2	

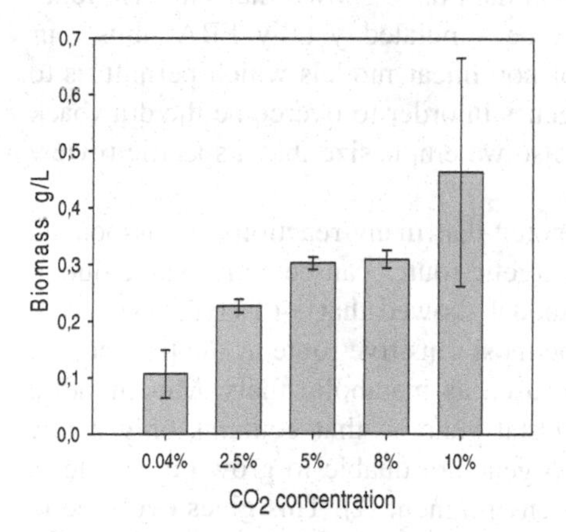

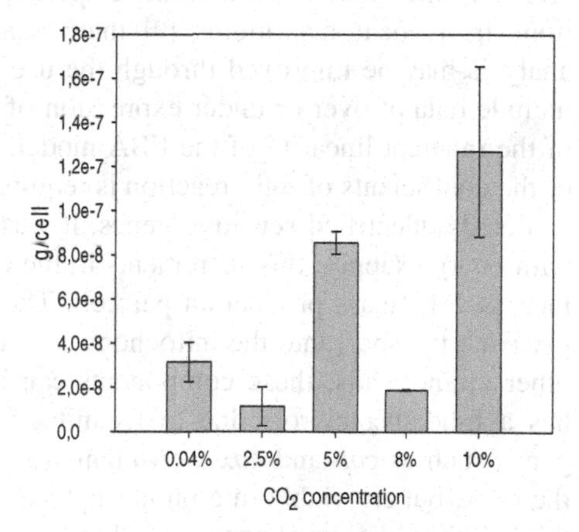

Fig. 1. Biomass results at different CO_2 concentrations in autotrophic condition. Left: dry weight (g/L). Right: biomass per cell (g/cell).

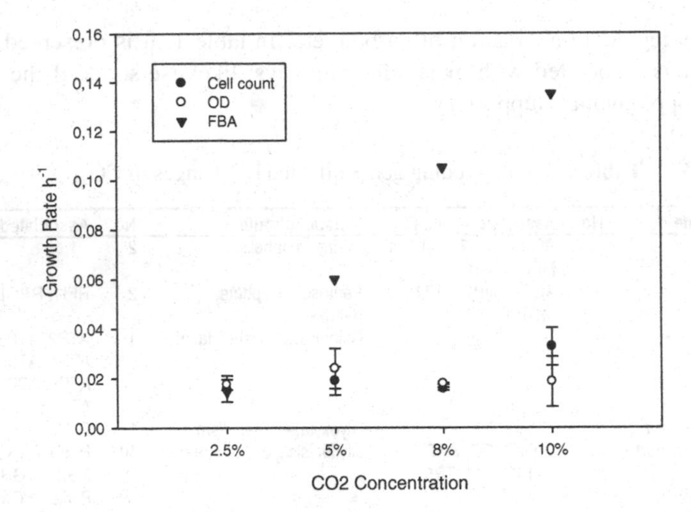

Fig. 2. FBA Results. Growth rate ($gDW{*}h^{-1}$) is compared with experimental data at different CO_2 concentrations in autotrophic condition.

4 Discussion

Our experimental results suggest that the cells are increasing its production capability towards biomass, while the growth rate has reached a boundary or achieved a saturation level. Although the FBA model does not describe this tendency, the magnitudes of the values are consistent. This can be explained because the model is resolved by linear optimization and a change in some parameters represents a proportional increase in the objective function. Experimental data have shown that biomass functions have local maximums [9] that cannot be simulated yet by FBA; thus, our analysis may be improved through the use of non-linear models which permitt us to include data of over or under expression of genes in order to overcome the drawback of the inherent linearity of the FBA model. Also we emphasize that a specific review of the coefficients of light reaction is required.

For the identified sensitive genes, it was noted that many reactions are associated with energy supply; this increments in the energetic routes can be conceivable due to increased biomass production per cell. The model showed that 3-(4-Hydroxyphenyl) pyruvate transport into the mitochondria is the most sensitive route as the transport of other amino acids; these components can be used as metabolic fuels. Mitochondria has a fundamental role due to its mitochondrial genome that contains only a few genes such as *cob* and *cox1*. Mutants to *cob* gene are unable to grow on acetate in the dark, but are viable in a phototrophically environment [2]. This gives evidence to think about the importance of these routes. The reactions related to glycolysis metabolism showed also high variations indicating possible arguments to the previous hypothesis.

5 Conclusion

FBA model has certain limitations but is still a robust tool to predict cellular components fluxes based on a mass balance and others constraints. Although our experimental conditions were not good represented by simulation data, it is suggested that there was saturation for the growth rate at working conditions indicating the need for nonlinear models. It is proposed to reevaluate the coefficients associated with the photons and sensitive genes to fully validate the results. Research has shown that the CO_2 fixation in Chlamydomonas is altered by the external condition and that a mechanism called the Carbon Concentrating Mechanism (CCM) is activated at low-CO_2 concentration as a storage system [11]. This process can be used at high-CO_2 to increase cellular performance towards biomass production. For this perspective, it is necessary to identify the mainly genes and to combine genetic manipulation techniques in order to generate more productive strains. Our sensitive genes may be great candidates on further studies about enhancement of biomass production.

References

1. Rosenberg, J., et al.: A green light for engineered algae: redirecting metabolism to fuel. Curr. Opin. Biotechnol. 19, 430–436 (2008)
2. Harris, E.H.: Chlamydomonas as a model organism. Annu. Rev. Plant Physiol. Plant Mol. Biol. 52, 363–406 (2001)
3. Savage, N.: Algae: The scum solution. Nature 474, S15–S16 (2011)
4. Matsumoto, M., et al.: Saccharification of marine microalgae using marine bacteriafor ethanol production. Appl. Biochem. Biotechnol. 105, 247–254 (2003)
5. Miller, R., et al.: Changes in transcript abundance in Chlamydomonas reinhardtii following nitrogen deprivation predict diversion of metabolism. Plant Physiol. 154(4), 1737–1752 (2010)
6. Orth, J., Thiele, I., Palsson, B.: What is flux balance analysis? Nat. Biotechnol., 245–247 (2010)
7. Boyle, N., Morgan, J.A.: Flux balance analysis of primary metabolism in Chlamydomonas reinhardtii. BMC Systems Biology 3, 4 (2009)
8. Kliphuis, A., Martens Dirk, E., Marcel, J., Wijffels, R.: Effect of O2:CO2 ratio on the primary metabolism of Chlamydomonas reinhardtii
9. Chang, R., et al.: Metabolic network reconstruction of Chlamydomonas offers insight into light-driven algal metabolism. Mol. Syst. Biol. 7, 1–13 (2011)
10. Carvalho, A., Meireles, A., Malcata, F.: Microalgal Reactors: A Review of Enclosed System Designs and Performances. Biotech. Progress, 1490–1506 (2006)
11. Badger, M.R., Kaplan, A., Berry, J.A.: Internal inorganic carbon pool of Chlamydomonas reinhardtii: Evidence for a carbon dioxide-concentrating mechanism. Plant Physiol. 3, 407–413 (1980)
12. Sorokin, C., Krauss, R.: The Effects of Light Intensity on the Growth Rates of Green Algae Plant. Physiol. 33(2), 109–113 (1958)
13. Kanehisa, M.: Post-genome Informatics. Oxford University Press (2000)

Analysis of Metabolic Functionality and Thermodynamic Feasibility of a Metagenomic Sample from "El Coquito" Hot Spring

Maria A. Zamora[1], Andres Pinzón[2], Maria M. Zambrano[3], Silvia Restrepo[4], Linda J. Broadbelt[5], Matthew Moura[5], and Andrés Fernando González Barrios[1]

[1] Grupo de Diseño de Productos y Procesos (GDPP), Departamento de Ingeniería química, Universidad de los Andes, Bogotá, Colombia
{ma.zamora72,andgonza}@uniandes.edu.co

[2] Centro de Bioinformática y Biología Computacional, Manizales, Colombia
andres.pinzon@cbbc.org.co

[3] GEBIX- Centro Colombiano de Genómica y Bioinformática, Bogotá, Colombia
mzambrano@corpogen.org

[4] Laboratorio de Micología y Fitopatología, Departamento de Ciencias biológicas, Universidad de los Andes, Bogotá, Colombia
srestrep@uniandes.edu.co

[5] Departament of Chemical and Biological Engineering, McCormick School of Engineering and Applied Sciences, Northwestern University, Evanston, Illinois, USA
broadbelt@northwestern.edu, MatthewMoura2015@u.northwestern.edu

Abstract. The study of metagenomic samples is crucial for understanding microbial communities. In this study, genomic samples of the "El Coquito" hot spring were analysed to identify their metabolic functionality, the thermodynamic restrictions and the influence of biogeochemical cycles. The metabolic functionality was determined assigning reactions and enzymes to the metabolic routes. To determinate the reversibility of the reactions we used the group contribution method. We also performed a topological analysis of the network. We found a total amount of 1930 reactions and 130 metabolic pathways. It was determined that at a pH of 3 there was 256 irreversible reactions and that the reactions involved in energy metabolism belonged to Carbon Fixation, Nitrogen and ammonia assimilation, and sulphur reduction. We found that the "El Coquito" metabolic network is a free scale network and that the clustering coefficients vary if the thermodynamic restrictions are included.

Keywords: Metabolic reconstruction, Thermodynamic restrictions, Network topology.

1 Introduction

Metagenomics consists on the Genome-Based analysis of entire communities on diverse ecological contexts, such as soil, ocean, rivers, hot springs [1, 2] or communities associated to the human or animal gut [3]. The metagenomic comparative

L.F. Castillo et al. (eds.), *Advances in Computational Biology*,
Advances in Intelligent Systems and Computing 232,
DOI: 10.1007/978-3-319-01568-2_41, © Springer International Publishing Switzerland 2014

approach has been used to study a wide variety of environments to elucidate the functional potential of nine biomes[4], also the Global Ocean Sampling (GOS) expedition focused in discovering the influence of marine microbes in the biogeochemical cycles in the planet [1]. The metabolic behavior of microbial communities has become interesting and requires identifying the metabolic pathways involved in the whole process, therefore metabolic reconstructions have been developed for this purpose. Then the appearing of algorithms capable of reconstructing metagenomic networks has been increasing over time with pipelines such as MGRAST [5] or methodologies such as the pipeline proposed by Pinzon in 2008 [6].

There are several factors that must be considered in a metabolic reconstruction, like variations in sample composition [2], extreme temperatures [7], extreme salinity conditions[2], variations in pH [8] compartmentalization of the system and thermodynamic feasibility of the reactions[9].

The thermodynamic feasibility of a metagenome derived from experimental data is highly restrained due to the lack of data available for biological systems. Nevertheless, there exist approximations applied for biological systems. Mavrovouniotis and collaborators [10] utilized the contribution method to calculate the standard Gibbs free energy change of reaction ($\Delta_r G'^{\circ}$) and the standard Gibbs free energy of formation ($\Delta_f G'^{\circ}$) of the compounds present in the Kyoto Encyclopedia of Genes and Genomes (KEGG) [11]. Moreover, there are expanded group contribution methods available that allow reducing the uncertainty [9]. On the other hand, considering the effect of pH on the metagenome is crucial as acidophilic microorganisms control biogeochemical cycling in a variety of environments; such as hot springs, and acid mine drainages[8].

Systems biology approaches under the network topology fundamentals analysis can provide important information when analyzing metagenomes reconstructions. The biochemical reactions in a cellular metabolism can be integrated into a metabolic network [12]. Topological features like shortest path, connectivity node degrees and node edge metrics have become common investigation tools; on the other hand knowing the directionality of the reactions allows a better reconstruction of the network reducing the number of reactions and restraining some pathways.

Then the aim of this work is to identify the metabolic functionality and analyze the thermodynamic restrictions of the reactions involved, taking into account pH variations of the system and perform a topological comparative analysis of the network considering the thermodynamic restrictions from existing genomic sequenced samples of the "El Coquito" hot spring located at the "Parque Nacional de los Nevados" in Colombia,

2 Methods

2.1 Metabolic Functionality

The genomic data already assembled in a previous work [13] was obtained from GeBiX (Colombian Center for Genomics and Bioinformatics of Extreme Evironments). The sequences were translated with the EMBOSS tool sixpack leaving the wider coverage ORFs. We determined the metabolic functionality with the methodology proposed by Pinzon in 2008 (mTools) [6], the basic principle of this profile reconstruction is HMMER a set of algorithms based on hidden Markov models, that finds a set of enzymes (E.C's) belonging to metabolism present in the metagenome and then compares this set with the KEGG database, the next step assigns a metabolic pathway to each E. C. . Finally it obtains a list of reactions associated to the metabolic pathways and enzymes.

2.2 Reaction Reversibility

To determine the reversibility of the reactions the method of group contribution was used as proposed by Henry [9, 11], in this method the molecular structure of a compound is decomposed into a set of smaller molecular substructures and using a linear model estimates the $\Delta_f G'^{\circ}$ and $\Delta_r G'^{\circ}$.

$$\Delta_r G' = \sum_{i=1}^{m} n_i \Delta_f G_i'^{\circ} + RT \ln\left(\prod_{i}^{m} x_i^{n_i}\right) \tag{1}$$

Where $\Delta_f G_i'^{\circ}$: standard Gibbs free energy of formation of compound i, R: universal gas constant, T: temperature, *m:* number of compounds involved in the reaction, $x_{i:}$ activity of the compound i, and $n_{i:}$ stoichiometric coefficient of compound i in the reaction

Since the cellular activity is on the order of 1mM the method employed uses a reference state of 1mM, leading to :

$$\Delta_r G_{est}'^{m} = \sum_{i=1}^{N_{gr}} n_{gr,i} \Delta_{gr} G_i'^{\circ} + RT \sum_{j}^{m} n_j \ln 0.001 \tag{2}$$

BNICE (Biochemical Network Integrated Computational Explorer) [9] was utilized to carry out the calculation of which of the reactions in the metagenome were reversible based on the group contribution method described above. To evaluate the effects of pH in the calculation, the values of pH were set at 2, 3, 4, 5, 7 and 8.

2.3 Topological Analysis

To perform the topological analysis we used the Cytoscape platform and the Network analyzer app [14], this app computes parameters such as clustering coefficient that represents the cohesiveness around a node, the average path length an important feature that, if short, ensures an efficient reaction to perturbations, and average number of neighbors related to path redundancy also known as the availability of multiple paths between a pair of nodes. To perform the comparison we took two network reconstructions, one taking into account the reversibility of the reactions and the other one assuming that all the reactions are reversible.

3 Results

3.1 Metabolic Functionality

On table 1 we found the most representative sub-metabolisms, which correspond to specific classifications of each metabolism category, evidencing the metabolic composition and functionality of the sample.

Table 1. Metabolic composition and functionality of the El Coquito sample

Type of metabolism	Representative type of sub metabolism	Percentage of reactions associated to the sub metabolism
Carbohydrate	Starch and sucrose	13%
Lipids	Glycerophospholipids and Glycerolipids	14%
Amino acid	Arginine and Proline	16%
Complex amino acids	Seleno Amino acids	28%
Glycans	Peptidoglycans biosynthesis	36%
Vitamins and cofactors	Porphyrin and chlorophyll	20%
Terpenoids and poliketids	Diterpenoids biosynthesis	33%
Synthesis of secondary metabolites	Isoquinoline alkaloid biosynthesis	25%
Xenobiotic degradation	Gama-Hexachlorocyclohexane	15%

The results for energy metabolism are shown in Fig. 1.

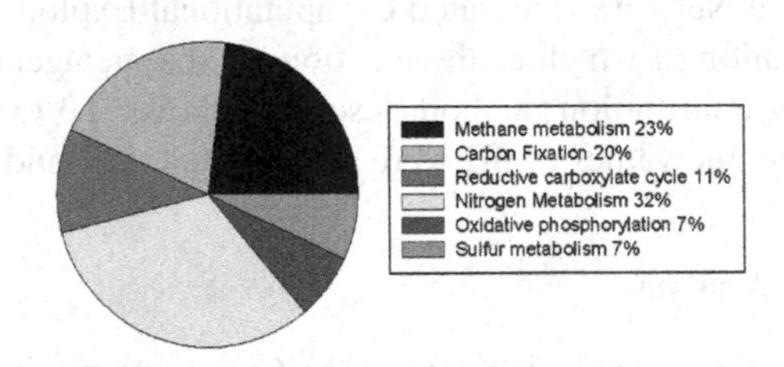

Fig. 1. Metabolic Frequencies in energy Metabolism

3.2 Reaction Reversibility

Out of the 1930 reactions it was found that that an average of 12.33% of the reactions was irreversible at different values of pH (Fig 2).

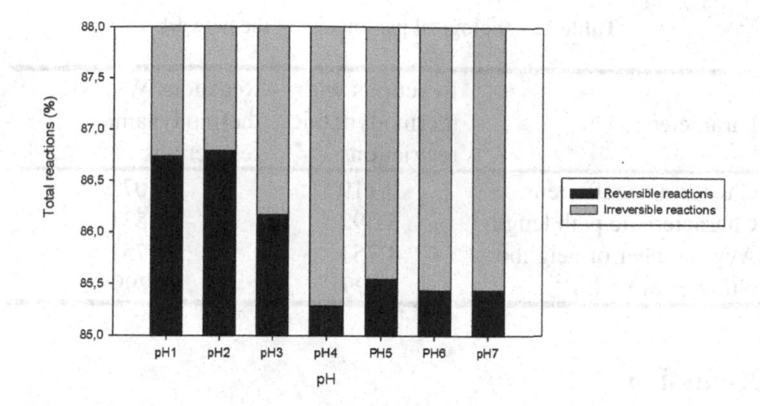

Fig. 2. Total percentage of irreversible reactions at different values of pH

It was found that for a pH of 3, 13 of the 65 reactions belonging to energy metabolism were irreversible, 21 for pH 3 and 16 for the rest of the values of pH.

3.3 Topological Comparison Including Thermodynamic Restrictions

For both cases, with and without thermodynamic restrictions of system we found the same neighbor connectivity as shown in Fig. 3.

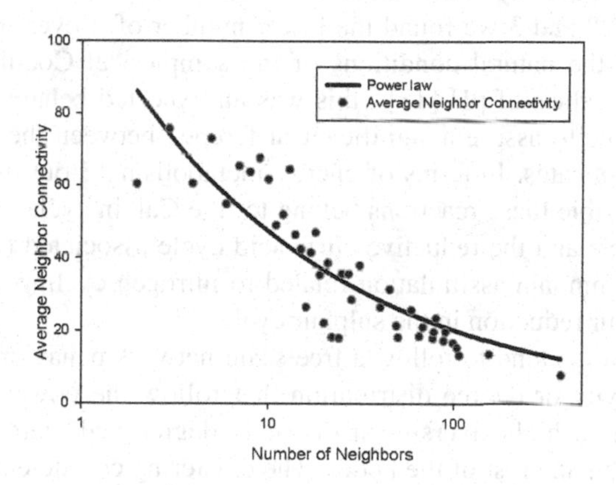

Fig. 3. Neighbor Connectivity

In table 2 there are the general topological parameters of the network.

Table 2. Topological parameters of the network

Parameter	Reactions with thermodynamic restrictions	Reactions Without thermodynamic restrictions
Clustering coefficient	0.119	0.072
Characteristic path length	3.392	3.531
Avg. number of neighbors	4.751	4.751
Number of nodes	1299	1299

4 Discussion

There is a bias in the metabolic identification of metagenomic samples, this is directly related to the fact that the comparisons are made against databases of already sequenced organisms and known proteins. Several studies have shown that despite of the strong environmental heterogeneity in metagenomic samples the metabolic composition frequency tends to remain the same, an example of this is the metabolic composition found in four metagenomic libraries derived from Coorong lagoon sediment [2] and it shows in "El Coquito" sample as well, we were able to found a wide variety of metabolisms from the essential like amino acids and glycans synthesis to xenobiotics degradation, interestingly we found the presence of the gama-hexachlorocyclohexane degradation pathway in the sample.

At pH values of 2 and 3 we found the lower number of irreversible reactions, this is possible due to the natural conditions of the sample, "el Coquito" hot spring is known for its low values of pH [13] , this was an expected behaviour of the system but it is not possible to assure a significant difference between the cases, due to the absence of data replicates. In terms of energy metabolism 13 out of the 56 reactions were found irreversible these reactions belong to: the Calvin cycle, the reductive pentose phosphate cycle and the reductive citric acid cycle associated to carbon fixation, the nitrogen and ammonia assimilation related to nitrogen cycling, and the reactions belonging to sulphur reduction in the sulphur cycle.

Metabolic Networks tend to follow a free scale network behavior represented by a neighbor connectivity or degree distribution that follow the power law (Fig. 3) this means that there is a high diversity in the node degree and there is not a "typical node" that represents the rest of the nodes. The clustering coefficient shows the cohesiveness of the neighborhood of a node, quantifies how close the local neighborhood is to being part of a clique, a region of the graph where every node is connected to every other node [12] in this case we found a higher clustering coefficient for the network with reversibility which could mean that this network possess a higher level of aggrupation between nodes and it is likely to find some nodes belonging to sub graphs that represent particular metabolisms in the ecosystem. Also we found a shortest path length (3.392) for the network reconstructed involving thermodynamic restrictions; this means that the network has a highest level of reaction to perturbations, therefore including the thermodynamic restrictions in the metabolic reconstructions improves the results obtained, giving as a result a network more curated and with an improvement in the parameters calculated (clustering coefficient and shortest path length).

References

1. Gilbert, J.A., Dupont, C.L.: Microbial Metagenomics: Beyond the Genome. Annual Review of Marine Science 3, 347–371 (2011)
2. Jeffries, T.C., Seymour, J.R., Gilbert, J.A., Dinsdale, E.A., Newton, K., Leterme, S.S.C., Roudnew, B., Smith, R.J., Seuront, L., Mitchell, J.G.: Substrate Type Determines Metagenomic Profiles from Diverse Chemical Habitats. PLoS ONE 6, e25173 (2011)
3. Karlsson, F.H., Nookaew, I., Petranovic, D., Nielsen, J.: Prospects for systems biology and modeling of the gut microbiome. Trends in Biotechnology 29, 251–258 (2011)
4. Dinsdale, E.A., Edwards, R.A., Hall, D., Angly, F., Breitbart, M., Brulc, J.M., Furlan, M., Desnues, C., Haynes, M., Li, L.L., McDaniel, L., Moran, M.A., Nelson, K.E., Nilsson, C., Olson, R., Paul, J., Brito, B.R., Ruan, Y.J., Swan, B.K., Stevens, R., Valentine, D.L., Thurber, R.V., Wegley, L., White, B.A., Rohwer, F.: Functional metagenomic profiling of nine biomes. Nature 452 (2008)
5. Logares, R., Haverkamp, T.H.A., Kumar, S., Lanzén, A., Nederbragt, A.J., Quince, C., Kauserud, H.: Environmental microbiology through the lens of high-throughput DNA sequencing: Synopsis of current platforms and bioinformatics approaches. Journal of Microbiological Methods 91, 106–113 (2012)
6. Pinzón, A., Rodriguez-R, L.M., González, A., Bernal, A., Restrepo, S.: Targeted metabolic reconstruction: a novel approach for the characterization of plant–pathogen interactions. Briefings in Bioinformatics 12, 151–162 (2011)
7. Lewin, A., Wentzel, A., Valla, S.: Metagenomics of microbial life in extreme temperature environments. Current Opinion in Biotechnology
8. Jones, D.S., Albrecht, H.L., Dawson, K.S., Schaperdoth, I., Freeman, K.H., Pi, Y., Pearson, A., Macalady, J.L.: Community genomic analysis of an extremely acidophilic sulfur-oxidizing biofilm. ISME 6 (2012)
9. Jankowski, M.D., Henry, C.S., Broadbelt, L.J., Hatzimanikatis, V.: Group Contribution Method for Thermodynamic Analysis of Complex Metabolic Networks. Biophysical Journal 95, 1487–1499 (2008)
10. Mavrovounotis, M.: Estimation of standard Gibbs energy changes of biotransformations. The Journal of Biological Chemistry 266, 14440–14445 (1991)
11. Henry, C., Jankowsky, M., Broadbelt, L., Hatzimanikatis, V.: Genomc-scale thermodynamic analysis of Escherichia coli metabolism. Biophysical Journal 90, 1453–1461 (2006)
12. Albert, R.: Scale-free networks in cell biology. J. Cell Sci. 118, 4947–4957 (2005)
13. Bohorquez, L.C., Delgado-Serrano, L., López, G., Osorio-Forero, C., Klepac-Ceraj, V., Kolter, R., Junca, H., Baena, S., Zambrano, M.M.: In-depth characterization via complementing culture-independent approaches of the microbial community in an acidic hot spring of the Colombian Andes. Microb. Ecol. 63, 103–115 (2012)
14. Doncheva, N.T., Assenov, Y., Domingues, F.S., Albrecht, M.: Topological analysis and interactive visualization of biological networks and protein structures. Nat. Protocols. 7, 670–685 (2012)

Identification of Small Non-coding RNAs in Bacterial Genome Annotation Using Databases and Computational Approaches

Mauricio Corredor[1] and Oscar Murillo[1,2]

[1] Genética y Bioquímica de Microorganismos (GEBIOMIC), Instituto de Biología, Universidad de Antioquia, Medellín, Colombia
mcorredor@matematicas.udea.edu.co
[2] Grupo Genética y Sociedad, Facultad de Medicina, Universidad Cooperativa de Colombia, Medellín
oscar.murillo@campusucc.edu.co

Abstract. RNA genes are unquestionable today, non-coding RNA is functional but its classification at the present is complex due to lack of computational tools. The vast progress in computer science for DNA and protein is not enough to resolve folding and function of RNA. Fortunately, computational tools for solving RNA concerns are in progress: web source as Centroid Homofold, CentroidFold (CBRC), Rfam (Sanger-HHMI, Janelia farm), sRNAdb (MGIL), RNApredator (Vienna RNA web server), TargetRNA2 (Wellesley College), Noncoding RNAdatabase (IBC), Mfold (CAS) and RNAcon (IMTC) are quickly supplying the bioinformatics gaps. In this work was used those tools to fill and appoint the intergenic annotation in the *Leuconostoc mesenteroides* bacterium, recently sequenced in 454 Roche. More than 2000 intergenic sequence were run on the mentioned tools. Various ncRNA were classified as Mir-(#)s, many T-Boxes, various L(#) leaders and some ones TPPs, yybp-ykoY and ykkC-yxkD, between others. Other interesting structures without matching in Rfam, ncRNA databases were annotated as hypothetical ncRNA.

Keywords: ncRNA, sRNA, fRNA, RNA folding, *Leuconostoc mesenteroides*.

1 Introduction

Secondary and tertiary structures prediction of RNA folding is a key point in molecular biology. There have been progresses in this area, but the accuracy of prediction from RNA sequence is still limited. In many cases, however, homologous RNA sequences are available with the target RNA sequence whose secondary structure is to be predicted [1].

In bacteria non-coding RNA (ncRNA) is a small functional RNA molecule (50-250 nucleotides), whose is transcribed but is not translated into a protein. The term small RNA (sRNA) is used for identify short bacterial ncRNAs. Less often used synonyms

L.F. Castillo et al. (eds.), *Advances in Computational Biology*,
Advances in Intelligent Systems and Computing 232,
DOI: 10.1007/978-3-319-01568-2_42, © Springer International Publishing Switzerland 2014

are functional RNA (fRNA) or small RNA (sRNA) o small non-coding RNA (sncRNA) [2].

Small non-coding RNA (sncRNA) genes play significant regulatory role about diversity of cellular processes, however, prediction of non-coding RNA genes is a great challenge, using either experimental or computational approach, due to the characteristics of sRNAs, which are short in nucleotides, are not translated into proteins and show variable stability [3]. The greater part of sRNAs known in bacteria, have been identified in *Escherichia coli*, which are conserved in closely related bacteria. Shapes of non-coding RNAs (ncRNA) with novel functions remain to be discovered [4].

Little information is published about ncRNA in Lactic Acid Bacteria (LAB), perhaps the most important bacteria in human gut. LAB modulates digestion, immune system and metabolism in human. However, the number of characterized small RNAs (sRNAs) in LABs has steadily increased at the same time with the new bioinformatics RNA tools capable to distinguish structure and function. Nevertheless, experimental evidence will be essential and some works goes in this way. For instance, *Corynebacterium glutamicum* ctRNA (counter-transcribed RNA) is a plasmid encoded non-coding RNA that binds to the mRNA of repB and causes translational inhibition. The ctRNA[5] is encoded by plasmids and have functions in rolling circle replication to hold up a low copy number. In this work was possible annotating some small ncRNA homologous with other bacteria species and found interesting new ncRNA structures unclassified yet.

2 Material and Methods

2.1 Databases

Leuconostoc mesenteroides GBM002 was sequenced in 454 Roche FLX. Initially intergenic regions were highlighted in Artemis 15.0.0 and each intergenic region was matched with Rfam Sanger (WT). Also the same sequences were aligned with Blast tool, in sRNAdb web (MGIL). More than 300 ncRNAs from Gram negative and positive bacteria were used as a pattern to be matched with each intergenic region (≅ 2000 total) from *L. mesenteroides* GBM002. Flanking non-coding regions (5' or 3') from each contig were not analyzed with those tools. Subsequently, in Rfam (Sanger) were selected the ncRNA matching each positive ncRNA sequence. The alignments were verified and each positive (matched) region was selected for drawing secondary structure using CentroidHomofold and CentroidFold (ncRNA.org belongs to CBRC), and RNAcon (IMTC) web tool.

2.2 Homology and Folding

RNA folding structures were predicted using the web software CentroidHomofold (version 1.0) and CentroidFold (version 2.0), BSRD (Bacterial Small Regulatory RNA Database version 1.0) and RNAfold web server (version 1.7). Over 2000 intergenic regions were analyzed to predict the RNA secondary structure. Other computational approaches like HMMER (version3.0) have been applied to develop structures and alignments. Subsequently, intergenic sequences were transfer to sRNAdb (version 1.0) tool with Blast software to be searched against the database and to find sRNA candidates in *L. mesenteroides* ATCC 8293 [6].

2.3 Predict Function

The new ncRNAs identified in database and those RNA hypothetic regions, were analyzed on the web servers TargetRNA2 (version 1.0) and RNApredator (RNAplex version 1.33) using as replicon *L. mesenteroides* genome ATCC 8293. Those programs find target genes where ncRNAs interact. Figure 1, characterizes the steps followed to approach ncRNA functions. Every databases and software tools have been integrated to validate hypothetical structures with possible functions.

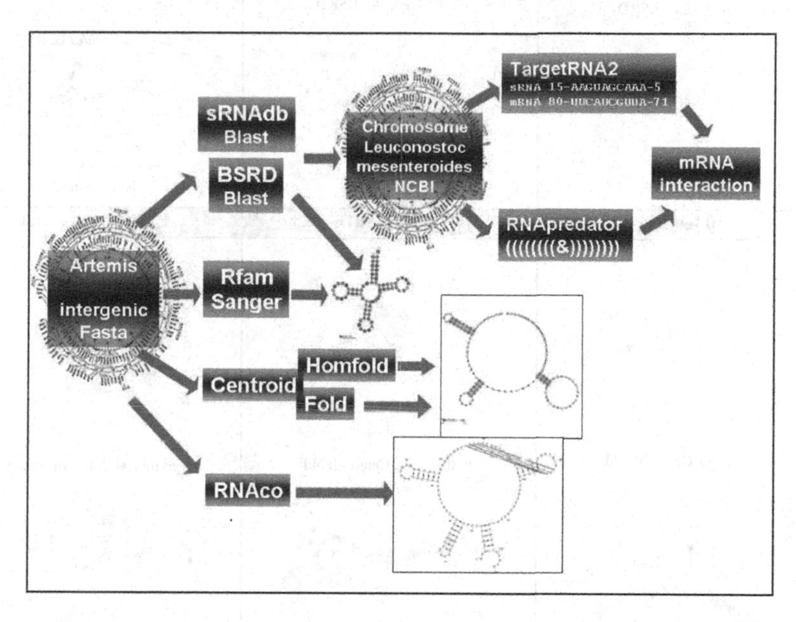

Fig. 1. The workflow chart followed to identify ncRNA folding and approaches theirs functions

3 Results

3.1 Rfam and BSRD Structures

In this work was possible characterize more than 50 sRNA homologous from Rfam database. Those ncRNA are well recognized and previously isolated in *Eschericia coli* and other bacteria. The *L. mesenteroide* GBM002 sequence homologous that hit Rfam ncRNAs, were as follow: Mir-3, Mir-273, Mir-283, L10 leader, L13 leader, yybp-ykoY, ykkC-yxkD, TPP, Bacterial Small SRP, ctRNA-pND324, Lysine riboswitch, and multiple T-Box, 5S, small SRP between others. The sequence from

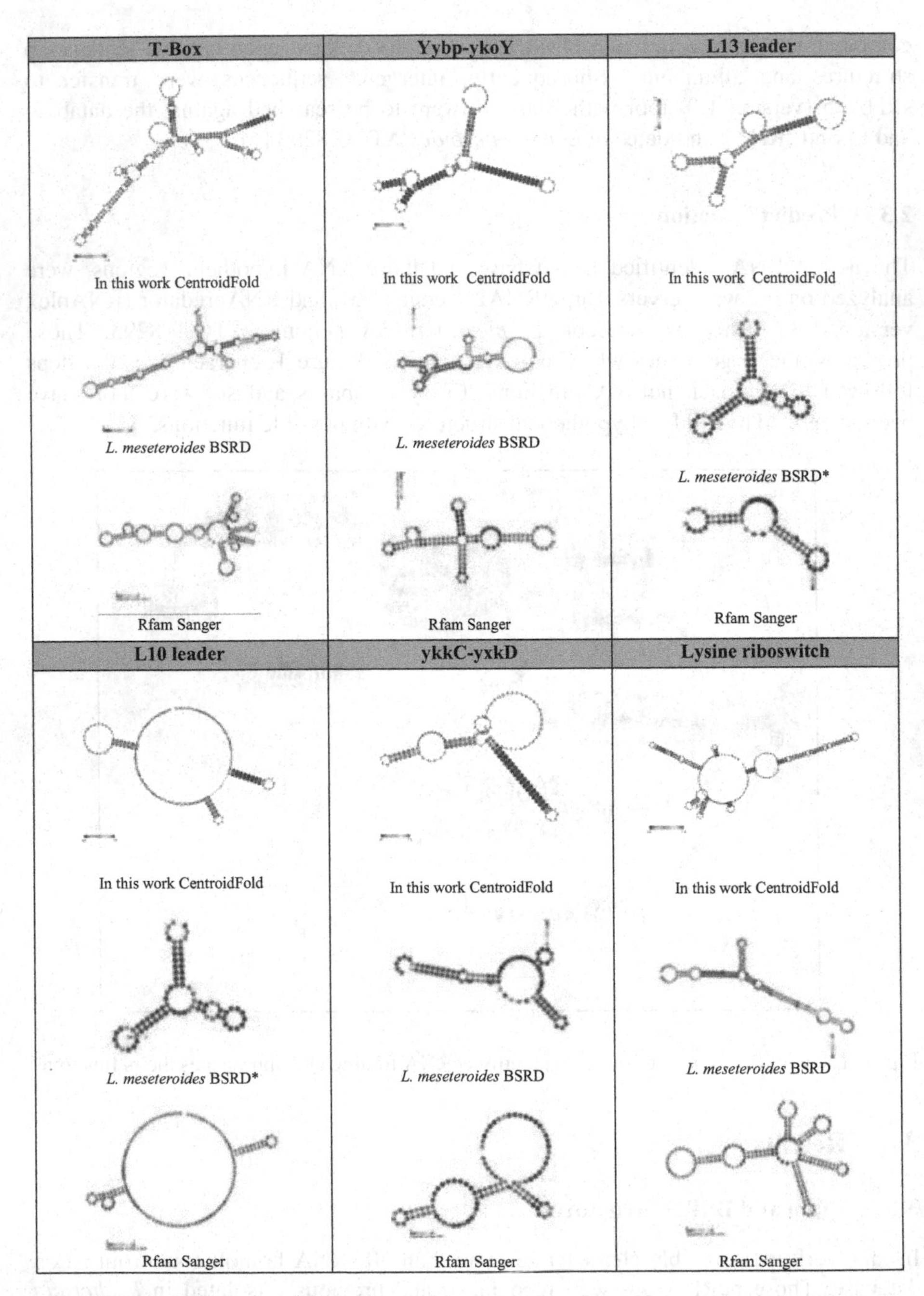

Fig. 2. Some examples of ncRNA structures analyzed in this work. The first structure (above) in each cell corresponds to an integenic sequences from *L. mesenteroides* GBM002 folding in CentroidFold. The second structure (bottom) corresponds to the same sequence running in the Blast BSRD that aligns with *L. mesenteroides* ATCC 8293. The third structure bottom is the same sequence matching Rfam. L10 and L13 leaders had the same structure in BSRD database.

each structure from Rfam was run over Blast BSDR. Figure 2, shows the examples T-Box, yybp-ykoY, L13 leader, L10 leader, ykkC-yxkD and Lysine riboswitch. We draw up more than 400 structures from those 50 well established with Rfam.

Some hypothetic ncRNA structures for *L. mesenteroides* ATCC 8293 as L10 leader and L13 leader are the same in BSRD database (Figure 2). In CentroidFold and CentroidHomolf was run different options (CONTAlign, probcons, different gamma and Rfam.full.99 between other), to get suitable structures, according to Rfam and BSDR. We analyze and compare the results with RNAcon to select the best scores with best folding. In some cases our structures are according with the structure of Rfam or BSDR, of course without any conclusions, because our structures are not always identical to Rfam, considering for us the master database.

3.2 ncRNA Function

The TargetRNA2 and RNApredator tools allow us to study, both, sequences well identified and those hypothetical structures. For example TargetRNA2 targeted between 2 and 10 different mRNA for *L. mesenteroides* ATCC 8293. At the other hand, RNApredator produced a large table with multiple alignments reaching more than 4000 possibilities for each sequence of *L. mesenteroides* ATCC 8293 drawing interaction (dot bracket). Those results could be post processed, throwing gene ontology ID with molecular function, biological process and cell component. It is impossible to report here those large results.

4 Discussion

In *E. coli* alone, have been identified more than 70 sRNA genes. Those bacterial sRNAs whose functions were characterized and can be sorted into three general categories: sRNAs that have intrinsic catalytic activity or are components of ribonucleoproteins, sRNAs that affect protein activity by structurally mimicking other nucleic acids and sRNAs that post-transcriptionally regulate mRNAs via basepairing interactions [7,8]. We characterized more than 2000 intergenic regions that could be fRNA, obviously without experimental evidence is impossible to confirm the function. However, we found that the different tools used in this work and our bioinformatics approach (Figure 1) allowed us annotate new sRNA genes in *L. mesesteroides* GBM002. In fact, allowed us setting up other relationships with annotate DNA genes, integrating them with new molecular cell functions.

The sRNA and ncRNA databases are still in construction and not always connected with the large NCBI, EBI and DDBJ databases. Interesting projects with reliable programs used in those works are in development and rapidly growing with the aim to supply these gaps in the three databases. Those tools are a value added to our *L. mesesteroides* GBM002 genome, because we include 50 or probably more genes in our genome annotation. At the other hand, it seems that some part of work with the *L. mesenteroides* ATCC 8293 ncRNA annotation is still in progress: such as, output structure for L10 and L13 in BSRD database was the same structure, well distinguish and different in Rfam, indicating that those analysis needs to be restored for improve a better prediction.

Genes that produce ncRNA cannot be detected by proteins genefindings algorithms. There is nothing in RNA genes as strong as the codon bias, hexamers frequency, and open reading frames signals exploited by protein genefinders [9]. But today this bioinformatics gap goes in the right way, with those tools applied in our work, and probably some ones could be the essential for new era: functionomics [10,11] or better known system biology.

Acknowledgments. We thank to Bioinformatics School, Gabi-CIDBIO, Bogota and Apolo server belongs to Eafit University, Medellin.

References

1. Hamada, M., Sato, K., Kiryu, H., Mituyama, T., Asai, K.: Predictions of RNA secondary structure by combining homologous sequence information. Bioinformatics 25(12), i330–i338 (2009)
2. Hershberg, R., Altuvia, S., Margalit, H.: A survey of small RNA-encoding genes in Escherichia coli. Nucleic Acids Res. 31, 1813–1820 (2003)
3. Tjaden, B., Goodwin, S.S., Opdyke, J.A., Guillier, M., Fu, D.X., Gottesman, S., Storz, G.: Target prediction for small, noncoding RNAs in bacteria. Nucleic Acids Res. 34, 2791–2802 (2006)
4. Hébrard, M., Kröger, C., Srikumar, S., Colgan, A., Händler, K., Hintons, J.: RNAs and the virulence of Salmonella enterica serovar Typhimurium. RNA Biol. 9(4), 437–445 (2012)
5. Venkova-Canova, T., Patek, M., Nesvera, J.: Control of rep gene expression in plasmid pGA1 from Corynebacterium glutamicum. J. Bacteriol. 185(8), 2402–2409 (2003)
6. Tjaden, T.: Biocomputational Identification of Bacterial Small RNAs and Their Target Binding Sites. In: Mallick, B., Ghosh, Z. (eds.) Regulatory RNAs: Basics, Methods and Applications. Springer (2012)
7. Gottesman, S.: The small RNA regulators of Escherichia coli: roles and mechanisms. Annu. Rev. Microbiol., 58303–58328 (2004)
8. Storz, G., Gottesman, S.: Versatile roles of small RNA regulators in bacteria. In: Gesteland, R.F., Cech, T.R., Atkins, J.F. (eds.) The RNA World, 3rd edn., pp. 567–594. Cold Spring Harbor Laboratory Press, Cold Spring Harbor (2006) (2006
9. Rivas, E., Eddy, S.: The greater part of sRNAs known in bacteria, have been identified in Escherichia coli, which are conserved in closely related bacteria. Bioinformatics 16(7), 583–605 (2000)
10. Amin, A.R.: A need for a "whole-istic functional genomics" approach in complex human diseases. Arthritis Res. Ther. 5, 76–79 (2003)
11. Neuman, E., Gay, R.E., Gay, S., Müller-Ladner, U.: Functional genomics of fibroblasts. Curr. Opin. Rheumatol. 16, 238–245 (2004)

Structural Modeling of *Toxoplasma gondii* TGME49_289620 Proteinase

Mateo Murillo León, Diego Mauricio Moncada Giraldo, Diego Alejandro Molina, Aylan Farid Arenas, and Jorge Enrique Gómez

Grupo de Estudio en Parasitología Molecular (GEPAMOL)
Gepamol2@uniquindio.edu.co

Abstract. Cysteine proteinases play key roles in host-parasite interactions, including host invasion, parasite differentiation, and intracellular survival. *Toxoplasma gondii* expresses five cysteine proteases, including one cathepsin L-like (TgCPL), one cathepsin B-like (TgCPB) and three cathepsin C-like (TgCPC1, 2 and 3) proteases. We performed the Structural modeling of catalytic domain of TgCPC1 with server I-TASSER, the template selected for homology modeling was Dipeptidyl peptidase I (Cathepsin C) (PDB code: 1JQP). The C-Score of structural modeling of catalytic domain was -0.5; in the L-domain there are nine α-helices and two β-strands and in the right domain there are four β-strands and two α-helices. Cys-440, His-651 and Asn-676, form the cysteine protease catalytic triad in the active site. Adjacent to active site there is a Tyr441 this residue may be involved in the binding of the N terminus of the peptide substrate. A tyrosine residue (Tyr 578) that binds a chloride ion in the crystal structures of rat and human Cathepsin C is also conserved.

Keywords: cathepsin, structural modeling, Toxoplasma gondii, protease.

1 Introduction

Peptidases play a critical role in protein catabolism by hydrolysis of peptide bonds in the polypeptides. Peptidases are classified into the seven categories based on the principal catalytic residue in the active site: Aspartic, Cysteine, Glutamic, Serine, Threonine, Metallo and Mixed, each of which can be further divided into clans and families [1].

Cathepsin peptidases belong to the C1 family clan CA of "Papain-like" cysteine peptidases are widely distributed in eukaryotic organisms. During catalysis, a basic amino acid in the catalytic triad, usually histidine, de-protonates the cysteine thiol group, which attacks the carbonyl carbon group in the substrate for hydrolysis [2].

Cysteine proteinases play key roles in host-parasite interactions, including host invasion, parasite differentiation, and intracellular survival [3]

Toxoplasma gondii expresses five members of the C1 family of cysteine proteases, including one cathepsin L-like (TgCPL), one cathepsin B-like (TgCPB) and three cathepsin C-like (TgCPC1, 2 and 3) proteases. *T. gondii* cathepsins Cs are exopeptidases

L.F. Castillo et al. (eds.), *Advances in Computational Biology*,
Advances in Intelligent Systems and Computing 232,
DOI: 10.1007/978-3-319-01568-2_43,

and remove dipeptides of unblocked N-terminal substrates of proteins or peptides, whereas cathepsin B and L are endopeptidases[4].

CPC1 was the most highly expressed cathepsin mRNA in tachyzoites (by real-time PCR). *T. gondii* cathepsins Cs are required for peptide degradation in the parasiphorous vacuole of transgenic tachyzoites was completely inhibited CPC1 was the most highly expressed cathepsin mRNA in tachyzoites (by real-time PCR). *T. gondii* cathepsins Cs are required for peptide degradation in the parasiphorous vacuole of transgenic tachyzoites was completely inhibited by the cathepsin C inhibitor [4].

Our aim was to determinate the tridimensional structure of catalytic region of TgCPC1 (TGME49_289620) by structural homology.

2 Materials and Methods

The Sequence of the TGME49_289620 protein from *T.gondii* was obtained from Genbank with the accession number: EEB01247. Secondary structure prediction was performed by web-based program namely Psipred [5] and the tertiary structure with the server I-Tasser [6]. The template selected for homology modeling was Dipeptidyl peptidase I (Cathepsin C) of crystallographic model deduced by Olsen *et al.*, 2001. (PDB code: 1JQP). The functional regions were determined with PROSITE [7], SMART [8], SOSUI [9] and NCBI CD [10]. FATCAT [11] was used for alignment of TgCPC1 with TgCPCL (PDB code: 3F75).

The structure was evaluated for its backbone conformation and stereochemical properties by PROCHECK Ramachandran Plot [12]. Visualization was performed with Chimera 1.8 [13].

3 Results

The TGME49_289620 are encoded in chromosome IX, this protein consist of 730 amino acids. The C-Score of structural modeling of catalytic domain was -0.5; the Catalytic domain consists of a L- domain and a R-Domain. The Left (L) domain (Residues: Leu 412 - Asp635) and the Right (R) domain (Residues: Leu 636- Met 730). There are nine α-helices and two β-strands in the L-domain and in the right domain there are four β-strands and two α-helices.

Cys-440, His-651 and Asn-676, form the cysteine protease catalytic triad in the active site. Adjacent to active site there is a Tyr441 this residue may be involved in the binding of the N terminus of the peptide substrate. A tyrosine residue (Tyr 578) that binds a chloride ion in the crystal structures of rat and human Cathepsin C is also conserved.

In our model the Asp1, which is a key residue that is involved in docking the substrate via interaction with the amino group and is conserved is all cysteine proteases, is far to the active site, but the Asp386 was found in this position and the propeptide occupied the deep active site cleft.

Results of FACAT showed that TgCPC1 and TgCPCL are significantly similar (P-value =3.41e-09), and they has 216 equivalent positions with an RMSD of 2.11. In this alignment we found that CPC1 have a stabilizing disulfide Bond (Cys-437 to Cys-508) that is highly conserved among other papain-like cysteine.

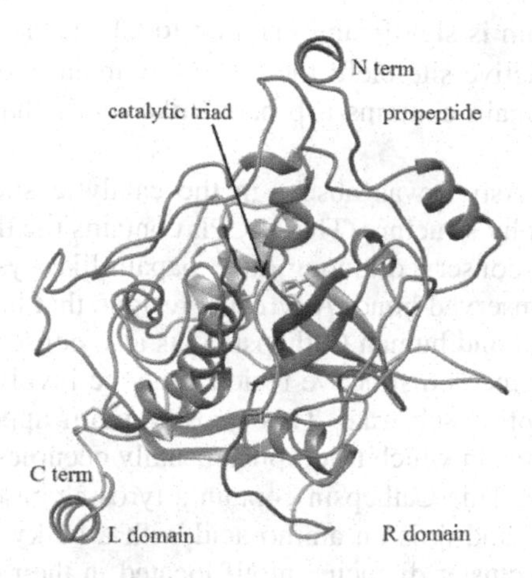

Fig. 1. Structure of TgCPC1 in complex with its propeptide. Stereoview looking into the active site cleft with the left (*L*) domain on the *left,* the right (*R*) domain on the *right*, and the propetide on the *top*. N- and C-terminal residues of each polypeptide are labeled. The catalytic triad (Cys_{440}, His_{651}, and Asn_{676}) with side chains shown as *sticks, disulfide bonds is also shown as sticks.*

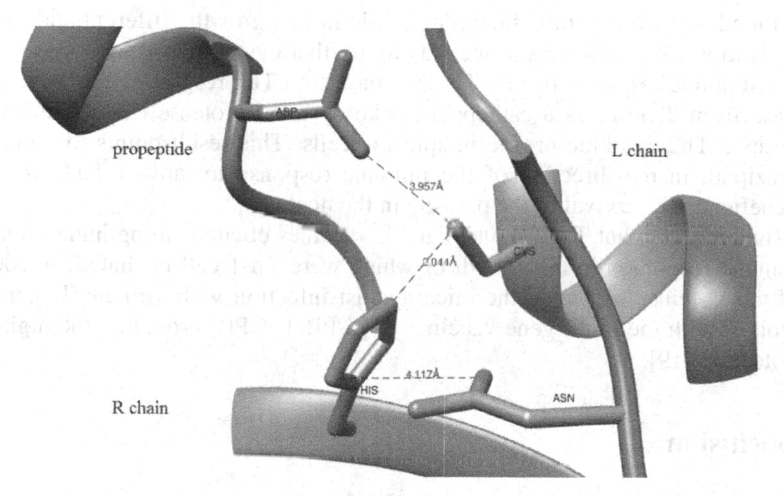

Fig. 2. View of catalytic triad. Cys_{440}, His_{651} and Asn_{676}, form the cysteine protease catalytic triad in the active site. Asp_{386} of the propeptide occupied the deep active site cleft.

4 Discussion

The amino acid sequences of the catalytic regions are similar to the mature papain family cathepsins, including cathepsins B and L from *T. gondii.* Several residues known to be important to activity are conserved in TgCPC: Cys, His and Asn, which form the cysteine protease catalytic triad in the active site [14].

The catalytic domain is significantly similar to CPCL that consist of two domains divided by the deep active site cleft; the left (L) domain is primarily α-helical, whereas the right (R) domain contains a β-barrel-like motif that is decorated by a few short α-helices [15].

In our model the Asp 1 was absent in the catalytic site, as well as results of TgCPCL crystallography structure. The TgCPL contains the three stabilizing disulfide bonds that are highly conserved among other papain-like cysteine [15] in our model we found only one conserved bond. A tyrosine residue that binds a chloride ion in the crystal structures of rat and human Cathepsin C is also conserved (Tyr 578). Adjacent to the active site cysteine, a distinctive residue may be involved in the binding of the N terminus of the peptide substrate. This tyrosine motif appears to be unique in the papain family proteases, in which tryptophan usually occupies the position adjacent to the active site residue. This Cathepsin contain a tyrosine based motif, YXXϕ, where X is any amino acid and ϕ is an amino acid with a bulky hydrophobic side chain, and/or an amino acid cluster dileucine motif located in their citoplasmatic tails know to participle in endosomal/lysosomal protein sorting in higher eukaryotes [16]. maturase for roptry proteins involved in modulation of the host cell [16]. TgCPL is predicted to be a type II membrane protein, it has an enzymatic activity with a low pH optimum and it occupies a membrane-bound structure in the apical region of extracellular parasites [17].

Studies found that Cathepsin C may play a role in cell growth, differentiation and protease activation [3].In other parasites was found that cysteine proteases may alter the adaptive immune response from a Th1 response to a Th2 response: The best studied protease from *T. cruzi* is a cathepsin L-like cysteine protease called cruzipain which induces a Th2 cytokine profile in spleenic cells. This result points to a major role for cruzipain in the direction of the immune response towards a Th2 profile, which is beneficial for survival of the parasite in the host [18].

Recent studies show that TgCPB and TgCPL vaccines elicited strong humoral and cellular immune responses in mice, both of which were Th-1 cell mediated. In addition, all of the vaccines protected the mice against infection with virulent T. gondii RH tachyzoites, with the multi-gene vaccine (pTgCPB/TgCPL) providing the highest level of protection [19].

5 Conclusion

TgCPC1 from *T.gondii* showed that Cys-440, His-651 and Asn-676, give the chemical environment for the cysteine protease catalytic triad activity, and the Asp386 on the propeptide is near to active site at ~3.9Å of distance to active Cys440 that was blocking the deep active site cleft.

References

1. Dou, Z., Carruthers, V.B.: Cathepsin proteases in toxoplasma gondii. In: Robinson, W.M., Dalton., J.P. (eds.) Cysteine Proteases of Pathogenic Organism, pp. 49–62. Landes Bioscience and Springer Science+Business Media (2011)

2. Rawlings, N.D., Barrett, A.J., Bateman, A.: MEROPS: the database of proteolytic enzymes, their substrates and inhibitors. Nucleic Acids Res. 40(D1), D343–D350 (2011)
3. Sajid, M., McKerrow, J.: Cysteine Proteases of Parasitic Organism. Mol. Biochem. Parasitol. 120(1), 1–21 (2002)
4. Que, X., Engel, J.C., Ferguson, D., Wunderlinch, A., Tomavo, S.: Cathepsin Cs Are Key for the Intracellular Survival of the Protozoan Parasite, Toxoplasma gondii. The Journal of Biological Chemistry 282(7), 4994–5003 (2007)
5. Buchan, D.W., Ward, S.M., Lobley, A.E., Nugent, T.C., Bryson, K., Jones, D.T.: Protein annotation and modelling servers at University College London. Nucl. Acids Res. 38, W563–W568 (2010)
6. Zhang, Y.: I-TASSER server for protein 3D structure prediction. BMC Bioinformatics 9, 40 (2008)
7. de Castro, E., Sigrist, C.J., Gattiker, A., Bulliard, V., Langendijk-Genevaux, P.S., Gasteiger, E., Bairoch, A., Hulo, N.: ScanProsite: detection of PROSITE signaturematches and ProRule-associated functional and structural residues in proteins. Nucleic Acids Res. 1(34)(Web Server issue), 362–365 (2006) PubMed PMID: 16845026; PubMed Central PMCID: PMC1538847
8. Ivica, L., Doerks, T., Bork, P.: SMART 7: recent updates to the protein domain annotation resourse. Nucleic Acids Research 40, 302–305 (2012)
9. Hirokawa, T., Boon-Chieng, S., Mitaku, S.: SOSUI: classification and secondary structure prediction system for membrane proteins. Bioinformatics 14, 378–379 (1998)
10. Marchler-Bauer, A.: CDD: a Conserved Domain Database for the functional annotation of proteins. Nucleic Acids Res. 39(D), 225–229 (2011)
11. Ye, Y., Godzik, A.: Flexible structure alignment by chaining aligned fragment pairs allowing twists. Bioinformatics 19(suppl. 2), ii246–ii255 (2003)
12. Lovell, S.C., Davis, I.W., Arendall III, W.B., de Bakker, P.I.W., Word, J.M., Prisant, M.G., Richardson, J.S., Richardson, D.C.: Structure validation by Calpha geometry: phi,psi and Cbeta deviation. Proteins: Structure, Function & Genetics 50, 437–450 (2002)
13. Pettersen, E.F., Goddard, T.D., Huang, C.C., Couch, G.S., Greenblatt, D.M., Meng, E.C., Ferrin, T.E.: UCSF Chimera–a visualization system for exploratory research and analysis. J. Comput. Chem. 25(13), 1605–1612 (2004)
14. Molgaard, A., Arnau, J., Lauritzen, C., Larsen, S., Petersen, G., Pedersen, J.: The crystal structure of human dipeptidyl peptidase I (Cathepsin C) in complex with the inhibitor Gly-Phe-CHN2. Biochem. J. 401, 645–650 (2007)
15. Larson, E.T., Parussini, F., Huynh, M.-H., Giebel, J., Kelley, A.M., Zhang, L., Carruthers, V.B.: Toxoplasma gondii Cathepsin L is the Primary Target of the Invasion- Inhibitory Compound Morpholinurea-leucyl-homophenyl-vinyl Sulfone Phenyl. Journal of Biological Chemistry 284(39), 26839–26850 (2009)
16. Que, X., Ngô, H., Lawton, J., Gray, M., Liu, Q., Engel, J., Reed, S.L.: The Cathepsin B of Toxoplasma gondii, Toxopain-1, Is Critical for Parasite Invasion and Rhoptry Protein Processing. The Journal of Biological Chemistry 277(28), 25791–25797 (2002)
17. Huang, R., Que, X., Hirata, K., Brinen, L.S., Lee, H.J., Hansell, E., Reed, S.: The Cathepsin L of Toxoplasma gondii (TgCPL) and its Endogenous Macromolecular Inhibitor, toxostantin. Mol. Biochem. Parasitol. 164(1), 86–94 (2009)
18. Giordanengo, L., Guinazu, N., Stempin, C., Fretes, R., Cerban, F., Gea, S.: Cruzipain, a major Tripanosoma cruzi antigen, Condictions the Host Immune Response in Favor of Parasite. Eur. J. Inmunol., 1003–1011 (2002)
19. Zhao, G., Zhou, A., Lv, G., Meng, M., Sun, M., Bai, Y., Sheyi, H.: Toxoplasma gondii cathepsin proteases are undeveloped prominent vaccine antigens against toxoplasmosis. BMC infectious Diseases 13(207) (2013)

Diversification of the Major Histocompatibility Complex (MHC) -G and -B Loci in New World Primates

Juan Sebastian Lugo-Ramos and Luis Fernando Cadavid

Departamento de Biología e Instituto de genética. Universidad Nacional de Colombia, Bogotá, Colombia
{jslugor,lfcadavidg}@unal.edu.co

Abstract. In primates, MHC class I family have evolved by continuous cycles of birth and death of genes leading to differential expansion of some loci in each major taxa. Humans, apes, and Old World Primates (*Catarrhini*) have expanded MHC-A and MHC-B loci, whereas New World Primates (*Platyrrhini*) have expanded MHC-G loci. Recent genomic studies in the Platyrrhini *Callithrix jachuus*, however, have showed several copies of MHC-B, some of which were predicted to be expressed. To inquire about the prevalence and expansion of MHC-B, and -G genes in Platyrrhini, we sequenced MHC-I cDNAs in 10 Platyrrhini species and evaluated polymorphism of MHC-G and -B in 11 individuals of *Cebus albifrons*. Phylogenetic analysis and genic distances showed an expansion of MHC-B with some functional copies in many species with the concomitant expansion of MHC-G previously identified. Polymorphism levels in MHC-B and MHC-G in *C. albifrons* were similar and both groups have been under diversifying selection at the Peptide Binding Region.

Keywords: Major histocompatibility complex, New World Primates, Platyrrhini, immunology, evolution.

1 Introduction

The Major Histocompatibility Complex class I molecules (MHC-I also named in humans IILA) play a central role initiating the adaptive immune response by presenting processed peptides to T lymphocytes. In humans, they are encoded by 6 main loci, HLA-A, -B, -C, E, -F, -G [1]. The classical MHC-I loci HLA-A, -B and -C are highly polymorphic with around 7000 alleles characterized, whereas the non-classical HLA-E, -F, -G have low polymorphism with about 83 alleles [2]. These loci are conserved in other hominoids [3], with the exception of MHC-C in orangutans which is present only in half of the populations studied [4], supporting the birth and death mode of evolution [5]. These features leads to a multiplicity of loci in other monkeys as Macaques with 19 loci of MHC-B and 7 of MHC-A with the psudogenization of G loci and emerge of MHC-AG [6]. In the process of Birth and death of genes, many of the loci are duplicated and rapidly pseudogenized maintaining some loci differentially expanded in different species.

L.F. Castillo et al. (eds.), *Advances in Computational Biology*,
Advances in Intelligent Systems and Computing 232,
DOI: 10.1007/978-3-319-01568-2_44, © Springer International Publishing Switzerland 2014

In Platyrrhini (i.e. marmosets, tamarins) the non-classical loci MHC-E and –F are maintained but in contrast to their Catarrhini counterparts, they have expanded the MHC-G loci, which appear to function as classical loci [7]. These MHC-G-like loci in Platyrrhini are polymorphic in most species, although it has limited polymorphism in *Saguinus oedipus* (cotton top tamarin) [8]. Interestingly, some species of Platyrrhini, like *Aotus sp, Pithecia pithecia*, and *Ateles belzebuth* transcribe genes related to MHC-B [9-10]. In addition, genome sequencing of *Callithrix jachuus* MHC-I region [11] showed the presence of at least 9 MHC-B related loci, 4 of which are potentially expressed. This indicates that the MHC-B loci has been overlooked in most studies in Platyrrhini, and that perhaps these genes are also expended in Platyrrhini, as it is the case for MHC-G-related loci. In this study we identified MHC-B and MHC-G secuences in 10 species of New World primates and analyzed their polymorphism levels in a single species, *Cebus albifrons*.

2 Methods

2.1 Blood Samples

Whole blood samples were obtained by venipuncture from one individual of 10 species of Platyrrhini: *Alouatta seniculus*, *Ateles belzebuth*, *Ateles hybridus*, *Lagothrix lagotricha*, *Saimiri sciureus*, *Cebus apella*, *Cebus capuchinus*, *Cebus albifrons*, *Saguinus oedipus*, and *Saguinus leucopus*. Samples from another 11 individuals of *C. albifrons* were obtained to analyze polymorphism levels. Animals were maintained at two wild life rescue centers in Colombia, (Corporación Autónoma Regional de Norte de Santander and Corporación Autónoma Regional del Alto Magdalena).

2.2 RNA Extraction, Reverse Transcriptase (RT)-PCR, and Sequencing

Total RNA was extracted with the TRIzol reagent following manufacturer indications. Complementary cDNA was synthetized with the RevertAid First strand cDNA Synthesis Kit (Fermentas). PCR was executed using forward primer: 5'TAACGGTCMTGGMGCCCCGAA3' and Reverse primer: 5'AATGAGAGACACATCAGAGCC 3'. Agarose gel-purified amplicons were cloned into the pGEM T Easy Vector (Promega) to transform chemically competent JM109 (Promega) E. coli cells. Colony PCR was performed with standard T7 and SP6 primers, and DNA products were sequenced in an ABI prism Automated sequencer. At least 24 clones per individual were sequenced in two directions. One sequence was considered for the study if it was represented in at least 3 clones.

2.3 Phylogenetic and Evolutionary Analysis

Sequences were aligned against each other with blastn and sequence pairs with an e-value higher than 10e-165 were used to perform a similarity network analysis. Novel and published Sequences were aligned with the MUSCLE algorithm and edited manually with MEGA 5.1. The substitution model for phylogenetic analyses was determined with the software Modeltest. Phylogenetic trees were constructed by Bayesian inference with MrBayes 3.2 using 12M generations. Neighbor Joining (NJ) analysis was performed with 2000 bootstrap replicates. Postitions under selection for *C. albifrons* sequences were detected with the Fast, Unconstrained Bayesian AppRoximation (FUBAR) algorithm including positions with a posterior probability >0.95. This analysis was performed in the Datamonkey server.

3 Results and Discussion

A total of 54 MHC class I cDNAs 1,032 bp long were obtained from the 10 different species of Platyrrhini. A similarity network analysis was performed assigning an edge if the e-value from blastn comparisons where higher than 10e-165. The sequences were clustered using the weighted-spring-embedded algorithm in Cytoscape which puts highly connected nodes with their neighbors close together (Fig. 1). The network showed a pattern with two main groups corresponding to MHC-B-like and MHC-G-like cDNAs (Fig. 1), as these sequences were closely related to MHC-B or MHC-G/A of Catarrhines, respectively. However, there were nodes falling outside from the main clusters, suggesting that there are sequences from different loci, including those related to MHC-E. No Platyrrhini sequences were found in the MHC-C subnetwork, indicating that this locus is unique of Catarrhini. There were also single nodes, composed by sequences with low similarity to the other nodes and included mostly pseudogenes. The nodes in the network analysis were reflected the phylogeny as there were two main clades corresponding to MHC-B-like MHC-G-like sequence (Fig. 2). These two clades, however, had various groups, indicating the presence of various loci within each of the B-like and G-like groups. These loci were not direct orthologous to those of Catarrhini (star-marked clades in Fig 2), suggesting that MHC-I in primates have evolved by a continuous process of gene duplication and diversification.

Platyrrhini MHC-G-like sequences were clustered into 6 main groups (MHC-G1 to G6), with two additional groups including the processed psudogenes MHC-G-PS1 and -PS2 (Fig 2). MHC-G1 clustered in a genus-specific fashion, and included alleles from the *Callitrichinae* subfamily (*Saguinus, Callithrix and Leontopithecus*). MHC-G2 had sequences from the *Atelidae* family that also were clustered in a genus-specific fashion, and included representative cDNAs from *Ateles sp.* and *Alouatta seniculus*. MHC-G3 sequences were restricted to Cebus species including *Cebus albifrons*, *C. capuchinus* and *C. apella*. In contrast to the previous groups, the MHC-G4 clade included sequences from all genera, except for those from the *Callitrhicinae* subfamily (tamarins and marmosets), indicating this clade contains ancestral MHC-G-like loci. MHC-G5 cDNAs were restricted to Aotus species, although it also included pseudogenes from *Saguinus oedipus*. The last MHC-G-like clade, MHC-G6, were restricted to *Saimiri sciureus* sequences. The MHC-I processed psudogenes PS1 and PS2 were exclusive of *Callithricinae* species with the single exception of a PS2 sequence from *Aotus.*

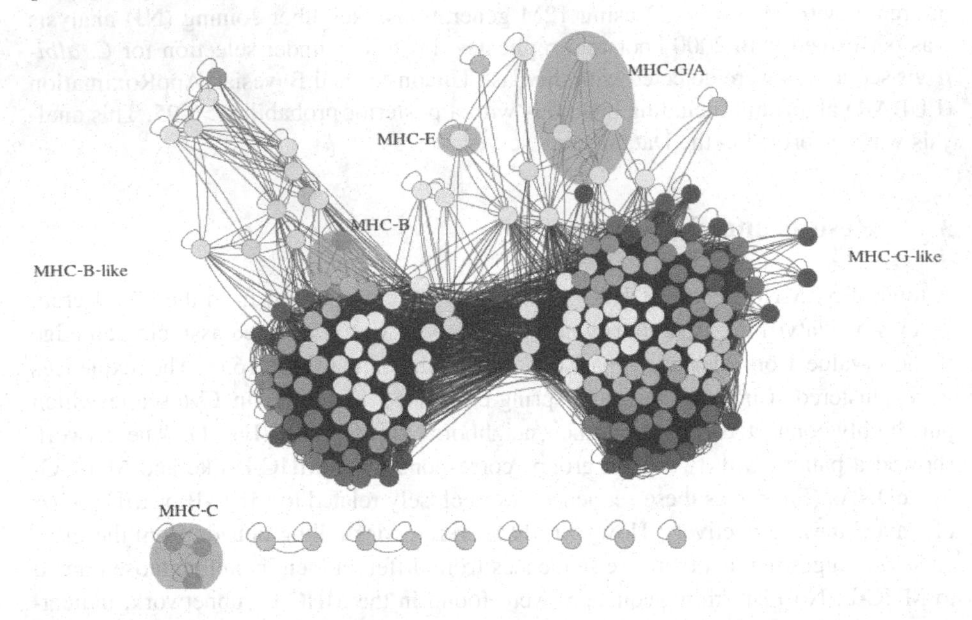

Fig. 1. Similarity *network of Primate MHC class I sequences*. Similarity network were obtained by BLAST alignments of exons 4-8 of all cDNAs and visualized in Cytoscape with spring embedded layout. Nodes indicate individual sequences, and edges similarity above threshold (see text). Node color indicate cluster coefficient among sequences. Shadowed nodes are Catarrhini sequences of indicated group (-A, -B, -C, -E, -G). All other nodes belong to Platyrrhini cDNA.

At least three MHC-B-like groups were identified in New World primates (Fig 2). MHC-B1 contained representative sequences from *Cebus, Ateles, Aotus Lagothrix* and *Alouatta* genera. This group tended to have sequences clustered by families, such that *Atelidae* and *Cebidae/Aotidae* clustered independently, with the exception of *Lagothrix lagotricha* MHC-B1 alleles that grouped with those of *Cebidae*.

The MHC-B2 clade had sequences from the same species found in B1 together with sequences from *Pithecia* and *Callithrix*, but it lacked those from *Cebus.* This clade had a clear pattern of trans-specis polymorphism, although some sequences from Callithrix clustered together likely as a result of recent duplications. Finally the MHC-B3 clade had sequences from Cebus, *Ateles* and *Saimiri* genera, each each forming an independent lineage. This clade is the closest to MHC-B in Catarrhini although they are not direct orthologous. Thus, in contrast to MHC-G-like cDNAs, MHC-B-like loci tended to be more conserved between Platyrrhini genera.

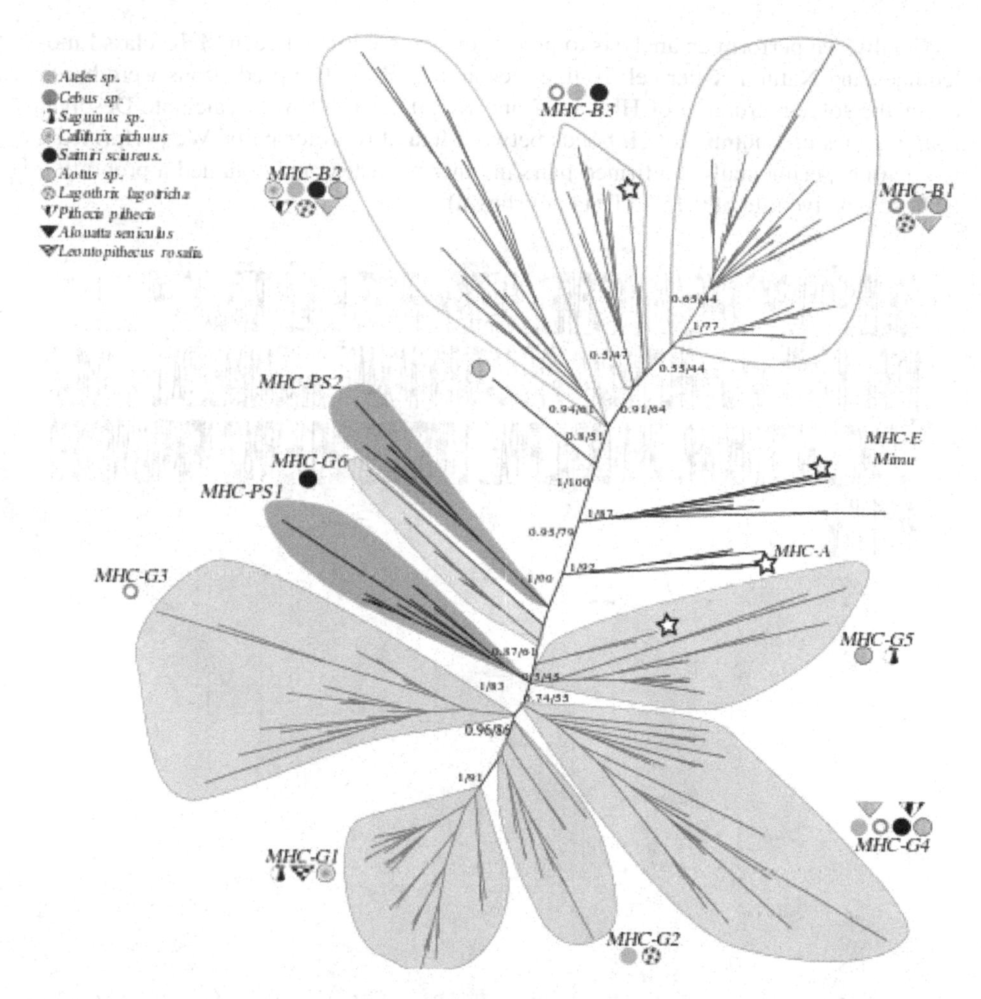

Fig. 2. Phylogenetic *relationships among Platyrrhini MHC-I.* Bayesian phylogenetic tree showing MHC-G-like and -B-like sequences. Posterior probabilities and Neighbor Joining bootstrap, are indicated in the main nodes (pp/NJ). Conventions next to groups indicate the species or genus towhich the sequence belong as indicated in the box. Starred clades indicate the place of HLA related sequences. Blue shadowing indicate B-like related groups, red G-like, and green others not related to G or B like.

To evaluate the polymorphism of MHC-G-like and MHC-B-like loci, cDNAs from 11 individuals of Cebus albifrons were cloned and sequenced. As expected, most variation occurred at the peptide binding region (fig 3A). This diversification is promoted by positive selection for peptide binding, as evidenced by dN-dS values significantly higher than 0 (dN=dS neutral evolution) (Fig 3B). The two groups of sequences seem to be equally polymorphic, suggesting that both are capable of presenting antigenic peptides.

Finally, we perform an analysis to predict the interaction between MHC class I moleculaes and Natural Killer cell Ig-like receptors (KIRs). The predictions were based on of the solved structure of HLA-B27 and KIR3DL1 (3VH8) to calculate Desolvation energies and number of H. bonds between ligand receptor pairs. We performed a test with experimentally confirmed pairs in other primates, and obtained a proportion of false positive rate of 0.15% (Data not shown).

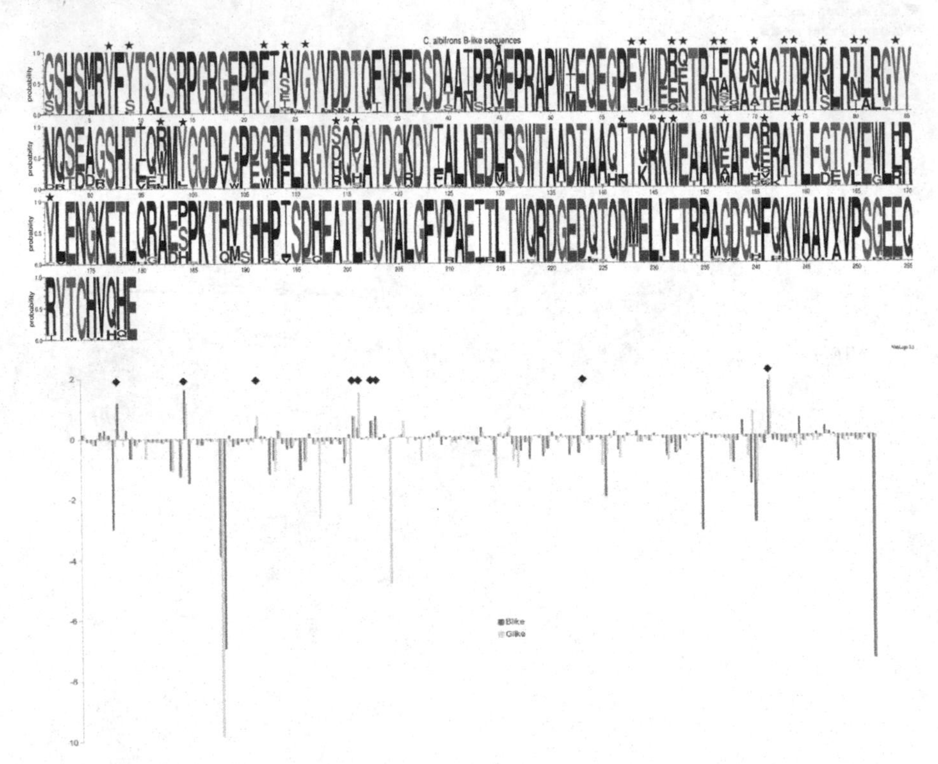

Fig. 3. Polymorphism *of MHC-G-like and MHC-B-like in C. albifrons.* Variation in amino acid positions in MHC-B and –G-like (A) sequences. Each position has a representation of the probability that the amino acid occurs in that position. Star-marked positions belong to the Peptide binding region (PBR). The histogram plot (B) shows the normalized dN-dS obtained with the FUBAR method as an indicator of Selective pressures for MHC-G-like (yellow) and MHC-B-like (red). If values are significantly above 0, then positive selection is acting to diversify alleles. . All values marked with a blue square have a posterior probability of positive selection >0.95.

References

1. Horton, R., Wilming, L., Rand, V.: Gene map of the extended human MHC. Nature Reviews 5 (December 2004)
2. Robinson, J., Halliwell, J.A., McWilliam, H., Lopez, R., Parham, P., Marsh, S.G.E.: The IMGT/HLA database. Nucleic Acids Research 41(Database issue), 1222–1227 (2013)

3. Adams, E.J., Parham, P.: Species-specific evolution of MHC class I genes in the higher primates. Immunological Reviews 183(650), 41–64 (2001)
4. Adams, E.J., Thomson, G., Parham, P.: Evidence for an HLA-C-like locus in the orangutan Pongo pygmaeus. Immunogenetics 49(10), 865–871 (1999)
5. Piontkivska, H., Nei, M.: Birth-and-death evolution in primate MHC class I genes: divergence time estimates. Molecular Biology and Evolution 20(4), 601–609 (2003)
6. Watanabe, A., Shiina, T., Shimizu, S., Hosomichi, K., Yanagiya, K., Kita, Y.F., Kimura, T., Soeda, E., Torii, R., Ogasawara, K., Kulski, J.K., Inoko, H.: A BAC-based contig map of the cynomolgus macaque (Macaca fascicularis) major histocompatibility complex genomic region. Genomics 89(3), 402–412 (2007)
7. Cadavid, L.F., Shufflebotham, C., Ruiz, F.J., Yeager, M., Hughes, A.L., Watkins, D.I.: Evolutionary instability of the major histocompatibility complex class I loci in New World primates. Proceedings of the National Academy of Sciences of the United States of America 94(26), 14536–14541 (1997)
8. Watkins, D.I., Garber, T.L., Chen, Z.W., Toukatly, G., Hughes, A.L.: Unusually limited nucleotide sequence variation of the expressed major histocompatibility complex class I genes of a New World primate species (Saguinus oedipus). Immunogenetics 33, 79–89 (1991)
9. Cardenas, P., Suarez, C.F., Martinez, P., Patarroyo, M.E., Patarroyo, M.A.: MHC class I genes in the owl monkey: mosaic organisation, convergence and loci diversity. Immunogenetics 56(11), 818–832 (2005)
10. Shiina, T., Kono, A., Westphal, N., Suzuki, S., Hosomichi, K., Kita, Y.F., Roos, C., Inoko, H., Walter, L.: Comparative genome analysis of the major histocompatibility complex (MHC) class I B/C segments in primates elucidated by genomic sequencing in common marmoset (Callithrix jacchus). Immunogenetics 63(8), 485–499 (2011)

A Methodology for Optimizing the E-value Threshold in Alignment-Based Gene Ontology Prediction Using the ROC Curve

Ricardo Andrés Burgos-Ocampo[1], Andrés Felipe Giraldo-Forero[1], Jorge Alberto Jaramillo-Garzón[1,2], and C. German Castellanos-Dominguez[1]

[1] Signal Processing and Recognition Group, Universidad Nacional de Colombia, Campus la Nubia, Km 7 vía al Magdalena, Manizales, Colombia
[2] Grupo de Máquinas Inteligentes y Reconocimiento de Patrones - MIRP, Instituto Tecnológico Metropolitano, Cll 54A No 30-01, Medellín, Colombia
{raburgoso,afgiraldofo,jaramillog,cgcastellanosd}@unal.edu.co

Abstract. The prediction of gene ontology (GO) terms is an important field of study in computational biology. With the advent of high-throughput experimental technologies, large quantities of sequenced proteins have emerged and, consequently, the number of computational tools that are used for analysing such data has also increased. In this field, methods based on sequence alignments like BLASTP are the most commonly used tools by biologists and bioinformaticians. However, an incorrect choice for the e-value threshold advised to identify homology and subsequently spread GO terms, may originate predictors with very low sensitivities and thus achieve poor prediction performances. In this work, a new methodology for optimizing the e-value threshold used in alignment-based predictors is proposed. The methodology is based on selecting a neighborhood for creating a veto scheme among proteins with similar e-values.

Keywords: Alignment sequences, Gene ontology, Proteomics, ROC curve.

1 Introduction

Protein function prediction has become one of the central problems in molecular biology due to the increasing need to associate novel protein sequences with its corresponding biological function. The Gene Ontology (GO) project [1] is an initiative for standardizing the representation of genes and gene products. This project has developed three controlled structures describing gene products in terms of biological processes, cellular components and molecular functions. Also, it has reduced sets named GOslims that are intended to provide descriptors for specific domains. One of them is the GOslim for plants, developed by The Arabidopsis Information Resource (TAIR) [2].

Experimental annotation is an almost impossible task due to the large amount of data to be processed. For this reason, during the last decade there have been

L.F. Castillo et al. (eds.), *Advances in Computational Biology*,
Advances in Intelligent Systems and Computing 232,
DOI: 10.1007/978-3-319-01568-2_45, © Springer International Publishing Switzerland 2014

several studies in the usage of computational methods to create protein function predictors and assist the biologists in this work. One of the most commonly used methods to succeed in annotation of proteins is *transfer approach*, where the protein to be annotated is searched against a database with already annotated proteins and, according to a similarity measure, the protein with the highest score transfers all its annotations to the query protein. It is important to say that this similarity measure can be a distance.

However, the rules based on transfer approach were designed to work in databases with high identity percentage and, consequently, they report very low sensitivities for databases with low identity percentage (tipycally under 40%. Further approaches are designed in order to implement geometric strategies for improving Blast results. For instance, the blast-knn [3], uses the BLASTP algoritm for extracting the k-nearest neighbours and in conjunction with a system of voting the annotation are transferred. Another example is blast2go [4], which uses an annotation rule that seeks to find the most specific annotations, but is unable to identify a high number of true positives.

This work proposes a new classification methodology based on finding the threshold that provides an optimal trade off between specificity and sensitivity, and for that purpose the receiver operating characteristic (ROC) curve is employed.

2 Materials and Methods

The notations that will be used throughout this paper are defined as follows. Consider a classification problem where each sequence (protein) ($\boldsymbol{x} \in \mathcal{X}$ can be associated with one or more labels from a finite set $\mathcal{Y} = \{1, 2, ..., Q\}$. In this setting, $\mathcal{Y}$ is associated to the set of categories defined by the GO. A labeled training set $\mathcal{S} = \{(\boldsymbol{x}_1, \mathbf{Y}_1), ..., (\boldsymbol{x}_s \mathbf{Y}_s), ..., (\boldsymbol{x}_S, \mathbf{Y}_S)\}(\boldsymbol{x}_s \in \mathcal{X}, \mathbf{Y}_s \subseteq \mathcal{Y})$ and an unlabeled test set $\mathcal{T} = \{x_1, ..., x_t, ..., x_T\}, x_t \in \mathcal{X}$ are given. The following approaches are used to infer the labels of the test set from the training set:

Blast ϵ-neighborhood: this is our proposed methodology and is based on a veto rule. Consider a ball $B(\boldsymbol{x}_t, \epsilon_i)$ with center in the sequence $\boldsymbol{x}_t$ and radius ϵ_i. Then, let $\mathcal{N}(\boldsymbol{x}_t)$ be the neighbors associated to the sequence $\boldsymbol{x}_t$, defined as:

$$\mathcal{N}(\boldsymbol{x}_t) = \{\boldsymbol{x}_s \in \mathcal{X} | d(\boldsymbol{x}_t, \boldsymbol{x}_s) < \epsilon_i\} \tag{1}$$

where $d()$ is e-values obtained from the BLASTP algorithm between sequences $\boldsymbol{x}_t$ and $\boldsymbol{x}_s$. The query sequence $\boldsymbol{x}_t$ will then be associated with the all the labels from its neighbors. that is:

$$\mathbf{Y}_t = \bigcup_{i=1}^{k} \{\mathbf{Y}_i | \boldsymbol{x}_i \in \mathcal{N}(\boldsymbol{x}_t)\} \tag{2}$$

A graphical description of the method is shown in Figure 1a.

Blast k-nearest neighbors: this approach is part of the GOPred method [5] and is based on the nearest neighbors algorithm. The k-nearest neighbors with the highest k blast scores are extracted. The output of BLAST-kNN, O_B for a target protein, is computed as:

$$O_B = \frac{P_p - P_n}{P_p + P_n} \tag{3}$$

where P_p is the sum of BLAST scores of proteins in the k-nearest neighbors in the positive training data of a given GO term. Similarly, P_n is the sum of scores of the k-nearest neighbor proteins in the negative training data of that GO term [3]. Positive values of O_B indicate membership while negative values indicate that the protein must not be associated to the gven term. A graphical description of the method is shown in Figure 1b.

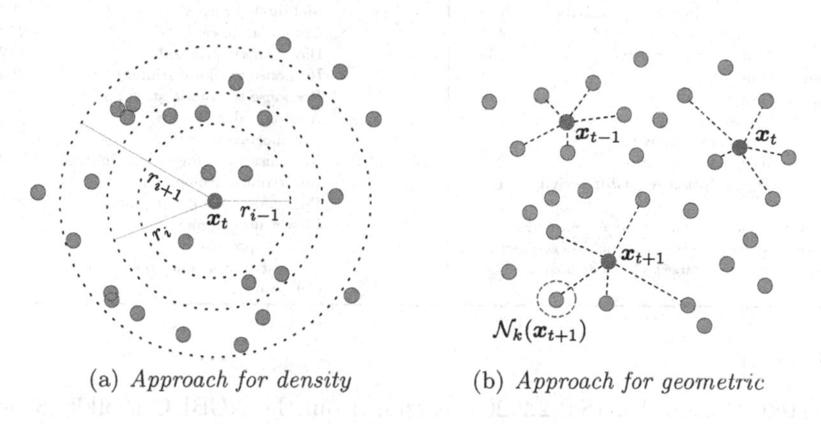

(a) *Approach for density* (b) *Approach for geometric*

Fig. 1. Methods to propagate the GO terms

3 Experimental Setup

3.1 Dataset

The database comprises all the available Embryophyta proteins at UniProtKB database [6], with at least one annotation in the Gene Ontology Annotation (GOA) project [1]. The resulting set comprises proteins from 189 different land plants. The dataset was filtered at 50% sequence identity using the Cd-Hit software [7], the class are defined according with the GOslim for plants developed by The Arabidopsis Information Resource - TAIR [2], are taken into account only the classes that are associated with at least 100 proteins. The final set is thus comprised by 10 GO terms in the molecular function ontology, 16 GO terms in the cellular component ontology and 29 GO terms in the biological process ontology. Table 1 shows the final list of categories.

Table 1. Number of protein for each class

Ontology	Class	Sample
Cellular Component	Extracellular region	123
	Cell wall	220
	Intracellular*	153
	Nucleus*	869
	Nucleolus	170
	Cytoplasm*	310
	Mitochondrion	476
	Vacuole	320
	Endoplasmic reticulum	218
	Golgi apparatus	106
	Cytosol	294
	Ribosome	118
	Plasma membrane	1112
	Plastid	1466
	Thylakoid	298
	Membrane*	879
Molecular Function	DNA binding	166
	Transcription factor activity	119
	Catalytic activity*	451
	Transporter activity	163
	Binding*	214
	Protein binding*	1288
	Kinase activity	128
	Transferase activity*	254
	Hydrolase activity	278
	Transcription regulator activity	178

Ontology	Class	Sample
Biological Process	Reproduction*	353
	Carbohydrate metabolic process	156
	Nucleobase, nucleoside, nucleotide and nucleic acid metabolic process*	193
	DNA metabolic process	148
	Transcription	538
	Protein modification process	184
	Cellular amino acid and and derivative metabolic process	165
	Lipid metabolic process	231
	Transport	292
	Response to stress	994
	Cell cycle	134
	Cell communication*	115
	Multicellular organismal development*	641
	Secondary metabolic process	173
	Cell differentiation	235
	Biological process*	587
	Metabolic process*	154
	Catabolic process	212
	Biosynthetic process*	532
	Response to biotic stimulus*	344
	Response to abiotic stimulus	747
	Anatomical structure morphogenesis	345
	Response to endogenous stimulus	542
	Embryonic development	206
	Post-embryonic development*	429
	Flower development	220
	Cellular process	1236
	Cellular component organization	432
	Cell growth	148

Classes marked with an asterisk (*) were redefined, the number of samples corresponds to the sequences associated to that class and no other of its also listed descendants

3.2 Blastp

The Protein-Protein BLAST 2.2.26+ version from the NCBI C toolkit [8] was used in this work with BLOSUM62 matrix and an e-value maximum of 40, for each ontology and each, e-value and number of neighbors considered, a cross-validation with ten fold was used. All the algorithms were developed on the R environment for statistical computing, using the *seqinr* package [9], available in RCRAN.

3.3 Annotation

The annotation process is carried out using both approaches described in section 2. For the reference method (k-nearest neighbors), the number of neighbors was set to one since the low sequence similarity among proteins in the dataset produced a very low number of blast hits, difficulting the implementation of this method for a higher number of neighbors.

Regarding our proposed methodology (the Blast $\epsilon-$neighborhood), a grid of twelve different e-values was chosen in order to analyze the behavior of the algorithm. The grid was obtained from the deciles corresponding to the increased number of neighbors, with the aim of estimating equally spaced values on the ROC curve.

4 Results and Discussion

The ROC curve is used to represent the optimal point, where the elbow region samples the best measures of sensitivity and specificity. Since the prediction of GO terms is a multi-label problem, a bipartition strategy is used because it is the most common framework for this kind of problems.

Figure 2 depicts the ROC curves constructed for each ontology in GO. The shaded region around each curve represents the standard deviation of the results and the best result appear in table 2 along with the results from the Blast k-nn method.

Table 2. Performance of annotation prediction over the three ontologies with the Blast k-nearest neighbors method

Ontology	Blast-knn				Blast ϵ-neighborhood			
	k	*Specificity*	*Sensitibity*	*G.Mean*	e-value	*Specificity*	*Sensitibity*	*G.Mean*
Cellular Component	1	0.940 ± 0.003	0.417 ± 0.015	0.626 ± 0.012	10	0.667 ± 0.006	0.711 ± 0.023	**0.689 ± 0.011**
Molecular Function	1	0.949 ± 0.003	0.614 ± 0.026	0.763 ± 0.017	1.2	0.856 ± 0.010	0.720 ± 0.025	**0.785 ± 0.012**
Biological Process	1	0.932 ± 0.004	0.386 ± 0.021	0.600 ± 0.016	1.2	0.708 ± 0.016	0.603 ± 0.018	**0.654 ± 0.009**

The best results for each of the comparison between the algorithms are highlighted in bold type.

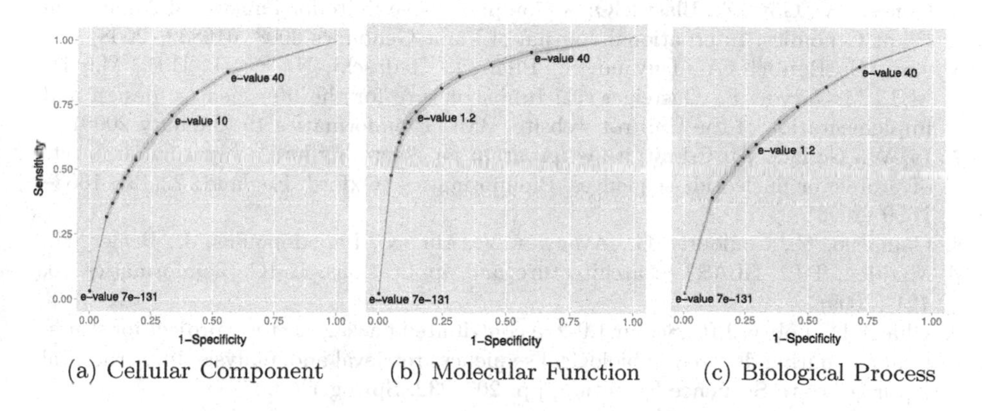

(a) Cellular Component (b) Molecular Function (c) Biological Process

Fig. 2. ROC curves for the three ontologies with the $\epsilon-$neighborhood method

5 Conclusions

The proposed methodology for optimizing the e-value threshold used in alignment-based protein function predictors, combined with the $\epsilon-$neighborhood approach, proved to work better than the GOPred method over the three ontologies in GO. The proposed methodology improves the sensitivity and balances the specificity in an specific e-value threshold. As a future work, it would be important to know to what extent is it possible to combine Blast $\epsilon-$ neighborhood with other strategies such as machine learning methods, obtaining a robust protein function predictor.

Acknowledgements. This work was partially funded by the Research office (DIMA) at the Universidad Nacional de Colombia at Manizales and the Colombian National Research Centre (COLCIENCIAS) through grant No.111952128388.

References

1. Gene, T., Consortium, O.: Gene Ontology: tool for the unification of biology. Gene Expression 25, 25–29 (2000)
2. Swarbreck, D., Wilks, C., Lamesch, P., Berardini, T.Z., Garcia-Hernandez, M., Foerster, H., Li, D., Meyer, T., Muller, R., Ploetz, L., Radenbaugh, A., Singh, S., Swing, V., Tissier, C., Zhang, P., Huala, E.: The Arabidopsis Information Resource (TAIR): gene structure and function annotation. Nucleic Acids Research 36(Database issue), D1009–D1014 (2008)
3. Saraç, O.: GOPred: GO Molecular Function Prediction by Combined Classifiers. PloS one 5(8), 1–11 (2010)
4. Conesa, A., Götz, S., García-Gómez, J.M., Terol, J., Talón, M., Robles, M.: Blast2GO: a universal tool for annotation, visualization and analysis in functional genomics research. Bioinformatics (Oxford, England) 21(18), 3674–3676 (2005)
5. Conesa, A., Götz, S.: Blast2GO: A Comprehensive Suite for Functional Analysis in Plant Genomics. International Journal of Plant Genomics 2008, 619832 (2008)
6. Jain, E., Bairoch, A., Duvaud, S., Phan, I., Redaschi, N., Suzek, B.E., Martin, M.J., McGarvey, P., Gasteiger, E.: Infrastructure for the life sciences: design and implementation of the UniProt website. BMC Bioinformatics 10 (January 2009)
7. Li, W., Godzik, A.: Cd-hit: a fast program for clustering and comparing large sets of protein or nucleotide sequences. Bioinformatics (Oxford, England) 22(13), 1658–1659 (2006)
8. Camacho, C., Coulouris, G., Avagyan, V., Ma, N., Papadopoulos, J., Bealer, K., Madden, T.L.: BLAST+: architecture and applications. BMC Bioinformatics 10, 421 (2009)
9. Charif, D., Lobry, J.R.: Seqinr 1.0-2: a contributed package to the r project for statistical computing devoted to biological sequences retrieval and analysis. In: Structural Approaches to Sequence Evolution, pp. 207–232. Springer (2007)

Hydrolytic Activity of OXA and CTX-M beta-Lactamases against beta-Lactamic Antibiotics

Ana Rosa Rodríguez Blanco, María Teresa Reguero Reza*, and Emiliano Barreto*

Bioinformatics Center, Biotechnology Institute, National University of Colombia,
Carrera 30 # 45-03, Edificio Manuel Ancízar, Bogotá DC
ebarretoh@unal.edu.co

Abstract. Beta-Lactamases OXA and CTX-M are highly disseminated and confer resistance to antibiotics. This project gathers, classifies and analyzes information from indexed articles and databases as UNIPROT, EMBL and PDB, in order to correlate molecular data with activity profile of beta-lactamases against beta-Lactamic antibiotics. Sequences of CTX-M and OXA enzymes are analyzed by ClustalW multiple alignment and distance trees are built by Neighbor – joining with default parameters and a Bootstrap of 1000, through MEGA 5.0.

Results from analysis were systematized with BLA.id system which documented 125 CTX-M, organized into four groups. Results show that CTX-M have punctual variations which change their hydrolytic activity profile. 310 OXAS grouped in 11 sets were analyzed and they show a high degree of conservation. Punctual changes in amino acids are correlated to changes in hydrolytic activity. BLA.id system fills a void in information about beta-Lactamases activity, which is disperse.

Keywords: beta-Lactamases OXA, beta-Lactamases CTX-M, antimicrobial sensitivity, Bioinformatics; Bla.id.

1 Introduction

Continuous rising in appearance and dissemination of antimicrobial drug-resistant microorganisms is a serious public health concern around the world. Inadequate use of antibiotics has applied selective pressure and this, added to diverse mechanisms of genetic transfer that bacteria can carry, contributes to multiresistant strains dissemination. [3]Design and development of new antibiotics have high costs, and as soon as a new antibiotic is created, microorganisms acquire new strategies to counteract its activity. [5] One of these strategies are beta-Lactamases, enzymes that hydrolyze amide bond of beta-Lactam ring. This makes the antibiotic loses its activity against cell wall synthesis [4] Currently, beta-Lactam antibiotics are the most viable alternative to treat bacterial infections and studying them is important and of a great relevance. It is especially important to focus on extended spectrum

* Corresponding authors.

L.F. Castillo et al. (eds.), *Advances in Computational Biology*,
Advances in Intelligent Systems and Computing 232,
DOI: 10.1007/978-3-319-01568-2_46,

beta-Lactamases as CTX-M and OXA enzymes [9,10], as they are of high dissemination in Colombia. Biological data systematization, correlation between hydrolysis profiles and structural data will allow predicting their responses against different antibiotics. This is also important for rational selection of antibiotics and design of new and better drugs and also to contribute with WHO strategies [11, 12].

Ambler et al [6] classified beta-Lactamases in four classes (A, B, C, and D), in terms of the mechanisms of interaction enzyme-substrate and their primary structure, [1,2]. Class A (serine-penicillinases), class B (metaloenzymes), class C (serine-cephalosporinases) and class D (serine-oxacillinases) [8]. CTX-M beta-Lactamases belong to class A. They are defined by efficiently hydrolyze cefotaxime. OXA beta-Lactamases belong to class D and are defined by their capacity to hydrolyze more efficiently oxacillin and cloxacillin than conventional penicillins [13]

2 Materials and Methods

Initially we searched for general information on beta-Lactamases in databases as PDB (Protein Data BanK) and EMBL and in scientific articles regarding primary structure and hydrolysis profiles of beta-Lactamases. Analysis of this information allowed us to select beta-Lactamases for the study, considering the following issues: dissemination, resistance level, sequences and general information. We selected OXA and CTX-M beta-Lactamases to follow with next stages.

2.1 Information Searches, Analysis and Selection: OXA and CTX-M beta-Lactamases Sequences, Structures and Antibiotic Hydrolysis Profiles

Information collection was made using database http://www.lahey.org/studies, which keeps global information on beta-Lactamases, such as number of reported OXA and CTX-M; primary and tertiary structure. Information of activity profile of CTX-M and OXA against antibiotics was gathered from indexed scientific articles in databases as PubMed. Other specialized databases were consulted as: UNIPROT (Universal Protein Resource), PDB (Protein Data Bank), Drugbank and EMBL (Nucleotide Sequence Database Integration of data on each beta-Lactamase was made by SRS system. Information was classified as: type of beta-Lactamase, microorganism in which it is found, sequence, origin, tertiary structure and antibiotic resistance profile for each class of OXA and CTX-M.

2.2 Correlation of Sequence vs. Activity against Antibiotic Profile

Data were organized in a new database within BLA.id system. Correlation of beta-Lactamases sequences with their respective resistance profile was made through an unsupervised clustering of type sequences of different OXA and CTX-M, considering their differences patterns and hydrolytic activity profiles. Changes in sequences were analyzed by multiple alignment using ClustalW in Mega 5.0. All OXA and CTX-M sequences were entered and phylogenetic trees were done using Neighbor-joining and maximum parsimony with 1000 replies each. The tree was validated with bootstrap method. Searches were made for the antibiotic susceptibility profile of each clade in the tree, taking into account the enzyme hydrolysis profile and MIC's against different antibiotics. Results were then interpreted according with CLSI 220430 guides.

3 Results

3.1 Search, Analysis and Selection of Information

Information search was focused on all OXA and CTX-M reported to date in different databases. With collected information we have a total of 256 OXA enzymes according Lahey Institute, from which, OXA-38, OXA-39, OXA-41, OXA-44, OXA-52, OXA-81, OXA-121 to OXA-127, OXA-135, OXA-140, OXA-151 to OXA-159, OXA-184, OXA-185, OXA-220 to OXA-222, OXA-226, OXA-227, OXA-232 to OXA-234, OXA-238, OXA-246, OXA-247, OXA-251 to OXA-255 have not yet been released but they were isolated, for this reason Lahey Institute classifies them as in assignation stage. It is worth noting that amino acids sequence of OXA-1 is the same as OXA-30 and amino acids sequences of OXA-24, OXA-40, OXA-46 and OXA-81 show not visible differences. (http://www.lahey.org/Studies/other.asp#table1)

There is a total 140 CTX-M, from which sequences of amino acids of CTX-M-14 and CTX-M-18; CTX-M-55 and CTX-M-57 are identical. There are reported as assigned: CTX-M-35, CTX-M-70, CTX-M-73, CTX-M-103, CTX-M-115, CTX-M-119, CTX-M-127, CTX-M-128, CTX-M-135, CTX-M-137, CTX-M-138, CTX-M-140 y CTX-M-141. CTX-M-118 is withdrown. For this project we worked with information of 125 CTX-M in total.

Information was selected and organized according with the following criteria: group, class, family, variant, gene name, organism, UNIPROT, PUbMEd and GenBank references, isoelectric point, number of amino acids, sequence, sensitivity, origin and bibliography. Most difficult to find was information about sensitivity and isoelectric points, as it is not available for all OXA and CTX-M enzymes or there is not information available for the same antibiotic. Information was stored in BLA.id system (http://bioinf-servicios.ibun.unal.edu.co/BLA.id/), which allows the crosschecking of clinical and molecular data, for the identification of beta-Lactamases from bacteria that are resistant to beta-Lactamic antibiotics. This information system BLA_ID was implemented on a PowerEdge M605 Dell server by the Bioinformatics group at Biotechnology Institute of the National University of Colombia. Operative system is Linux SUSE 10, web server Apache 2.0; PHP version 4.3.4; MySQL version 4.0.18 and Perl version 5.8.3. It is available in http://bioinf.servicios.ibun.unal.edu.co/BLA.id/ [7].

3.2 Correlation of Sequence Data vs Antibiotic Resistance Profile

OXA. From information gathered about activity against antibiotics is concluded that on a general level it is few information about susceptibility profiles against different groups of antibiotics, except for carbapenems (imipenem and meropenem).

In all OXA's were identified structural motifs as 108STFK111, except for some changes in OXA-9 T109 S, OXA-62 F110Y, OXA-85 T109 S, OXA-50 F110Y, OXA-186 T109P and motif 185YGN187 typical of class D beta-Lactamases.

Distance tree groups OXA's in 11 clades or groups. (http://bioinf-servicios.ibun.unal.edu.co/BLA.id/arbolOXA.pdf). Clade 1 is largest with 79 OXA. Of them, OXA-51 confers resistance to oxacillin and OXA-68 and 86 to piperacillin. Regarding monobactamics, presence of OXA-68 and 86 confers resistance to aztreonam and there is scarce information about oximino-cephalosporins. Against carbapenems in general, there is information on meropenem and imipenem in 24 OXA's that produce profiles from susceptible to resistant. In the group of aminoglycosides, we found that OXA-194 confers to the microorganism resistance to kanamycin. OXA's belonging to clade 3, in general, confers resistance to carbapenems imipenem and meropenem, with exception of OXA-23 that confers resistance to imipenem. This group presents a change of conserved motif YGN, which is typical of class D beta-Lactamases by FGN, but as stated by some authors this does not modify imipenem hydrolysis.

12 OXA's belong to clade 6; OXA-60 hydrolyzes imipenem more efficiently than oxacillin, however, there is not enough information to correlate hydrolysis with susceptibility and its sequence. Clade 8 contains 23 OXA and they have the tetrad STFK and conserved motifs YGN, KTG which are typical of this class. In this clade OXA-16, OXA-14 and OXA-17 are the closest of the tree and they are very similar in sequence. These three OXA's do not confer resistance to carbapenems and they are present in Pseudomonas aeruginosa. OXA-17 does not have aspartate substitution in position 148, keeping glycine for which it does not confer significative resistance to ceftazidime, compared with OXA-14 and OXA-16. OXA's in this clade show changes in structural motifs KTG by KSG and in third conserved motif YGN by FGN. It seems this change affects NaCl inhibition. It is a great diversity in profiles and sequences between enzymes in this group.

Enzymes belonging to a same clade show few changes in amino acids, but, when comparing sequences between clades, there are several changes in amino acids. However, there is low information to correlate in accurate way susceptibility with sequences.

CTX-M. We used a total of 125 sequences out of 140 reported CTX-M. Enzymes were grouped in a dendrogram obtained by neighbor joining (http://bioinf-servicios.ibun.unal.edu.co/BLA.id/arbolCTXM.pdf). Alignment and dendogram show four groups. Comparison of these shows similitudes in amino acids sequences inside each group. There are highly conserver sequences as 73STSK76, 132SDN135 (with exception of CTX-M 81 which shows H135 by N) and 238KYGS241 (with exception of CTX-M 106 and CTX-M 107 238R por K), features associated with class A beta-Lactamases to which CTX-M belong.

Distance tree reveals four large clades. Clade 1 is formed by 46 CTX-M highly related. All enzymes show leucine 122 that seems to confer a higher resistance to ceftriaxone. CTX-M-11 shows proline at this position but there is not information about hydrolysis or susceptibility against ceftriaxone,. For this reason it is not possible to affirm if this change affects resistance to this antibiotic.

CTX-M-62 and CTX-M52 show the change P170S, which contributes to efficiently hydrolyzing ceftazidime. Clade 2 has 21 enzymes with amino acid glycine 126, compared with other CTX-M that have serine in this position, its presence in microorganisms confer resistance to cefotaxime. Clade 3 is formed by 12 enzymes, of which CTX-M 40, CTX-M 63 and CTX-M 8 are the most distant from clade 4. CTX-M-40 confers to microorganism resistance to gentamicin and tobramycin but there is not enough information to correlate changes of sequence with those of susceptibility profile of microorganism that possess it. In this clade, leucine 122 is changed by phenylalanine, except for CTX-M 94, CTX-M 78, CTX-M 40, CTX-M 63 and CTX-M 8 which have the change alanine by glycine in position 123. Clade 4 contains 43 CTX-M with the most distant CTX-M-44, probed by alignment because it does not have amino acids R194 and N195. All enzymes in this clade have changed A65 by G and A70 by P. Some authors affirm that resistance to ceftriaxone is due to leucine 122, except for CTX-M-85 that has proline at this position, but it is not available information about microorganisms' susceptibility or hydrolysis of antibiotic to correlate this change with ceftriaxone resistance. CTX-M 93, 121,27, 98, 102, 105 and 16 change aspartate 242 by glycine, which increases hydrolytic activity against ceftazidime. CTX-M enzymes confer resistance to cefuroxime, cefotaxime and cefepime.

4 Conclusions

Available information on CTX-M and OXA enzymes is broad but scattered and it is mostly recorded in individual way. BLA.id system allows to systematize information in a dynamic way, trying to fill a void that currently exists as information about activity of beta-Lactamases is scattered and difficult to correlate. Multiple alignment generates information that allows to deduct punctual differences between different CTX-M and OXA. There is lack of information regarding their hydrolysis profile, or the susceptibility of microorganisms owing them; just some articles refer to enzymatic hydrolysis rate, which gives a higher reliability about enzymatic activity. In other publications minimal inhibitory concentration (MIC) is informed for a given microorganism which is less reliable information as activity against an antibiotic might be affected by other factors as simultaneous presence of other beta-Lactamase, efflux pumps, modifications in penicillin-binding proteins (PBP), among others. Given CTX-M and OXA diversity, their hydrolytic activity may be explained, in part, by amino acid sequence, but it is necessary to complement it with three-dimensional and functional information.

References

1. Bush, K., Jacoby, G., Medeiros, A.: A Functional Classification Scheme for b-Lactamases and its Correlation with Molecular Structure. Antimicrobial Agents and Chemotherapy 39(6), 1211–1233 (1995)

2. Adriana, C.Y.: Descripción del ambiente genético del gen blaCTX-M_12 detectado en aislamientos de Klebsiella pneumoniae en hospitales de tercer nivel de (Tesis Maestría en Mi-crobiología), Universidad Nacional de Colombia. Facultad de Ciencias. Bogotá, 73 p. (2008)
3. Ângela, N., Iñaki, C., Fernando, B., Rafael, C., Teresa, M.C., Andrés, M., Fernando, G.-C., Juan, C.G.: Evolutionary Trajectories of Beta-Lactamase CTX-M-1 Cluster Enzymes: Predicting Antibiotic Resistance. PLoS Pathogens 6(1), 1–16 (2006)
4. del Carmen Rocha Gracia, R., Lozano P., Laguna, M.Y.: Mecanismos de Patogenicidad e Inter-acción: Parásito-Hospedero II, 245p. Benémerita Universidad Autónoma de Puebla, México (2006)
5. Sánchez, L., Ríos, R., Máttar, S.: Detección de beta-lactamasas de espectro exten-dido en Escherichia coli y Klebsiella pneumoniae aislados en una clínica de Villavicencio, Colombia. Asociacion Colombiana De Infectologia 12(3), 193–200 (2008)
6. Ambler, R.P., Coulson, A.F.W., Frére, J.-M., Ghuysen, J.M., Joris, B., Forsman, M., Lévesque, R.C., Tiraby, G., Waley, S.G.: A standard numbering scheme for the class A -lactamases. Biochem. J. 276, 269–272 (1991)
7. Emiliano, B.H., Reza, M.T.R.: Diseño de un modelo bioinformático para la detección, identificación y clasificación de genes codificantes de β-lactamasas,como herramienta para el cruce de datos moleculares con datos clínicos. Rev. Colomb. Cienc. Quím. Farm 36(2), 109–126 (2007)
8. Helena, H.Á.: Escherichia coli. productores de Blee aislados de urocultivo: implicaciones en el diagnóstico y tratamiento de la infección urinaria. Tesis. Universidad Complutense de Madrid (2010)
9. Villegas, M.V., Kattan, J.N., Correa, A., Lolans, K., Guzman, A.M., Woodford, N., Livermore, D., Quinn, J.P., The Colombian Nosocomial Bacterial Resistance Study Group.: Dissemination of Acinetobacter baumannii Clones with OXA-23 Carbapenemase in
Colombian Hospitals. Antimicrobial Agents and Chemotherapy 51(6), 2001–2004 (2007)
10. Villegas, M., Correa, A., Pérez, F., Miranda, M., Zuluaga, T., Quinn, J.: Colombian Nosocomial Resistance Study Group. Prevalence and characterization of extended-spectrum beta-lactamases in Klebsiella pneumoniae and Escherichia coli isolates from Colombian hospital. Diagn. Microbiol. Infect. Dis. 49, 217–222 (2004)
11. ONU: Resistencia a los antimicrobianos: el drama del abuso. Boletín de la Organización Mun-dial de la Salud. 88, 797–876 (2010)
12. Resistencia a los antimicrobianos (RAM). Nota descriptiva N°194 (2012), `http://www.who.int/mediacentre/factsheets/fs194/es/index.html`
13. Sun, T., Nukaga, M., Mayama, K., Braswell, E.H., Knox, J.R.: Comparison of-lactamases of classes A and D: 1.5-Å crystallographic structure of the class D OXA-1 oxacillinase. Protein Science 12, 82–91 (2003)

Analysis in Silico of 5'-Terminal Secondary Structures of Hepatitis C Virus Sequences Genotype 1 from Colombia

Luisa Fernanda Restrepo, Johanna Carolina Arroyave,
and Fabian Mauricio Cortés-Mancera

Grupo de Investigación e Innovación Biomédica (GI^{2}B). Facultad de Ciencias Exactas y Aplicadas–Programa Ingeniería Biomédica, Instituto Tecnológico Metropolitano (ITM), Institución Universitaria Adscrita a la Alcaldía de Medellín, Colombia
fabiancortes@itm.edu.co

Abstract. Hepatitis C Virus (HCV) genome consists in a single strand RNA molecule that includes in its 5'-untranslated region (5'-UTR) a secondary (2D) structure, designated IRES (Internal Ribosome Entry Site), which has four structural domains (I-IV). Specifically domain III presents a regulatory element involved in RNA stability and translation efficiency, in a 2D structure-dependent manner. Sequences from HCV genotype 1 isolated from Colombian Blood Donors and patients with Hepatocellular Carcinoma were analyzed for genetic variability (*Bioedit v7.1.11*) and RNA conformation (*Assemble 2.0* ™ and *Mfold v2.3*). A total of 25 substitutions were observed between sequences, and some of them showed an effect in RNA 2D structure. Interestingly, most substitutions were located in Stems, altering the topology of IRES domain III with a high broad potential of biological impacts. The *in silico* analysis showed here reveals new substitutions located in important structural motifs, thus epidemiological and experimental studies will be necessary to confirm their biological and clinical implications.

Keywords: HCV, IRES, RNA secondary structure, *Assemble 2.0* ™, *Mfold v2.3*.

1 Introduction

Hepatitis C virus (HCV) is a member of the *Flaviviridae* family, which is considered one of the most important risk factor for Hepatocellular Carcinoma (HCC) development; it is estimated more than 170 million people infected for this agent [1].

HCV genome consists in a positive strand RNA molecule flanked for two untranslated region in 5' and 3'- termini (5'UTR and 3'UTR, respectively). 5'UTR segment has been extensively used for genotyping studies in HCV but it is also a topic of interest because includes RNA secondary (2D) structures that play a critical role in translation of viral polyprotein. This function is essential for HCV replication and it is related to the conformational folding Stem-Loops called IRES (Internal Ribosome Entry Site), which is formed by four structural domains (I-IV). On this line, domain III has a particular importance derived of its interaction with 40S Ribosomal subunit and Eukaryotic Initiation Factor 3 (eIF3) recruiting before protein synthesis initiation [2], thus the study of HCV IRES structure by *in silico* analysis can contribute to understand the biological significance in RNA 2D alterations (Topology alterations)

L.F. Castillo et al. (eds.), *Advances in Computational Biology*,
Advances in Intelligent Systems and Computing 232,
DOI: 10.1007/978-3-319-01568-2_47,

and their possible roles in viral pathogenesis. Furthermore, *in silico* prediction approaches have been strengthened with the improvement of their algorithms and accuracy that allow a high correlation with molecular structures obtained through Nuclear Magnetic Nuclear and X-Ray Crystallography methodologies [3], becoming in a strong tool for studying sequence-structure-function relations in RNA molecules.

At present, six genotypes have been described for HCV (1-6), with a broad and heterogeneous worldwide distribution. In the Americas and specifically in Colombia, HCV genotype 1 is the most prevalent; significantly, this genotype has been related with higher risk of HCC development and antiviral treatment resistance [4]. In this paper we analyzed HCV available sequences from Colombia and described the presence of nucleotide substitution in relation to IRES domain III structure by *in silico* approaches. Biological implications of these findings are discussed.

2 Material and Methods

Dataset and sequence analysis. Presence of substitutions were studied in 46 sequences of the HCV 5'UTR (5'-termini), previously characterized in three genotyping studies developed in Colombia [5–7]. Sequences included in the present study corresponded to subtypes 1a (22) and 1b (24), previously determined in two different Blood Donor populations (Accession number on GanBank: GQ379738-69; AM69927-37) and patients with diagnosis of Hepatocellular Carcinoma (JF693486-89). HCV 5'-termini analysis was limited only to HCV IRES domain III (119-302 and 316-324), subdomains IIIa-IIe, because adjacent segments (domain I, domain II and IV) were partially incomplete in most of the dataset. Sequences were aligned using the ClustalW algorithm with 1000 repetitions for bootstrapping (*Bioedit v7.1.11*) [8], and substitutions were analyzed in comparison to prototype references in the alignment (Accession number: 1a-D29815 and 1b-D90208). Recombination events were also discarded by Bootscanning (*SimPlot V3.5.1*) [9], applying Neighbor Joining method and K2 as a substitution model, conducting the analysis on 120bp sliding windows with steps of 10bp, under strict consensus and 80% as a cut value for recombination point determination.

Analysis of RNA secondary structure. HCV IRES domain III sequences were used to search for possible associations between Stem-Loops topology alterations and substitutions observed during multiple alignments. For RNA 2D analysis, two bioinformatics tools were selected according to their high values on Sensitivity (**S**) and Positive Predictability (**PPV**) values, previously determined using HCV IRES as a model [10]: *Assemble 2.0* ™ and *Mfold v2.3* Web-server platform; the former tool delegates prediction to *RNAfold* based on Minimum Free Energy (MFE) with the Nearest-Neighbor thermodynamic algorithm and the Maximizing Expected Accuracy (MEA) [11], while the latter platform implements folding at 37°C with ionic conditions in1M NaCl (non-divalent ions)based on Zuker and Turnerv method [12]; the core algorithm of *Mfold v2.3* predicts a Minimum Free Energy (**ΔG**), as well as minimum free energies for folding that must contain any particular base pair [13].

3 Results

Sequence analysis of HCV IRES domain III. Recombination analysis of HCV 5'-termini sequences characterized in Colombia was performed using *Simplot* software (figure 1). Once groups of reference including all genotypes of HCV were constructed, five sequences per genotype, there were not evidence of recombination events in any sequence analyzed by bootscaning; this analysis was consistence even with systematic modifications in the established consensus level and modifications of sliding windows and steps values. In figure 1 is shown 1 out of 46 results, with high percentage of similarity between query and subtype 1a group sequence under strict consensus and 80% as a cutoff value for determine recombination events point; after confirming the absence of recombination events, HCV sequences were aligned next to prototypes for substitution analysis development.

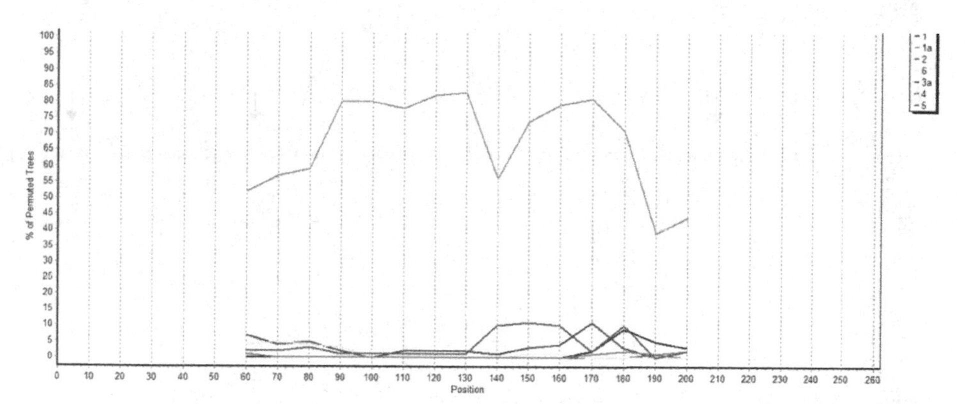

Fig. 1. Representative result of 5'-termini HCV sequence recombination analysis by Bootscanin (Simplot), showing the percentage of permuted tree (Y axis) by position of the sequence (X axis) with a sliding window of 120bp. In the figure is observed the analysis performed with one of the sequence of subtype 1a. The event of recombination was set up at 80%, but any query sequence showed a recombinant breakpoint.

In turn, the multiple alignment of HCV IRES domain III sequences (nucleotides 119-330 of HCV 5'-termini) showed a total of 25 substitutions, with higher number of changes in sequences belonging to subtype 1a than 1b (figure 2). In this context subtype 1a presented more transitions than transversión while subtype 1b sequences had an opposite tendency; briefly, the alignment of 22 sequences subtype 1a (subdomains IIIa-e) showed 16 substitutions in relation to the reference. Among these sequences were characterized transversions that involved the change of Adenine (A) to Cytosine (C) at position 119 (A119C), also A204C, G258U, and G286U, and nine transitions: G135A, U175C, A243G, G245A, C247U, U248C, G271A G273A and G283A; one insertion was also detected, specifically Uracil (U) at 126 position (126U) and two deletions characterized for nucleotide C absence at position 126 (C126Δ) and G268Δ. Among them A204C was present in sixteen sequences, while transitions C247U and U248C were observed in ten members of this dataset; sequences from nucleotide 119 to 256 and substitutions are shown in the alignment (figure 2a).

On the other hand, 9 substitutions were identified during the analysis of the 24 sequences of HCV subtype 1b (figure 2b). Within them, transitions A185G and C204U were found, six transversions A134C, G135U, A136C, A205U, U287G and C289G, and one deletion C209Δ. Interestingly, C204U, A205U and C209Δ were presented in sequences derived from HCC patients.

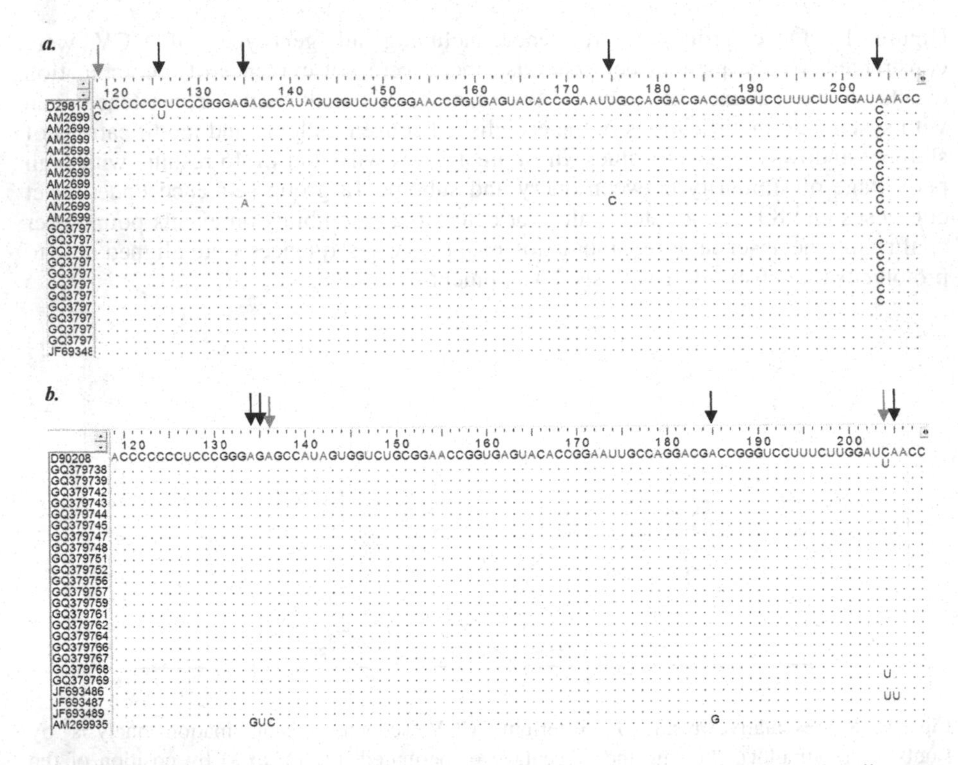

Fig. 2. Alignment of HCV IRES Domain III. Top and bottom panels are shown the alignment (*Bioiedit*) of subtypes 1a and 1b sequences isolated in Colombia, respectively. Upper numbers in each panel represents the nucleotide position in relation to each reference sequence; accession numbers of every sequence are also indicated on left side of the figure. Arrows indicates substitution localization; gray arrows are highlighting substitutions with report in previous studies. Although his figure only shows a portion of domain III (119-208nts) the following substitutions are presented for subtype 1a (**a**) A119C, C126U, C126Δ, G135A, U175C and A204C; for subtype 1b (**b**) A134G, G135U, A136C, A185G, C204U, and A205U.

RNA secondary structure prediction. In silico structure of IRES Domain III was generated by *Mfold v2.3* web server platform and *Assemble 2*™ program in order to relate the RNA sequence substitutions with 2D structure topology, either the stems or loops present in the assessed region. Although both bioinformatics tools have a high **PPV** and **S** values [10], *Mfold* was used for secondary RNA structure prediction and *Assemble 2*™ for modeling the output (figure 3), having into account that both programs generate similar structures, but *Assemble 2*™ has a friendly visualization interface. As observed in figure 3, predicted RNA 2D structures showed important mispairing sites affecting Stems I, II, III, V, VI and VIII. Interestingly, when IRES domain III topology was analyzed based on subtype identity, it was observed a clustering of substitutions in Stem III and II for subtype 1a and 1b, respectively.

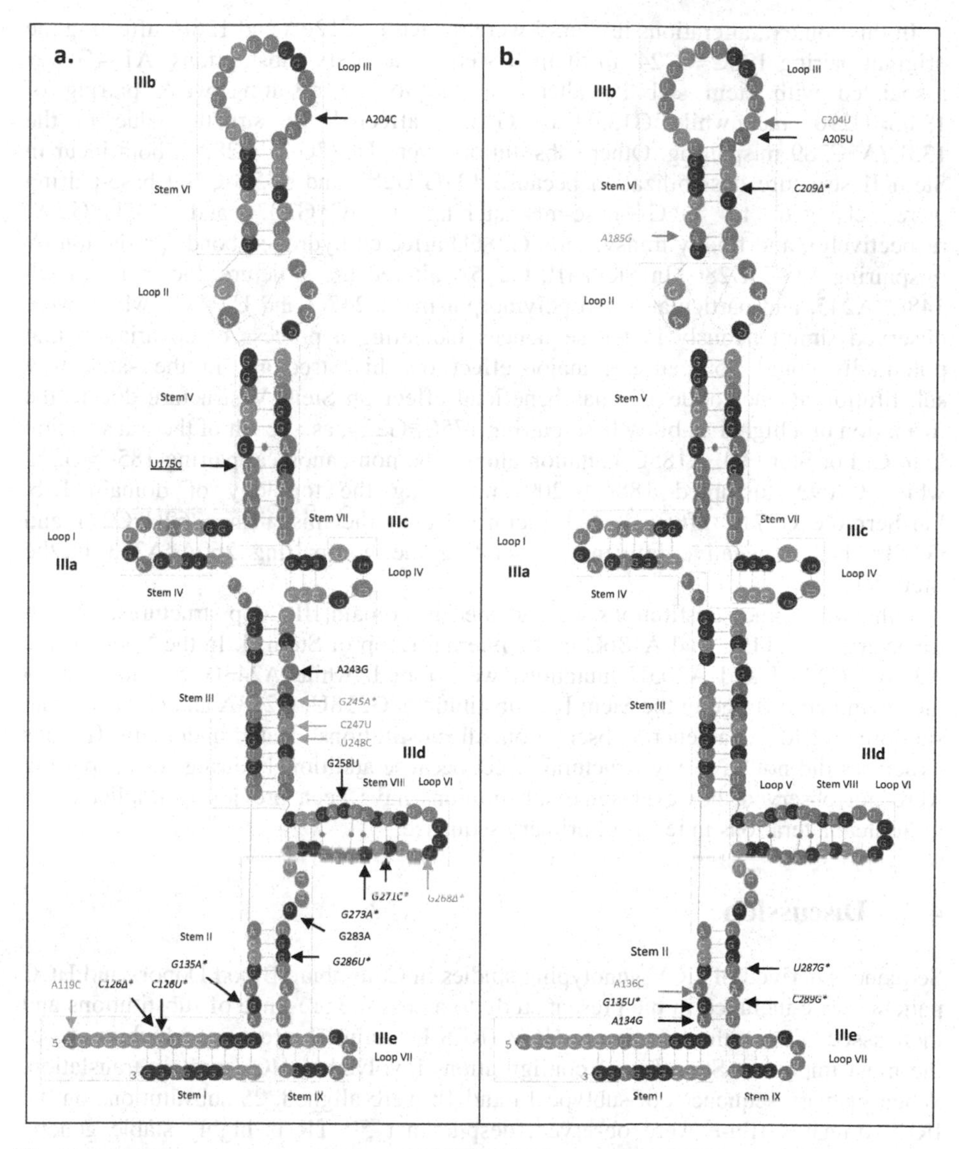

Fig. 3. RNA 2D structure prediction of HCV IRES domain III. This analysis was conducted by *Assemble 2* using nucleotides 119 to 324 of the references sequences, showing a representative secondary structure of IRES for subtype 1a (**a.**) and 1b (**b**). In the structures are indicated the subdomains IIIa-IIIe, the Stem (I-IX) and Loops (I-VII) conformation. Arrows highlight variable position on the IRES structure. Δ indicates a nucleotide deletion. The underlined substitution remark a position of stability stem increment, while italicized letter indicates mispairing. Asterik symbol designates a mismatch and gray letters indicate variable positions previously reported in others studies. Localization of variable positions in secondary structure of HCV IRES domain III Domains are numbered as described by *Brown et al*; and *Honda et al* [14,15].

In this context, alterations in Stem I were related to C126Δ and 126U, affecting the original pairing 126C≡G324. In Stem II were located six substitutions: A134G was associated with stem stability alteration due to the resulting weak pairing of 134G=U290, meanwhile G135U or G135A affected the structure due to the 135U/A*C289 mispairing. Other substitutions were U287G or C289G, both incur in Stem II structure destabilization because 137G-U287 and 135G-C289 base-pairing were changed to G*G base-mismatching (137G*G287 and 135G*G289, respectively), and finally transversion G286U affected hydrogen bonds prediction by mispairing 138C*U286. In Stem III, G245A altered the structure due to mismatch 148C*A245, and particularly two polymorphisms (C247U and U248C), which were observed simultaneously in ten sequences indicating a process of covariation that potentially could not cause a major effect on this structure; in the same way substitution at nucleotide 175 has beneficial effect on Stem V structure due to the formation of a higher stability base-pairing 175C≡G224, as a result of the transversion U to C. For Stem VI, A185G mutation altered the non-canonical pairing 185G=U212 while C209Δ disrupted 188G≡C209, impacting the topology of domain IIIb. Furthermore G271C affected the structure due to the mispairing 262U*C271 and G273A had a negative influence generating the non-pairing 260A*A273 in the Stem VIII.

Otherwise, nine substitutions were located in domain III Loop structures; briefly, transversions A119C and A136C in the internal Loop of Stem II. In the Loop III the A204C, C204U and A205U mutations were found, while A243G was located on the asymmetric Bulge in the Stem III; substitutions G258U, G283A and G268Δ in the subdomain IIId. As a general observation, all substitutions located in domain III loops structures did not have any structural effect because additional pairings or mispairing were not observed, however some substitutions have been previously implicated in biological alterations in terms of primary sequence.

4 Discussion

Sequences derived of HCV genotyping studies in Colombian Blood Donors and HCC patients were included in the present study to assess the presence of substitutions and their association with alteration in HCV IRES Domain III structure, which is one of the most important Stem-Loops configurations involved in HCV protein translation. When primary sequences of subtype 1a and 1b were aligned, 25 substitutions on the IRES domains IIIa-e were observed, despite that 5'UTR is highly stable genetic region used for HCV detection and genotyping, where mutations are generally compensatory, preserving the base-pairing pattern in order to conserve structural properties related to translation efficiency [16, 17]. However, in this *in silico* analysis we could observe that not all substitutions conserved the base-pairing on Stems; indeed some of them implicate alterations in translational activity IRES-dependent.

Among substitutions observed, 17 were localized in Stem structures, and 9 in Loops. For the case of substitutions located on Stems, three mutations presented available experimental data; specifically transition A185G in Stem VI has been related with activity translation decreasing in more than 70%, while the polymorphism G245A (Stem III) has shown a marginal effect on translation (95.8%)

[18]; in these *in vitro* studies, IRES activity was assessed by functional luciferase reporter assay [18, 19]. In contrast, A136C has demonstrated a translational activity similar to wild type [20].

For Loops structure, three variable positions were previously reported in experimental studies. The insertion 126U that simultaneously involved transversion A119C, presented a 5-fold increase of translational activity [21]. Additionally, C204U localized on Loop III, were reported by *Barría et al.* as a mutation that did not have effect on proteins synthesis in comparison to the reference subtype 1b [22].

As shown in figure 3a, among deletions the G268Δ was found. Although this substitution was present in only one sequence of subtype 1a, G268Δ has a deleterious effect according to experimental analysis. Indeed, this deletion localized in the third position of the conserved triplet GGG in the loop VIII (subdomain IIId), affects an essential element for translational IRES activity in all HCV genotypes, as well as others *Flaviviridae*-members family, such as Pestiviruses and Pegiviruses [23, 24]. In other study, *Romero-López et al.* indicated that this triplet is required for long-range RNA–RNA interaction between the 5′ and 3′ ends of the HCV genome, and mutants harboring substitutions in the apical loop of domain IIId are capable of disrupting this complex, indicating that these regions are essential in the kissing interaction initiation [25].

Previous studies have described that HCV IRES structural domain III contains three crucial Stem-Loops structures (subdomains IIIc, IIId, and IIIe), which contribute to the high affinity interaction with the 40S ribosomal subunit, indispensable for viral polyprotein translation [2]. In this context, eight substitutions observed here are located in the basal domain IIId, specifically A134G, G135A/U, G271C, G273A, G286U, U287G and C289G, however in most cases there were not experimental evidence to support their biological implications. How these mutations could implicate alterations in viral replication and fitness, further experimental and clinical studies will be necessary to confirm their relevance.

5 Conclusions

In silico analysis conducted by *Mfold v2.3* and *Assemble 2*™ allows to determine substitutions in HCV sequences from Colombia with biological implications associated to RNA 2D structure alteration. There were differences in substitutions per each HCV subtype in relation to their classification (Transition/Transversion), number and structural localization.

Acknowledgements. Authors thanks to Dirección de Investigación of Instituto Tecnologico Metropolitano (ITM) for funding (Grants P08204-P10245)

References

1. Szabo, S., Bibby, M., Yuan, Y.: The epidemiologic burden of hepatitis C virus infection in Latin America. Annals of Hepatitis 11(5), 623–635 (2012)
2. Lukavsky, P.J.: Structure and function of HCV IRES domains. Virus Research 139(2), 166–171 (2009)

3. Moretti, S.: In Silico Experiments in Scientific Papers on Molecular Biology. Science Studies 24(2), 23–42 (2011)
4. Pawlotsky, J.M., Germanidis, G., Neumann, A.U., Pellerin, M., Frainais, P.O., Dhumeaux, D.: Interferon resistance of hepatitis C virus genotype 1b: relationship to nonstructural 5A gene quasispecies mutations. Journal of Virology 72(4), 2795–2805 (1998)
5. Cortes-Mancera, F., Loureiro, C.L., Hoyos, S., Restrepo, J.-C., Correa, G., Jaramillo, S., Norder, H., Pujol, F.H., Navas, M.-C.: Etiology and Viral Genotype in Patients with End-Stage Liver Diseases admitted to a Hepatology Unit in Colombia. Hepatitis Research and Treatment 2011, 1–10 (2011)
6. Mora, A., Gomes-gouve, M.S.: Molecular Characterization, Distribution, and Dynamics of Hepatitis C Virus Genotypes in Blood Donors in Colombia. Journal of Medical Virology 1898, 1889–1898 (2010)
7. Moratorio, G., Martínez, M., Gutiérrez, M.F., González, K., Colina, R., López-Tort, F., López, L., Recarey, R., Schijman, A.G., Moreno, M.P., García-Aguirre, L., Manascero, A.R., Cristina, J.: Evolution of naturally occurring 5'non-coding region variants of Hepatitis C virus in human populations of the South American region. Virology Journal 4(1), 79 (2007)
8. Hall, T.: BioEdit: a user-friendly biological sequence alignment editor and analysis program for Windows 95/98/NT. Nucleic acids symposium series (1999)
9. Lole, K., Bollinger, R.: Full-length human immunodeficiency virus type 1 genomes from subtype C-infected seroconverters in India, with evidence of intersubtype recombination. Journal of Virology 73(1), 152–160 (1999)
10. Restrepo, L.F., Arroyave, J.C., Cortes, F.M.: Evaluation of in silico tools for RNA secondary structures determination, using the Hepatitis C Virus IRES sequence as a model. In: Pan American Health Care Exchanges-V Congreso Colombiano de Bioingeniería e Ingeniería Biomédica, p. 310 (2013) ISBN: 978-1-4673-6256-6
11. Lorenz, R., Bernhart, S.H., Höner Zu Siederdissen, C., Tafer, H., Flamm, C., Stadler, P.F., Hofacker, I.L.: ViennaRNA Package 2.0. Algorithms for Molecular Biology: AMB 6, 26 (2011)
12. Mathews, D.H., Sabina, J., Zuker, M., Turner, D.H.: Expanded sequence dependence of thermodynamic parameters improves prediction of RNA secondary structure. Journal of Molecular Biology 288(5), 911–940 (1999)
13. Zuker, M.: Mfold web server for nucleic acid folding and hybridization prediction. Nucleic Acids Research 31(13), 3406–3415 (2003)
14. Brown, E.A., Zhang, H., Ping, L.H., Lemon, S.M.: Secondary structure of the 5' nontranslated regions of hepatitis C virus and pestivirus genomic RNAs. Nucleic Acids Research 20(19), 5041–5045 (1992)
15. Honda, M., Rijnbrand, R., Abell, G., Kim, D., Lemon, S.M.: Natural variation in translational activities of the 5' nontranslated RNAs of hepatitis C virus genotypes 1a and 1b: evidence for a long-range RNA-RNA interaction outside of the internal ribosomal entry site. Journal of Virology 73(6), 4941–4951 (1999)
16. Soler, M., Pellerin, M., Malnou, C.E., Dhumeaux, D., Kean, K.M., Pawlotsky, J.-M.: Quasispecies Heterogeneity and Constraints on the Evolution of the 5′ Noncoding Region of Hepatitis C Virus (HCV): Relationship with HCV Resistance to Interferon-α Therapy. Virology 298(1), 160–173 (2002)
17. Araújo, F.M.G., Sonoda, I.V., Rodrigues, N.B., Teixeira, R., Redondo, R.A.F., Oliveira, G.C.: Genetic variability in the 5' UTR and NS5A regions of hepatitis C virus RNA isolated from non-responding and responding patients with chronic HCV genotype 1 infection. Memórias do Instituto Oswaldo Cruz 103(6), 611–614 (2008)

18. Tang, S., Collier, A.J., Elliott, R.M.: Alterations to both the primary and predicted secondary structure of stem-loop IIIc of the hepatitis C virus 1b 5' untranslated region (5'UTR) lead to mutants severely defective in translation which cannot be complemented in trans by the wild-type 5'UTR s. Journal of Virology 73(3), 2359–2364 (1999)
19. Collier, A.J., Gallego, J., Klinck, R., Cole, P.T., Harris, S.J., Harrison, G.P., Aboul-Ela, F., Varani, G., Walker, S.: A conserved RNA structure within the HCV IRES eIF3-binding site. Nature Structural Biology 9(5), 375–380 (2002)
20. Berry, K., Waghray, S., Mortimer, S., Bai, Y., Doudna, J.: Crystal structure of the HCV IRES central domain reveals strategy for start-codon positioning. Structure 19(10), 1456–1466 (2011)
21. Zhang, J., Yamada, O., Ito, T., Akiyama, M., Hashimoto, Y., Yoshida, H., Makino, R., Masago, A., Uemura, H., Araki, H.: A single nucleotide insertion in the 5'-untranslated region of hepatitis C virus leads to enhanced cap-independent translation. Virology 261(2), 263–270 (1999)
22. Barría, M.I., González, A., Vera-Otarola, J., León, U., Vollrath, V., Marsac, D., Monasterio, O., Pérez-Acle, T., Soza, A., López-Lastra, M.: Analysis of natural variants of the hepatitis C virus internal ribosome entry site reveals that primary sequence plays a key role in cap-independent translation. Nucleic Acids Research 37(3), 957–971 (2009)
23. Kolupaeva, V.G., Pestova, T.V., Hellen, C.U.: An enzymatic footprinting analysis of the interaction of 40S ribosomal subunits with the internal ribosomal entry site of hepatitis C virus. Journal of Virology 74(14), 6242–6250 (2000)
24. Jubin, R., Vantuno, N.E., Kieft, J.S., Murray, M.G., Doudna, J.A., Lau, J.Y., Baroudy, B.M.: Hepatitis C virus internal ribosome entry site (IRES) stem loop IIId contains a phylogenetically conserved GGG triplet essential for translation and IRES folding. Journal of virology 74(22), 10430–10437 (2000)
25. Romero-López, C., Berzal-Herranz, A.: A long-range RNA–RNA interaction between the 5′ and 3′ ends of the HCV genome. RNA, 1740–1752 (2009)

In Silico Hybridization System for Mapping Functional Genes of Soil Microorganism Using Next Generation Sequencing

Guillermo G. Torres-Estupiñan and Emiliano Barreto-Hernández

Bioinformatics Center, Biotechnology Institute, Universidad Nacional de Colombia, Bogotá D.C. – Colombia
{Ggtorrese,ebarretoh}@unal.edu.co

Abstract. Nowadays a widely production and low cost of sequencing has allowed the extension of metagenomics in order to explore the genomic information from diverse environments. This offers the opportunity to examine new approaches for sequence binning and functional assignment. Driving of metagenomic studies using high throughput sequencing, usually follows the same pipeline used to analyze single genomes: sequence assembling, gene prediction, functional annotation and phylogenetic classification of reads, contigs or scaffolds, nevertheless, the accuracy of this approach is limited by the length of the reads or resulting contigs.

In Silico Hybridization System is an approach of functional and taxonomical assignment. It has default assignment at gender taxonomic level binning. This tool works with two general pipelines: Probes Creator (CrSo), ordered to design DNA probes (fingerprints) to gender taxonomic level and Sequential In silico Hybridator (HISS) which use the probes to make the hybridization with the community reads.

This bioinformatics tool allows characterization of the microbial metabolism in charge of biogeochemical cycles, tracking their key stages using debugged reference information. This strategy resulted in an increasing of binning accuracy. The simulated and real scenarios were better described using one probe and selection threshold fitted to a logarithmic distribution, with mean sensitivity of 85% and mean specificity of 83%.

Keywords: Metagenomics, binning, metatranscriptomics, soil, nitrogen, phosphorus, biogeochemical cycles.

1 Introduction

High-throughput sequencing (HTS) technologies currently have offered an unprecedented opportunity to examine the microbial ecology on a wide scale by three general approaches: 1) 18S/16S ADNr, fast and effective way to characterize communities structure [10, 23]; 2) Whole genomic sequencing (metagenomics), this approach is able to deal with the structure of communities along with the functional potential of them [12, 26] and 3) Whole transcriptome sequencing (metatranscriptomics), in order

L.F. Castillo et al. (eds.), *Advances in Computational Biology*,
Advances in Intelligent Systems and Computing 232,
DOI: 10.1007/978-3-319-01568-2_48, © Springer International Publishing Switzerland 2014

to examine the gene potential and metabolic capabilities of the microbial communities under certain conditions [17, 27]. Metagenomic analysis follows a relative similar pipeline as working with a single organism: sequence assembling, gene prediction, functional annotation and phylogenetic classification (binning) of reads, contigs or scaffolds [15]. Nevertheless, the metagenomic assembling approach has had not so good result so far [16]. In the binning process we attempted phylotyping the sequences through composition-based or similarity-based strategies.

The composition-based strategy extracts information related with GC content [7], codon usage [19] or k-mer frequency [22] from the metagenome sequences and compares them with features calculated of reference sequences with known taxonomic origin. This strategy allowing achieves optimal assignments to sequence length larger than 800bp.

On the other hand, the similarity-based strategy relies on homology information obtained by database searches. The databases could contain nucleotide sequences (i.e. complete genome) or protein sequences with known taxonomic origin. However, commonly the bioinformatics tools use protein sequences as reference for metagenomic analysis, since protein sequences are more conserved than nucleotide sequences, they are better suited for detection of remote homologies in order to explore an ecosystem which mainly consists of uncultured microorganisms [2]. But, the usage of protein sequences as reference has the disadvantage that the metagenomic DNA fragments have to be translated into all six reading frames, which increases computation time of the homology search. This strategy can be sub-divided in two general methods: those to use Hidden Markov Models (HMM) [5] or BLAST-based [1] homology searches.

Despite of all the efforts, these bioinformatics tools have not achieved both the efficiency and accuracy level required by current metagenomics high-complexity data sets because of computational limitations, unable good accurate assignments for short DNA fragments (<400bp). Therefore, we present a new strategy to evaluate the taxonomic and functional features of soil microbial communities, an integrated approach that combines bioinformatic algorithms to probe (fingerprint) design (CrSo) from debugged coding gene sequences database and in silico hybridization (HISS), that allows to characterize the soil biogeochemical metabolism. Finally we discuss initial experimental results, which help evaluate our both fingerprint specificity and hybridization selection criteria.

2 Implementation

A specialized bibliographic review allowed identifying the metabolic processes, subprocesses and the reactions that compose them, in charge of Nitrogen and Phosphorus biogeochemical cycles. In order to distinguish microbial communities, the enzyme gene markers selected were not housekeeping genes, but them were differentially distributed in microbial communities. A PHP script was developed to retrieve the nucleotide, protein sequence and taxonomy identification via SOAP protocol (Simple Object Access Protocol) on EMBL -EBI database, this information was storage in local database named as SPH.

2.1 Probe Design

CrSo has been split the probe design problem in three general steps, however, these are not independent at all and indeed the input parameters in the first step have logical relationship with second and third step criteria. The first step reduces the redundancy of input information by clustering in an attempt to bring together DNA sequences of the same biological sequence, then extracting the consensus sequence. The second step, a probe design phase extracts from candidate consensus sequences only those subsequences that satisfy the experimental specificity conditions. In the third step a probe denoising process is executed, creating clusters with DNA probes to select just the singletons.

Step 1: Redundancy reduction. We start with large enzyme CDSs information allocated in SPH. However, this database presented two features: 1) several sequences covering the same biological sequence, and 2) sequence fragments of one biological sequence are globally alignable, so, it will be impractical testing each sequence from database for probe design because it will result in repeated probes. For this, we exploit the sequence similarities, clustering them, in order to estimate a biological sequence from consensus sequence derived from a multiple alignment using UCLUST [6]. But, clustering configuration depends on homology features of the target genes, so that we determine the level of taxonomic resolution of these functional genes. The gene sequences were grouped according to their taxonomic classification and aligned, later the sequence similarities were calculated with p-distance model using MEGA 5 [24].

Step 2: Probe design. At this stage we established a set of constraints to extract probes from candidate sequences. We have selected a probe design tool, OligoWiz 2.0 [28], that implements a complex score model to select the best probe. This model allows specificity constraints like: minimum homology, minimum length of homology stretch, maximum similarity, probes length and database. Probes satisfying these constraints are extracted and passed to the next step.

Step 3: Probe denoising. In this step, all probe set candidate are clustered [6] as a filter process to discard duplicates or close related ones, selecting just the singleton clusters for the hybridization process.

2.2 *In silico* Hibridization

The hybridization process follows a general rule: each read aligned significantly with a probe, should be considered originating or homology of the enzyme gene that probe represents, therefore HISS performs a nucleotide BLAST search for each probe against HTS reads database, and reads with non-significant alignments with target probes were not taken into account.

Due to the limitations of in silico hybridization models that determine which DNA alignments are significant, HISS examines alignment features to determine homology relationship between probe and reads. As the probes are designed to identify multiple target genes we have to consider multiple criteria to determine a homology selection threshold, such as overall sequence identity, contiguous matches, mismatches, gaps and alignment length [13]. Analytically HISS incorporate a threshold selection function generated from series of continuous matches thresholds Xi with i mismatches. From basis on empirical threshold [14], we define 21 to X0, the longest stretch of contiguous matches between a probe and metagenomic sequence, the following thresholds were calculated such as Xi = X0 + W(i). Where W is BLAST input parameter, termed word size, involved in BLAST heuristic approach. In blastn program used for BLAST searches the default value of W is 11 and the smallest value is 7. Because every BLAST result has to include an exact match of length W, it becomes a bound on values of specificity thresholds, therefore, in order to increase the sensitivity of HISS, it implements a default W value of 7 [29].

3 Results

The entire pipeline was implemented on a HPC environment at Bioinformatics Center server of Biotechnology Institute of Universidad Nacional de Colombia, the cluster consists of 10 X 2.8 GHz Quad Core processors, running with SUSE10 with 32 GB of shared memory. The computational time of the algorithm depends on the number of probes and metagenomic sequences for the hybridization, hence the computational time of processing is directly dependent on the speedup achieved by BLAST, so the execution of HISS could be improved by using parallel versions of BLAST like mpiBLAST [4] or pioBLAST [9].

The SPH database, lodge 33536 sequences for Nitrogen cycle, associated with 39 reactions of nitrogen fixation, nitrification, mineralization, assimilation and denitrification process, and 13883 sequences for Phosphorus cycle, associated with 34 reactions of mineralization process of phosphomonoesters, phosphodiesteres and inositol phosphates.

The taxonomic resolution suggests a high variability, and confirms that amoA gene is a species specific marker but not higher taxonomical levels, furthermore, we found that species of the same gender clustering at an identity mean of 82% and 73% for genders of the same family (Table 1). Therefore, for the first stage of CrSo, the redundancy parameter was settled to 82% of identity to clustering, in order to make gender lever bins. The probes were designed with 75% of identity for minimum homology, 30 of minimum length of holomoly stretch [14], 83% of identity for maximum homology, probes length of 100bp and local database named as ProEnvFun compounded by EMBL sequences from Prokaryotes, Environmentals and Fungy groups to calculate cross hybridization (deeper information in [28]). The denoising was configured with 82 % of identity clustering.

CrSo triggers five sets of three best probes according to their length (25bp, 40bp, 60bp, 80bp and 100bp) that were evaluated with simulated metagenomes made with Grinder [3]. Ten Illumina sequencing metagenomes were simulated mimic to a low complexity community [25], 20 soil representatives genomes were selected from NCBI RefSeq. With the best probes we fitted the selection HISS parameters, that were started as linear function traced using a series of thresholds X_0, X_1, X_2, X_3, X_4

and X_5, analytically calculated, such as: 21, 28, 35, 42, 49 and 56 respectively. However we acquire better results with a logarithmic function ($y = 9.9581\ln(x) - 32.067$), achieving a mean sensitivity of 85% and mean specificity of 83% for all five length sets of probes (The sensitivity was evaluated as the ratio of true positives and all Blast hits. The specificity was calculated as the ratio of true positives and the value of true positives plus false positives).

Table 1. Taxonomic resolution of functional genes. Data show the mean (±sd) of similarity values of the gene groups at different phylogenetic levels. N.A. not applicable

Cycle	EC number	Gene	Phylogenetic Hierarchies (% sequence similarity)		
			Strain	Specie	Gender
Nitrogen	1.14.99.39	*amo*A	0.98 ± 0.01	0.82 ± 0.01	0.78 ± 0.03
	1.18.6.1	*nif*D	0.95 ± 0.01	0.87 ± 0.01	0.82 ± 0.01
		*nif*K	0.91 ± 0.01	0.76 ± 0.01	0.73 ± 0.01
		*nif*H	0.92 ± 0.02	0.86 ± 0.01	0.83 ± 0.01
	1.7.2.1	*nir*K	0.91 ± 0.01	0.82 ± 0.01	0.72 ± 0.01
		*nir*S	0.93 ± 0.03	0.75 ± 0.02	0.66 ± 0.01
	1.7.99.4	*nap*A	0.96 ± 0.01	0.87 ± 0.01	0.76 ± 0.01
		*nar*G	0.93 ± 0.01	0.83 ± 0.02	0.79 ± 0.03
		*nar*H	0.90 ± 0.02	0.80 ± 0.01	0.71 ± 0.02
		*nar*I	0.92 ± 0.03	0.73 ± 0.03	0.66 ± 0.05
		*nar*Z	0.98 ± 0.02	0.90 ± 0.01	N.A.
		*nar*Y	0.98 ± 0.01	0.81 ± 0.01	0.81 ± 0.01
	1.7.1.4	*nas*B	0.95 ± 0.01	0.83 ± 0.02	0.72 ± 0.01
	1.7.7.1	*nas*A	0.92 ± 0.01	0.84 ± 0.01	0.69 ± 0.05
	1.7.2.5	*nor*B	0.94 ± 0.01	0.82 ± 0.02	0.70 ± 0.03
		*nor*C	0.93 ± 0.01	N.A.	N.A.
	6.3.1.2	*gln*A	0.98 ± 0.01	0.84 ± 0.01	0.73 ± 0.03
Phosphorus	3.1.3.11	fbp	0.95 ± 0.01	0.81 ± 0.02	0.76 ± 0.01
	3.1.3.8	phy	0.95 ± 0.01	0.87 ± 0.01	N.A.
	3.1.3.15	hisB	0.90 ± 0.02	0.75 ± 0.02	0.70 ± 0.02
		hisJ	0.95 ± 0.01	0.79 ± 0.02	N.A.
	3.1.3.18	cbbZ	0.92 ± 0.01	0.82 ± 0.02	0.64 ± 0.02

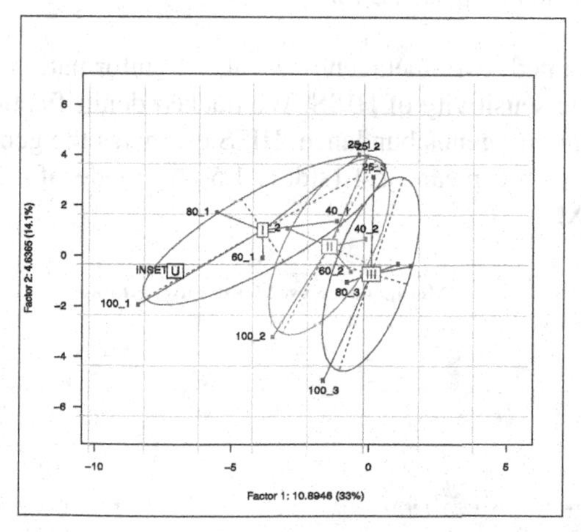

Fig. 1. Comparison of community characterization of different probe sets. We use an ACP to represent the synthetic metagenome INSET (U) as a dot and comparing with probe length and number of probes for gene descriptors.

The effect of multiple probes for gene was evaluated in order to figuring out the community structure. The ACP result suggests that the better way of characterize the community is with one probe of 100bp, 80bp or 60bp in length (Fig. 1), and there is not evidence to determine which is the best. With 100bp probes, we performed an evaluation of probes usage strategy against full-length gene to mapping the community. The full-length gene approach consider BLAST results with bitscore bigger than 55 and 100bp alignment length to assign reads [11]. The results suggest that using full-length genes overestimates the number of genes in the community (Fig. 2), similar scenery have been reported [8].

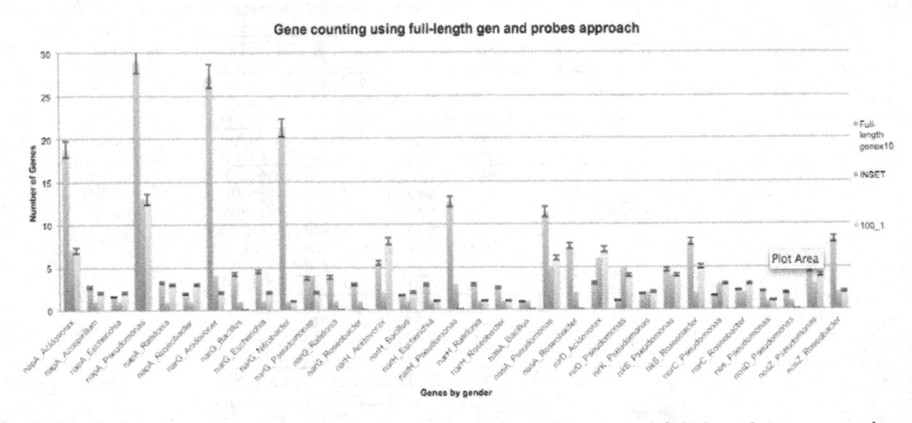

Fig. 2. Number of reads assigned with HISS one 100bp probe set and full-length gene mapping. INSET represents the expecting gen number. Full-lenth gene BLAST hits are represented 10 times less for representation simplicity. Error 5%.

There were generated two metagenomes at 1:5 information ratio (Lib1X and Lib5X) for evaluating sensitivity of HISS. We tracked denitrification process by characterizing the community gene abundance. HISS estimates the genes number close to the expected value, with a mean of 4 (stdev. 1.5 and mode of 4.3) times greater in Lib5X against Lib1X.

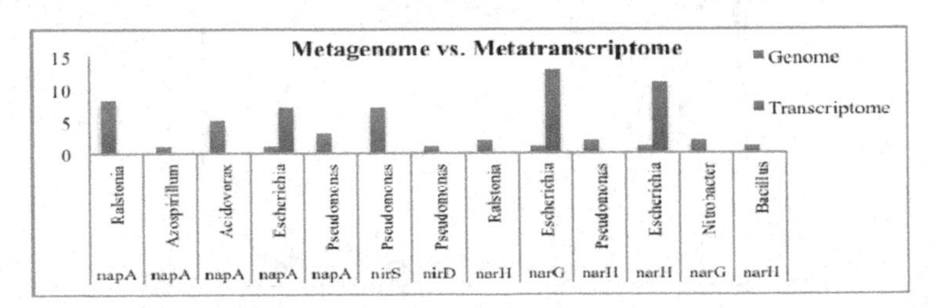

Fig. 3. Denitrification genes with different abundances detected in the metagenome and the metatranscriptome

Real Scenario. It was explored the functionality of HISS over real soil community sequences collected from potato (*Solanum phrueja*) crop, located in Cundinamarca department - Colombia. The system detects denitrification genes in metagenome and some of these were confirmed on metatranscriptome but with different abundances (Fig. 3). HISS founds a high richness of functional genes associated to denitrification process of different genders, however just few were expressed. That was expected, since the ecosystem functional richness possesses an independent dynamics regarding to functional homogeny, nevertheless there is a close relationship with taxonomical diversity, such as taxa richness and abundance [18].

4 Discussion

CrSo probes and HISS detection system were developed for gene detection in short reads, they provide a versatile method to predict functional genes in soil metagenomes using probes. The number of probes for gene is critical to characterize them in the samples; the results exhibit a better characterization with one probe for gene. It is due to the hard task of retrieve multiple subsequences from the gene with low cross hybridization with no overlapping.

The HISS selection function is stringent enough to predict gene fragments of at least 21bp at a gender taxonomic level. In contrast to other approaches that are able to assign as short as 60bp [20]. CrSo and HISS have showed a high gene abundance sensitivity using simulated metagenomic data sets, in contrast to similarity-based binning approaches that have indicated that a significant amount of reads get unclassified or even misclassified [8, 11].

In the current study, a novel functional gene assignation with gender taxonomical level has been devised that attempts to characterize soil metagenomes according to its functional groups. The evaluation of real samples suggests that it is possible to track the behavior of soil community in order to develop comparable functional and metabolic pathway profiles of communities.

This work has calculate the similarity of Nitrogen and Phosphorus functional genes at different phylogenetic levels, it confirms amoA gene as specie-specific marker [21], however its behavior differs at higher taxonomical levels.

References

1. Altschul, S.F., et al.: Basic local alignment search tool. J. Mol. Biol. 215(3), 403–410 (1990)
2. Amann, R.I., et al.: Phylogenetic identification and in situ detection of individual microbial cells without cultivation. Microb. Rev. 59(1), 143 (1995)
3. Angly, F.E., et al.: Grinder: a versatile amplicon and shotgun sequence simulator. Nucleic Acids Res. 40(12), 94 (2012)
4. Darling, A.E., et al.: The design, implementation, and evaluation of mpiBLAST. Presented at the Conference on Linux Clusters: The HPC Revolution 2003 in ClusterWorld Conference & Expo and the 4th International (2003)
5. Eddy, S.R.: Profile hidden Markov models. Bioinfor. 14(9), 755–763 (1998)
6. Edgar, R.C.: Search and clustering orders of magnitude faster than BLAST. Bioinformatics 26(19), 2460–2461 (2010)
7. Foerstner, K.U., et al.: Comparative Analysis of Environmental Sequences: Potential and Challenges. Philosophical Transactions: Biological Sciences 361(1467), 519–523 (2006)

8. Haque, M., et al.: SOrt-ITEMS: Sequence orthology based approach for improved taxonomic estimation of metagenomic sequences. Bioinformatics 25(14), 1722–1730 (2009)
9. Heshan Lin et al.: Efficient Data Access for Parallel BLAST. Presented at the 19th IEEE Internat. Parallel and Distributed Processing Symposium (2005)
10. Hugenholtz, P., Tyson, G.W.: Microbiology: metagenomics (2008)
11. Huson, D.H., et al.: MEGAN analysis of metagenomic data. Genome Res. 17(3), 377–386 (2007)
12. Huson, D.H., et al.: Methods for comparative metagenomics. BMC Bioinformatics 10(suppl. 1), S12 (2009)
13. Kane, M.D., et al.: Assessment of the sensitivity and specificity of oligonucleotide (50mer) microarrays. Nucleic Acids Res. 28(22), 4552–4557 (2000)
14. Liebich, J., et al.: Improvement of oligonucleotide probe design criteria for functional gene microarrays in environmental applications. Appl. Environ. Microbiol. 72(2), 1688–1691 (2006)
15. Mavromatis, K., et al.: Use of simulated data sets to evaluate the fidelity of metagenomic processing methods. Nat. Methods 4(6), 495–500 (2007)
16. Mende, D.R., et al.: Assessment of metagenomic assembly using simulated next generation sequencing data. PLoS ONE 7(2), e31386 (2012)
17. Moran, M.: Metatranscriptomics: eavesdropping on complex microbial communities. Microbe (2009)
18. Mouillot, D., et al.: Functional regularity: a neglected aspect of functional diversity. Oecologia 142(3), 353–359 (2004)
19. Noguchi, H., et al.: MetaGene: prokaryotic gene finding from environmental genome shotgun sequences. Nucleic Acids Res. (2006)
20. Rho, M., et al.: FragGeneScan: predicting genes in short and error-prone reads. Nucleic Acids Res. 38(20), e191 (2010)
21. Rotthauwe, J.H., et al.: The ammonia monooxygenase structural gene amoA as a functional marker: molecular fine-scale analysis of natural ammonia-oxidizing populations. Appl. Environ. Microbiol. 63(12), 4704–4712 (1997)
22. Sandberg, R.R., et al.: Capturing whole-genome characteristics in short sequences using a naïve Bayesian classifier. Genes Dev. 11(8), 1404 (2001)
23. Schloss, P.D., Handelsman, J.: Introducing SONS, a Tool for Operational Taxonomic Unit-Based Comparisons of Microbial Community Memberships and Structures. Appl. Environ. Microbiol. 72(10), 6773–6779 (2006)
24. Tamura, K., et al.: MEGA5: Molecular Evolutionary Genetics Analysis Using Maximum Likelihood, Evolutionary Distance, and Maximum Parsimony Methods. Mol. Biol. E 28(10), 2731–2739 (2011)
25. Tringe, S.G., et al.: Comparative metagenomics of microbial communities. Science 308(5721), 554–557 (2005)
26. Urich, T., et al.: Simultaneous assessment of soil microbial community structure and function through analysis of the meta-transcriptome. PLoS ONE 3(6), e2527 (2008)
27. Vila-Costa, M., et al.: Transcriptomic analysis of a marine bacterial community enriched with dimethylsulfoniopropionate. ISME J., 1–11 (2010)
28. Wernersson, R., Nielsen, H.B.: OligoWiz 2.0–integrating sequence feature annotation into the design of microarray probes. Nucleic Acids Res. 33(Web Server issue), W611–W615 (2005)
29. Ye, J., et al.: Primer-BLAST: A tool to design target-specific primers for polymerase chain reaction. BMC Bioinformatics 13(1), 134 (2012)

Identification of Differently Expresed Proteins Related to Drillings Fluids Exposure in Hydractinia Symbiolongicarpus by Mass Spectrometry

Iván Aurelio Páez-Gutiérrez and Luis Fernando Cadavid

Departament of Biology and Institute of Genetics, Universidad Nacional de Colombia, Cra. 30 # 45-08, Bogotá, Colombia
iapaezg@unal.edu.co

Abstract. Due to the interest of increasing the national oil production in Colombia, seawater exploration it is a real option. During well production, drilling fluids are used to support drilling boreholes into the earth or sea platform. Water-based drilling fluid (WBF) can be composed of several materials as clay, barite, emulsifier additives and metal ions. Although there are studies that evaluate mid and long-term effects to exposure to this fluids there is no way to monitor any potential effect on seawater organisms. *Hydractinia symbiolongicarpus* was used to evaluate the effect of WBF exposure in order to identify possible biomarkers that may aid to monitor seawater organism's health and prevent any ecosystem disruption. *Hydractinia* was exposed to WBF and 30 Up-Regulated spot proteins were determined by 2D-DIGE. All spots proteins were identified by mass spectrometry using MASCOT and X! Tandem search engine. Glutathione-S-transferase and peroxiredoxin VI were recognized and proposed as possible biomarker since they are related to cell detoxification and redox homeostasis, respectively.

Keywords: Mass spectrometry, Hydractinia, Drilling fluids, biomarkers.

1 Introduction

Colombia is ranked 28th in the world as crude producer with close to a million barrels per day annually. With the goal of increasing the total national oil production, exploration the seawater platform has became a real option. Seawater exploration involves offshore platforms, which would support the excavation machinery to drill boreholes. The adequate well production depends on using drilling fluids that contribute to the hydrostatic pressure, cooling the system and cleaning during the excavation process from drill cuttings. Those fluids are mainly composed by clay, polymers, barite ($BaSO_4$), several metal ions and emulsifier additives [1, 2].

Toxicological evaluation of drilling fluids has been done in few marine organisms. In Artemia salina and the diatoms Pheadactylum, it has shown a variable lethality [3], whereas in stony corals such as Montastrea annularis it reduces the growth rate [4], produces retraction of polyps and in some cases cause the organism's death [5].

L.F. Castillo et al. (eds.), *Advances in Computational Biology*,
Advances in Intelligent Systems and Computing 232,
DOI: 10.1007/978-3-319-01568-2_49, © Springer International Publishing Switzerland 2014

Although these studies have evaluated the effects of mid and long-term exposure to drilling fluids, little is known about the acute effects of these pollutants on the cellular physiology of marine organisms. Evaluation of drilling fluids exposure result in a general idea of how some organism responds to those xenobiotic fluids in a long period of time. However this is not an alternative right now, it is vital to monitor any potential effect in seawater organisms that could damage ecosystem homeostasis. Here we identified proteins differentially expressed as a consequence of an acute exposure to drilling fluids in the cnidarian model Hydractinia symbiolongicarpus (Cnidaria: Hydrozoa).

Environmental stress factors affect physiological and immunology functions of several seawater invertebrates and chemical pollutants like this present in drilling fluids. Thus, components as heavy metals and aromatic hydrocarbons could be modifying their immunocompetence and survival, affecting the ecosystem. In particular, cnidarian organisms are very susceptible to any physical o chemical change, for example, in Acropora grandis changes only in temperature generate altered expression of some conserved proteins as Hsp70 and heme-oxygenases [6]. Those proteins are conserved across species, so cnidarians, as basal organisms, may share many genetic elements with other phyla. In that way those conserved proteins could be used to deduce the physiological effect in other organisms [7]. H. symbiolongicarpus is a seawater cnidarian organism, which could be used in order to determine any possible effect generated by drilling fluids and to extrapolate to other organisms since it is the only seawater cultivable cnidarian. It is a dioicus colonial organism that lives on the hermit crab shells of Littorina gastropods species in the US east cost [8].

Colombia is becoming an important crude oil productive country and as part of a national legacy, local government is venturing in offshore exploration. In order to make this venture sustainable, an initiative to determine possible toxicity by drilling fluids would be assessed by proteomic identification. The main goal is to identify possible acute toxicity gene biomarkers in H. symbiolongicarpus, using 2D-DIGE and MALDITOF/MSMS to recognize differentially expressed proteins. Certainly, it will be necessary to determine the exact genes that code these proteins by transcriptome analyses qPCR.

2 Materials and Methods

2.1 Animal Cultures and Water-Based Drilling Fluid Preparation

A wild type H. symbiolongicarpus colony (HWB-29) was collected from Woods Hole (Massachusetts, USA) in 2009, cultured in artificial seawater at 20°C and fed three times week with Artemia salina nauplii. The water-based drilling fluids were prepared by mixing a 500 ml sample of mud with artificial seawater (distilled water) in a volumetric mud-to-water ratio of 1:9. The suspended particulate phase (SPP) was decanted into a container and a solution of 50% v/v SPP (1:1 seawater:mud) was used for the exposure.

2.2 Exposure and Protein Extraction

Colonies were incubated in 50% v/v SPP for 1.5 h and sampled 0, 3, 6, 12, 24, 48 hours post-exposure (hpe). Control subclones were sampled at the end of the experiment, 48 hours. Samples were immersed in 15 ml buffer containing 30 mM Tris-HCl pH 8.8, 7 M urea, 2 M thiourea, 4% CHAPS and 1X protease inhibitor cocktail (cOmplete ULTRA Tablets, Mini EDTA-free, ROCHE) and stored at -80°C. Tissue was later lysed with a glass tissue grinder and by pipetting at 4°C. Then, the homogenates were centrifuged for 30 minutes at 6000 g at 4°C and the supernatants were collected. Cold phenol was added to the supernatants in a 1:1 proportion, mixed by vortex, and submerged in hot water 70°C for ten minutes. The samples were then place on ice for 5 minutes and centrifuged for 30 minutes at 6000 g. After the organic phase was collected, a solution of 3:1 of ice-cold acetone:methanol was added and the mix was stored at -80°C overnight. Total protein was precipitated by centrifugation for one hour at 13 000 g and the pellet was washed with the same solution once.

2.3 2D-DIGE Mass Spectrometry

To identify proteins differentially expressed in Hydractinia pellets were run on 2D-DIGE (3-10 pH range, 12% v/v acrylamide), procedure performed by Applied Biomics (Hayward, CA, USA). 30 ug of each sample and the control sample were labeled with cyanine dyes, Cy5 and Cy3 (GE Healthcare) respectively, and the same amount of a pooled standard, which contained equal amounts of control and post-exposure samples, was labeled with Cy2. The standard was used to normalize the Cy3 and Cy5 samples. All three labeled samples were combined and solubilized in a buffer containing 8 M urea, 4% CHAPS, 20 mg/ml DTT, 2% pharmyltes and bromophenol blue. The immobilized pH gradient strips (pH 3-10, linear range 18 cm) were rehydrated overnight with destreak rehydration buffer which contained 7 M urea, 2 M thiourea, 4% CHAPS, 20 mg/ml DTT, 1% pharmalytes and bromophenol blue. For isoelectric focusing, the manufacturer's protocol was followed (GE Healthcare). Then, the IPG strips were equilibrated in buffer 1 (50 mM Tris-HCl pH 8.8, 6 M urea, 10mg/ml DTT, 2% SDS, 30% glycerol and a trace of bromopehol blue), later they were rinsed in equilibration buffer 2 (same as buffer 1 but 45 mg/ml iodoacetamide instead). The strips were then rinsed in SDS-gel running buffer before they were transferred into gradient SDS-gel (12% acrylamyde). Image scans were performed immediately using Typoon TRIO (GE Healthcare) according to manufacturer's recommendations. The images of Cy2-, Cy3- and Cy5- labeled samples, were acquired by using excitation/emission values of 488/520, 523/580 and 633/670 nm, respectively. The scanned images were analyzed by ImageQuantTL software (GE-Healthcare). The normalization of each spot amount and determination of spots size, abundance and statistics for the samples were performed using data analysis software DeCyder software version 6.5 (GE-Healthcare). The differentially expressed spots were detected by using the internal standard for normalization and then, matching the same spot across the gels. Proteins from post-exposure samples that showed increases compared to the control were selected for further mass spectrometry. The differentially expressed proteins or spots were subjected of mass spectrometry using MALDI-TOF/MS analysis performed by Applied Biomics (Hayward, CA, USA).

2.4 Protein Identification

The mass spectra of selected Up-Regulated proteins were analyzed in MascotDetiller version 2.4.3.3 evaluation software (Matix Science). The spectra were identified against a costumed database of predicted proteins from the H. symbiolongicarpus transcriptome (unpublished data). The protein database was modified with DBToolkit [9] and peptide matching was performed with two different approaches: searching with MASCOT by Applied Biomics (Hayward, CA, USA) and with X! Tandem [10]. Both cases with fixed carbamidomethylation of cysteine and variable oxidation of methionine residue. Then, the identified Up-Regulated proteins of H. symbiolongicarpus by X! Tandem were aligned against the non-redundant protein and nucleotide NCBI database in order to predict their function using BLASTp algorithm [11].

3 Results

3.1 Differentially Expressed Proteins

Protein samples were run on nine different 2D-DIGEs and each sample was run at least twice as is show in Table 1. The scanned image and statistical analysis with DeCyder detected and differentially quantified protein spots in the scanned images after matching, quantitation and statistical analysis across gels, and provided the average ratio of the spot relative abundance of hpe-samples/control-sample. A total of 109 spots were detected, where 56 were always Up-Regulated, 24 were always Down-Regulated and 29 showed variable protein expression. 30 Up-Regulated protein spots were selected for identification using mass spectrometry by MALDI-TOF MS/MS.

Table 1. Experimental design for 2D-DIGE differential profile

Gels	Labeled samples			Gels	Labeled samples			Gels	Labeled samples		
	Cy2	Cy3	Cy5		Cy2	Cy3	Cy5		Cy2	Cy3	Cy5
1	ST	C	0 hpe	4	ST	C	12 hpe	7	ST	0 hpe	3 hpe
2	ST	C	3 hpe	5	ST	C	24 hpe	8	ST	6 hpe	12 hpe
3	ST	C	6 hpe	6	ST	C	48 hpe	9	ST	24 hpe	48 hpe

Cy2-, Cy3-, Cy-5: Cyanine dyes corresponding to each sample. ST: pooled of all samples, including control. C: Control sample. 0-48 hpe: Post-exposure samples.

3.2 Peptide Identification

The peptide identification of those 30 proteins was carried out following two main approaches. The first approach involved searching the non-redundant database using MASCOT (Matrix Science) search engine, performed by Applied Biomics (Hayward, CA, USA) and showed homologous sequences identified in other organisms. In order to confirm the sequence identity in Hydractinia symbiolongicarpus, the second approach involved the use of a custom database to determine the exact sequence

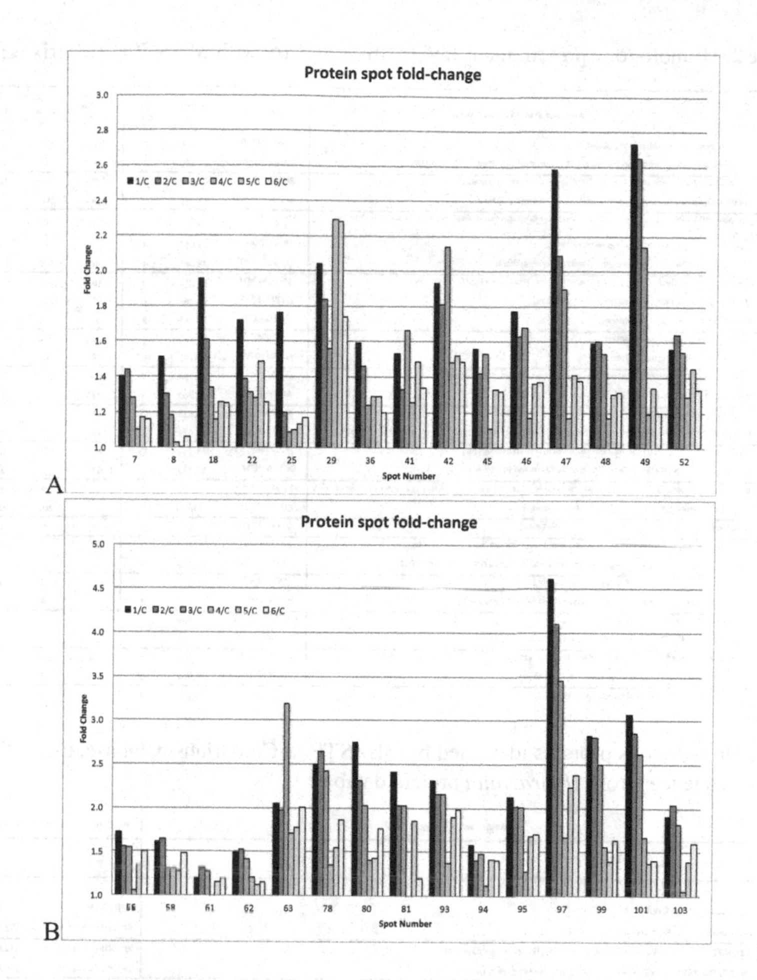

Fig. 1. Protein spot fold change of each differential Up-Regulated protein spot compared to the control

with X! Tandem [10]. Translating RNA sequences from a RNA-seq previous procedure generated this custom database. Proteins identified with each procedure are shown in Table 2 and 3.

Searching with MASCOT resulted in the homologous identification of all 30 proteins, whereas X! Tandem only identified 21 spot proteins as it is showed in table 2 and 3. Thought it is not possible to relate directly each MASCOT result with the correct sequence in Hydractinia, it is plausible to match peptide mass fingerprinting with related proteins in the closest organisms [12, 13]. From the 30 spots identified, 22 belong to Cnidarian organisms and 8 to other species. Even though 8 identified sequences did not belong to the Cnidarian phylum (18, 42, 48, 52, 58, 63, 80 and 95 spots), they would be useful for functional prediction (Table 2).

The proteins identified by X! Tandem [10] were used for database searching with BLASTp [14, 15] against the NCBI non-redundant database in order to predict their

Table 2. Homologous protein identified by mass spectra with MASCOT (MatrixScience)

Spot number	Top Ranked Protein Name [Species]	Accession No.	Protein MW	Pep. Count	Protein Score	Protein Score C.I.%	Total Ion Score	Total Ion C.I.%
7	Phospholipase A2 OS=Urticina crassicornis PE=1 SV=1	PA2_URTCR	17,308	2	22	0	14	96
8	Leucine-rich repeat protein soc-2 homolog OS=Nematostella vectensis GN=v1g189306 PE=3SV=1	SHOC2_NEMVE	64,199	3	22	0	21	99
18	PREDICTED: gelsolin-like protein 2-like [Strongylocentrotus purpuratus]	gi|115891439	41,202	1	129	100	129	100
22	PREDICTED: similar to Actin, non-muscle 6.2 [Hydra magnipapillata]	gi|221122242	41,770	12	514	100	427	100
25	Probable tRNA threonylcarbamoyladenosine biosynthesis protein osgep OS=Nematostella vectensis GN=os	OSGEP_NEMVE	36,964	2	15	0	15	96
29	Tropomyosin-2 OS=Podocoryne carnea GN=TPM2	TPM2_PODCA	29,044	4	18	0	8	84
36	PREDICTED: similar to 14-3-3 protein B [Hydra magnipapillata]	gi|221116303	28,081	4	246	100	235	100
41	Lipoyl synthase, mitochondrial OS=Nematostella vectensis GN=v1g225637 PE=3SV=1	LIAS_NEMVE	33,111	2	16	0	11	94
42	alanyl-tRNA synthetase [Arthrospira maxima CS-328]	gi|209523613	96,284	6	67	0	57	95
45	peroxiredoxin VI [Laternula elliptica]	gi|190360997	25,741	8	126	100	85	100
46	peroxiredoxin VI [Laternula elliptica]	gi|190360997	25,741	7	113	100	79	100
47	peroxiredoxin VI [Laternula elliptica]	gi|190360997	25,741	5	83	96	64	99
48	3-hydroxyacyl-CoA dehydrogenase [Cupriavidus taiwanensis LMG 19424]	gi|194291333	25,757	5	114	100	92	100
49	Mitrocomin OS=Mitrocoma cellularia GN=MI17 PE=2 SV=1	MYTR_MITCE	22,700	3	23	0	14	96
52	chaperonin subunit gamma CCTgamma [Trichomonas vaginalis G3]	gi|123448445	60,625	11	84	97		
55	Branched-chain-amino-acid aminotransferase OS=Nematostella vectensis GN=v1g246094 PE=3 SV=1	BCAT_NEMVE	45,629	3	26	2	19	99
58	PREDICTED: keratin, type I cytoskeletal 10 isoform 3 [Pan troglodytes]	gi|114667511	58,244	15	207	100	128	100
61	Dynactin subunit 6 OS=Nematostella vectensis GN=dctn6 PE=3 SV=1	DCTN6_NEMVE	20,141	2	21	0	14	96
62	Branched-chain-amino-acid aminotransferase OS=Nematostella vectensis GN=v1g246094 PE=3 SV=1	BCAT_NEMVE	45,629	5	36	91	27	100
63	PREDICTED: neurocalcin homolog [Nasonia vitripennis]	gi|156538104	21984.8	5	158	100	135	100
78	Annexin-B12 OS=Hydra vulgaris GN=ANXB12 PE=1 SV=1	ANX12_HYDVU	35,087	4	25	0	13	94
80	hypothetical protein GCWU000342_00030 [Shuttleworthia satelles DSM 14600]	gi|229827982	70,466	5	68	0	58	96
81	Elongation factor Ts, mitochondrial OS=Nematostella vectensis GN=v1g215604 PE=3 SV=1	EFTS_NEMVE	31,881	9	44	98		
93	Luciferin-binding protein OS=Renilla reniformis PE=1	LBP_RENRE	20,527	3	29	51	19	98
94	Heat shock 70 kDa protein OS=Hydra vulgaris GN=HSP70.1 PE=3 SV=1	HSP70_HYDVU	71,424	4	18	0	10	93
95	unnamed protein product [Ostreococcus tauri]	gi|308812722	58,288	3	64	0	58	97
97	Dynactin subunit 6 OS=Nematostella vectensis GN=dctn6 PE=3 SV=1	DCTN6_NEMVE	20,141	2	24	0	18	98
99	Costars family protein v1g158749 OS=Nematostella vectensis GN=v1g158749 PE=3SV=1	COSA_NEMVE	9,345	1	98	100	93	100
101	Toxin CqTX-A OS=Chiropsalmus quadrigatus PE=2 SV=1	CTXA_CHIQU	51,769	5	21	0	10	91
103	Dynactin subunit 6 OS=Nematostella vectensis	DCTN6_NEMVE	20,141	3	18	0	6	85

Table 3. Homologous proteins identified by BLASTP in Cnidarians database, using X!Tandem predicted sequences from *Hydractinia* protein database

Spot number	Proteins identified by BLASTP	Accession Number	Score	e-value
7	-	-	-	-
8	-	-	-	-
18	-	-	-	-
22	non-muscle actin II [Hydractinia echinata]	ADR10434.1	785.0	0.0E+00
	predicted protein [Nematostella vectensis]	XP_001639332.1	353.0	2.0E-116
25	PREDICTED: D-arabinitol dehydrogenase 1-like [Hydra magnipapillata]	XP_002164947.1	311.0	2.0E-102
29	PREDICTED: tropomyosin-1-like isoform 1 [Hydra magnipapillata]	XP_002165211.1	278.0	2.0E-90
36	PREDICTED: 14-3-3 protein zeta-like [Hydra magnipapillata]	XP_002161942.1	349.0	3.0E-118
	14-3-3 protein A [Hydra vulgaris]	AAN87349.1	349.0	3.0E-118
41 42	PREDICTED: Williams-Beuren syndrome chromosomal region 27 protein-like isoform 2 [Strongylocentrotus purpuratus]	XP_781570.1	93.6	4.0E-20
45	PREDICTED: glutathione S-transferase-like [Hydra magnipapillata]	XP_002158301.1	129.0	2.0E-34
46 47	PREDICTED: peroxiredoxin-6-like [Hydra magnipapillata]	XP_002170014.2	352.0	4.0E-120
48	predicted protein [Nematostella vectensis]	XP_001636473.1	351.0	3.0E-119
	PREDICTED: 3-hydroxyacyl-CoA dehydrogenase type-2-like [Amphimedon queenslandica]	XP_003385670.1	317.0	1.0E-105
49	PREDICTED: uncharacterized protein LOC100207073 [Hydra magnipapillata]	XP_002166352.1	70.5	1.0E-12
52	proteasome subunit beta type-2 [Ictalurus punctatus]	NP_001188165.1	241.0	1.0E-77
	predicted protein [Nematostella vectensis]	XP_001624954.1	240.0	2.0E-77
55	PREDICTED: glutathione S-transferase-like [Hydra magnipapillata]	XP_002158301.1	175.0	3.0E-51
58	PREDICTED: uncharacterized protein LOC100208799 [Hydra magnipapillata]	XP_002167799.2	229.0	1.0E-69
61	predicted protein [Nematostella vectensis]	XP_001637520.1	184.0	3.0E-55
	PREDICTED: transcription factor BTF3 homolog 4-like [Hydra magnipapillata]	XP_002156589.1	267.0	6.0E-89
62	PREDICTED: uncharacterized protein LOC101238714 [Hydra magnipapillata]	XP_004206074.1	152.0	2.0E-41
63	PREDICTED: neurocalcin homolog [Hydra magnipapillata]	XP_002159500.2	380.0	2.0E-132
78	PREDICTED: phospholipase A2-like [Hydra magnipapillata]	XP_002157393.1	95.5	3.0E-21
80	hypothetical protein BRAFLDRAFT_67403 [Branchiostoma floridae]	XP_002614013.1	51.2	1.0E-05
	predicted protein [Nematostella vectensis]	XP_001638232.1	115.0	2.0E-25
81	PREDICTED: uncharacterized protein LOC100197130, partial [Hydra magnipapillata]	XP_002161711.2	565.0	0.0E+00
	PREDICTED: chitotriosidase-1-like [Hydra magnipapillata]	XP_002163945.2	523.0	0.0E+00
93	PREDICTED: NHP2-like protein 1-like [Hydra magnipapillata]	XP_002160406.1	214.0	3.0E-69

identity. Comparing both results, some of them match each other. The corresponding spot identification of 22, 29, 36, 45, 46, 47, 48 and 63 spots, showed that the Hydractinia protein sequence could be definitely associated to the predicted protein, as the following: non-muscle actin II, tropomyosin, 14-3-3 protein, glutathione S-transferase, peroxiredoxin 6, 3-hydroxyacil-CoA dehydrogenase type 2, neurocalcin (Tables 2 and 3).

Other proteins, which were identified with MASCOT were not identified with X!Tandem and vice versa. Nine spots were assigned to any sequence in the predicted proteome, but their function could be associated to proteins in Hydractinia. Spots 7, 8, 18, 94, 95, 99, 101 and 103, were identified as homologous to phospholipase A2 in the sea anemone Urticina crassicornis; leucine-rich repeat protein soc-2, dynactin subuit 6 and costars family protein v1g158749 in the sea anemone Nematostella vectensis; gelson-like protein 2-like in the purple sea urchin Strongylocentrotus purpuratus; Heat shock 70 kDa protein in Hydra vulgaris (Table 2).

4 Discussion

Any component included in drilling fluids could be potentially toxic to H. symbiolongicarpus, and that could be recognized in the differentially expression of proteins of exposure samples related to the control sample. One of the processes that could be affected is the oxidative homeostasis represented in the up-regulation of antioxidant proteins as peroxiredoxin 6-like protein. This protein has been related to the "defensome" in N. vectensis contributing to prevent altered gene expression or damage to biomolecules [16]. The expression of this gene has been considered as a possible biomarker related to cytotoxicity due to hyperoxidative stress in HeLa cells [17] and to disturbed ecosystem in the oyster Crassostrea gigas [18].

Other protein that was affected by the WBF exposure was gluthathione S-transferase-like (GST) protein. It has been determined that this enzyme is crucial for detoxification process as it catalyzes the addition of gluthathione to electrophilic residues decreasing the reactivity of xenobiotic compounds [19]. Exposure to organic compounds as benzo(a)pyrene has showed that GST activity increased so it could be a biomarker in cnidarian larvae of Porites astreoides and Pocillopora damicornis [20, 21].

Heat shock 70kD protein-like, could be also considered a good biomarker since it has been suggested that this protein is related to the "defensome" in N. vectensis and it is overexpressed [16]. This chaperone has been identified as part of the defense mechanisms to protect against synthetic based drilling fluids in pink snapper Pagrus auratus [22] or exposure to copper and oil dispersant in the coral Montastrea franksi [23].

The recognition of proteasome subunit ß-like and BTF3 (Basic Transcription Factor 3)-like suggests that could be an overexpression those proteins related to protein degradation and the correct assembly of proteasome [24] and apoptosis [25], respectively. Also, overexpression of 14-3-3-like protein could be relevant due to that it is involved in a vast number of cellular processes as protein synthesis, protein interaction with Bcl2 member in apoptosis, metabolism, cytoskeletal organization and calcium signaling [26]. Some other identified proteins, as non-muscle actin II-like, tropomyosin-like and dynactin-like, could be just playing a structural role in Hydractinia.

Certainly, mass spectrometry allows us to identify individual proteins that could be involved in a wide variety of phenotypes. A correct determination of proteins depends on the organism database or the possibility to make a cross species searching. Thus, the function or identity assigned to those predicted proteins must be validated by the definition of the genes in the Hydractinia genome as the validation of differently expressed genes by qPCR in exposed organisms.

Acknowledgements. Thank you to personnel from Institute of Genetics of Universidad Nacional de Colombia and people from Group of Evolutionary Immunology and Immunogenetics for personal and technical support. This work was supported by grants from Instituto Colombiano del Petróleo.

References

1. Caenn, R., Chillingar, G.V.: Drilling fluids: State of the art. Journal of Petroleum Science and Engineering 14, 221–230 (1996)
2. NRC NRC (U. S. C on ADT, Drilling and Excavation Technologies for the Future. 175 (1994)
3. Terazaghi, C., Buffagni, M., Cantelli, D., et al.: Physical-chemical and ecotoxicological evaluation of water based drilling fluids used in Italian off-shore. Chemosphere 37, 2859–2871 (1998)
4. Dodge, R.E.: Effects of drilling mud on the reef-building coral Montastrea annularis. Marine Biology 71, 141–147 (1982)
5. Thompson Jr., J.H., Shinn, E.A., Bright, T.J., Thompson, J.: Effects of Drilling Mud on Seven Species of Reef-Building Corals as Measured in the Field and Laboratory. In: Marine Environmental Pollution, 1 Hydrocarbons. Series RAGBT-EO, ch. 16, pp. 433–453. Elsevier (1980)
6. Mydlarz, L.D., Jones, L.E., Harvell, C.D.: Innate Immunity, Environmental Drivers, and Disease Ecology of Marine and Freshwater Invertebrates. Annual Review of Ecology, Evolution, and Systematics 37, 251–288 (2006)
7. Ball, E.E., Hayward, D.C., Saint, R., Miller, D.J.: A simple plan–cnidarians and the origins of developmental mechanisms. Nature Reviews Genetics 5, 567–577 (2004)
8. Frank, U., Leitz, T., Muller, W.A.: The hydroid Hydractinia: a versatile, informative cnidarian representative. BioEssays 23, 963–971 (2001)
9. Martens, L., Vandekerckhove, J., Gevaert, K.: DBToolkit: processing protein databases for peptide-centric proteomics. Bioinformatics (Oxford, England) 21, 3584–3585 (2005)
10. Craig, R., Beavis, R.C.: TANDEM: matching proteins with tandem mass spectra. Bioinformatics (Oxford, England) 20, 1466–1467 (2004)
11. Altschul, S.F., Gish, W., Miller, W., et al.: Basic local alignment search tool. Journal of Molecular Biology 215, 403–410 (1990)
12. Champagne, A., Boutry, M.: Proteomics of nonmodel plant species. Proteomics 13, 663–673 (2013)

13. Liska, A.J., Shevchenko, A.: Expanding the organismal scope of proteomics: cross-species protein identification by mass spectrometry and its implications. Proteomics 3, 19–28 (2003)
14. Altschul, S.F., Madden, T.L., Schäffer, A.A., et al.: Gapped BLAST and PSI-BLAST: a new generation of protein database search programs. Nucleic Acids Research 25, 3389–3402 (1997)
15. Altschul, S.F., Wootton, J.C., Gertz, E.M., et al.: Protein database searches using compositionally adjusted substitution matrices. The FEBS Journal 272, 5101–5109 (2005)
16. Goldstone, J.V.: Environmental sensing and response genes in cnidaria: the chemical defensome in the sea anemone Nematostella vectensis. Cell Biology and Toxicology 24, 483–502 (2008)
17. Kim, S.Y., Jo, H.-Y., Kim, M.H., et al.: H2O2-dependent hyperoxidation of peroxiredoxin 6 (Prdx6) plays a role in cellular toxicity via up-regulation of iPLA2 activity. The Journal of Biological Chemistry 283, 33563–33568 (2008)
18. David, E., Tanguy, A., Moraga, D.: Peroxiredoxin 6 gene: a new physiological and genetic indicator of multiple environmental stress response in Pacific oyster Crassostrea gigas. Aquatic Toxicology (Amsterdam, Netherlands) 84, 389–398 (2007)
19. Eaton, D.L., Bammler, T.K.: Concise review of the glutathione S-transferases and their significance to toxicology. Toxicological Sciences: An Official Journal of the Society of Toxicology 49, 156–164 (1999)
20. Farina, O., Ramos, R., Bastidas, C., García, E.: Biochemical responses of cnidarian larvae to mercury and benzo(a)pyrene exposure. Bulletin of Environmental Contamination and Toxicology 81, 553–557 (2008)
21. Vijayavel, K., Downs, C.A., Ostrander, G.K., Richmond, R.H.: Oxidative DNA damage induced by iron chloride in the larvae of the lace coral Pocillopora damicornis. Comparative biochemistry and physiology Toxicology & Pharmacology: CBP 155, 275–280 (2012)
22. Bakhtyar, S., Gagnon, M.M.: Toxicity assessment of individual ingredients of synthetic-based drilling muds (SBMs). Environmental Monitoring and Assessment 184, 5311–5325 (2012)
23. Venn, A.A., Quinn, J., Jones, R., Bodnar, A.: P-glycoprotein (multi-xenobiotic resistance) and heat shock protein gene expression in the reef coral Montastraea franksi in response to environmental toxicants. Aquatic toxicology (Amsterdam, Netherlands) 93, 188–195 (2009)
24. De, M., Jayarapu, K., Elenich, L., et al.: Beta 2 subunit propeptides influence cooperative proteasome assembly. The Journal of Biological Chemistry 278, 6153–6159 (2003)
25. Kusumawidjaja, G., Kayed, H., Giese, N., et al.: Basic transcription factor 3 (BTF3) regulates transcription of tumor-associated genes in pancreatic cancer cells. Cancer Biology & Therapy 6, 367–376 (2007)
26. Pauly, B., Lasi, M., MacKintosh, C., et al.: Proteomic screen in the simple metazoan Hydra identifies 14-3-3 binding proteins implicated in cellular metabolism, cytoskeletal organisation and Ca2+ signalling. BMC Cell Biology 8, 31 (2007)

In silico Modificiton of Cathelicidins Generates Analogous Peptides with Improved Antimycobacterial Activity

Sandra Chingaté, Carlos Yesid Soto, and Luz Mary Salazar*

Departamento de Química, Facultad de Ciencias, Universidad Nacional de Colombia, Carrera 30 No. 45-03. Bogotá, Colombia
{smchingatel,Cysotoo,lmsalazarpu}@unal.edu.co

Abstract. Tuberculosis is a bacterial infection that annually produces approximately 1.3 million of deaths around the world; this infectious disease has become an important public health problem due to the emergence of multidrug-resistant *Mycobacterium tuberculosis* strains. In this study, bioinformatics tools were used to design 15 aminoacid analogous sequences from the cathelicidin LL37, CAP 18 and PMAP 36 that *in silico* displayed improved helical structure and antibacterial activity in comparison to the native sequences. The analogous and native aminoacid sequences were synthesized using the Fmoc strategy, purified by RP-HPLC and characterized by MALDI-TOF and circular dichroism. According to an *in vitro* assay of viability on *M. smegmatis* mc^{2}155 cells, an analogous peptide from CAP-18 displayed a minimal inhibitory concentration 10-fold lower that the showed for the native aminoacid sequence. In addition, the designed analogous peptides showed human erythrocytes hemolysis lesser than 10% and significant inhibition of the basal ATPase activity of mycobacterial plasma membrane vesicles. The results obtained suggest that the designed peptides diminish the mycobacterial viability and could be useful as antituberculous compounds.

Keywords: Antimicrobial peptides, tuberculosis, cathelicidins, ATPase activity.

1 Introduction

Tuberculosis (TB) is an important infectious disease caused by *M.* tuberculosis. In 2011, there were an estimated 8.7 million new cases of TB (13% co-infected with HIV) and 1.4 million people died from TB. Treatment for drug-susceptible TB consists of a 6-month regimen of four first-line drugs: isoniazid, rifampicin, ethambutol and pyrazinamide; the emergence of multidrug resistant *M. tuberculosis* strains and the co-infection with HIV has aggravated the prevalence of TB [1-2]. Mycobacteria have unique characteristics which endow them with natural resistance to many commonly used antibacterial agents. The highly hydrophobic cell envelope provides mycobacteria a barrier to many antimicrobial compounds. In addition,

L.F. Castillo et al. (eds.), *Advances in Computational Biology*,
Advances in Intelligent Systems and Computing 232,
DOI: 10.1007/978-3-319-01568-2_50, © Springer International Publishing Switzerland 2014

mycobacteria also have transporters which flush out toxic compounds and enzymes that are able to hydrolyze or modify the antimicrobial compounds [3].

There is an urgent need for the development of new molecules that can reduce the problem of mycobacterial resistance. In recent years, cationic antimicrobial peptides (AMPs) have drawn much attention as a promising antibacterial agent that could overcome the resistance problem. AMPs display ability to disrupt bacterial membranes via non-specific electrostatic interactions with the phospholipids [4,5]. Extensive structure-activity relationship studies have revealed that net charge, amphipathicity and structural tendency are among the most important physicochemical and structural parameters that dictate the ability of AMPs to interact with and disrupt membranes [6,7].

Cathelicidins are a group of structurally diverse AMPs found in several mammalian species, including humans, monkeys, horses, cattle, sheep, goats, pigs, rabbits, mice and guinea pigs [8]. The cathelicidin LL37 is a multifunctional molecule that may mediate various host responses, including bactericidal action, chemotaxis, ephitelial cell activation, angiogenesis, epithelial wound repair and activation of chemokine secretion [9]. Porcine cathelicidins were among the first groups of mammalian AMPs isolated, including the three porcine myeloid AMPs PMAP 23, PMAP 36 and PMAP 37 [10]. The cathelicidin CAP 18 also displays a potent antimicrobial activity against both Gram-negative and Gram-positive bacteria [11].

In a previous work Santos et al 2012, demonstrated that the modified peptide MIAP inhibits the F_1F_0 ATPase activity of mycobacterial plasma membrane suggesting alternative targets involved in cell viability that could be altered by antimicrobial peptides against mycobacteria [12]. In this work, *in silico* modifications of 15 aminoacids sequences from cathelicidins produced analogous peptides with improved structural and antimycobacterial characteristics, that also inhibited the basal ATPase activity of mycobacterial plasma membrane.

2 Materials and Methods

2.1 Mycobacterial Strains and Culture

The fast-growing *M. smegmatis* mc^{2}155 strain was used in this work. Mycobacteria were grown in LB broth with agitation at 70 rpm, 37 ° C until an OD 600: 0.4-0.5 for 4-6 days. This strain was used to estimate the minimum inhibitory concentration (MIC) of peptides and to isolate mycobacterial plasma membrane.

2.2 Bioinformatic Design of Antimicrobial Peptides

The server (http://www.imtech.res.in/raghava/antibp/index.html) and database (APD, http://aps.unmc.edu/AP/main.php) were used to identify positively charged and alpha-helical 15 amino acid sequences from LL37, CAP 18 and 36 displaying the best antimicrobial score [13]. The same analysis was used to modify the primary structure of the selected sequences to increase the antimicrobial score of the selected peptides. The prediction of structural and chemical characteristics were determined by PSSpred for secondary structure, and (http://zhanglab.ccmb.med.umich.edu/PSSpred and http://web.expasy.org/protparam/) for primary structure respectively.

2.3 Chemical Peptide Synthesis and Characterization

15 amino acid sequences of peptides derived from LL37, PMAP CAP 18 and 36 (peptides called A) and its analogs (peptides called B) were synthesized by the Fmoc solid phase methodology proposed by Houghten [14], using as support Rink amide solid resin with a substitution of 0.52 mmol / g. The peptides were analyzed by RP-HPLC in a column of ODS, in which 10.0 μL of sample eluted in a gradient A / B from 0 to 70% B (A: Deionized water 0.05% TFA, B: acetonitrile 0.05% TFA, flow 1.0 mL / min) for 45 minutes and then purified by preparative technique. The molecular weight was determined by MALDI mass spectrometry. Circular dichroism (CD) Assays were performed in a Jasco spectropolarimeter performer to determine the pattern of secondary structure. Spectra were registered in a 190-260 nm wavelength interval from average of three scans at a scan rate of 20nm/min. 0.30% aqueous 2,2,2-trifluoroethanol (TFE) in a volume of 200 uL end. Peptides at 2 mM in TFE-water mixture are stabilized but do not induce secondary structure.

2.4 Antimycobacterial Activity

The MIC of peptides was estimated on *M. smegmatis* $mc^2$155 cells using resazurin as an indicator of cells viability. 100 μL of serial dilutions of peptides (from 50 mg / mL) were mixed in 96-well microtiter plates with 100 μL of a 1:100 dilution of *M. smegmatis* $mc^2$155 cells growth until an OD_{600}= 0.4-0.5. Cells were incubated at 37 °C for 4 days, then 30μL of resazurin (5mg/mL) were added and cells were subsequently incubated for additional 24 hours. Bacterial growth was indicated by a change in color of broth from yellow to violet, and the MIC was defined as the lowest peptide concentration preventing this color change. Kanamycin (25 μg/mL) was used as control of growth inhibition and determinations were done in triplicate.

2.5 Hemolytic Activity

The hemolytic activity was determined in V-bottom 96-well plates (Corning, USA); 0,1 mL of red blood cells (RBC) of O Rh+ healthy human donors suspended in 0.9% saline solution (Baxter, Colombia) to achieve a final hematocrit concentration of 2% were mixed with 0,1 mL of serial dilutions of peptide (from 50μg/mL). The assay included control wells containing erythrocytes without peptide in saline solution (corresponding to 0%of hemolysis), erythrocyte in sterile distilled water (control for 100% of hemolysis), insulin as unrelated peptide and kanamycin at inhibitory concentration. After 2 h of incubation at 37°C with 5% of CO_2, plates were centrifuged at 3500 rpm and the supernatants were collected to determine the concentration of hemoglobin by measuring the absorbance at 540 nm in a digital spectrophotometer. Absorbance units from saline-diluted RBC that were not exposed to any substance were used as negative control (100% of viability). Absorbance readings from the supernatants from RBC solutions exposed to each peptide and Kanamycin were normalized with regards to the absorbance reading obtained for the RBC solutions exposed to sterile water, which were considered as 100% of hemolysis. The percentage of hemolysis was calculated in a MS Excel Worksheet were analyzed with the GraphPad Prism 5.0 Trial Version® program (LA Jolla, CA, USA) and calculate the median hemolytic concentration (HC_{50}), which is equivalent to the concentration of peptide that causes 50% of RBC lysis.

2.6 ATPasa Activity

M. Smegmatis mc^2155 plasma membrane was isolated based on the modified method by Santos and colleagues [12]. In brief, centrifuged bacteria was resuspended in a pH 7.4 buffer solution (10 mM MOPS, 1 mM EDTA and 0.3 mM phenylmethylsulfonyl fluoride) and lysed in a Mini Beadbeater-16 (Biospec, USA) for 7 min. Mycobacterial cell walls were isolated by centrifugation at 25000 rpm for 30 min, and the supernatants containing the plasma membranes were centrifuged at 35000 rpm for 90 min. The resulting pellets were frozen at -80°C in pH 7.4 buffer containing 10 mM MOPS and 0.08 g/mL sucrose, until further use. The ATPasa activity of plasma membrane vesicles was described by Cariani et al [15], who modified the Fiske & Subbarow method incorporating bismuth citrate to stabilize the molybdenum complexes formed as consequence of the Pi released in the enzymatic reaction. The ATPase assay was performed in 96-well microtiter plates, in which reactions including 3 mM $MgCl_2$ and 40 μg/mL mycobacterial plasma membrane (final concentration) were initiated by adding 3 mM Tris-ATP (final concentration) and stopped with sulfuric acid. The complex concentration was determined at 690 nm in a Microplate absorbance iMark ™ Reader (BioRad). Enzymatic reactions were separately supplemented with different peptide concentrations to estimate the effect of peptides on the basal ATPase activity of mycobacterial plasma membrane.

3 Results and Discussion

3.1 Bioinformatic Design Analog Peptides

AntiBP server and database APD2 were used to ascertain the antibacterial score of analogue peptides and native aminoacid sequences derived from cathelicidins LL37, CAP 18 and PMAP36. Thus, the 15 aminoacid fragment with the best antibacterial and helical structure from each amino acid sequence was identified (peptides A). The replacement of some aminoacids from peptides A allowed to find analogue amino acid sequences (peptides B) with an *in silico* improved helical structure and antibacterial activity. The Table 1 shows the results obtained for the LL37 peptide; in this case, the peptide A displayed an antimicrobial score of 2.108 (peptide A) that was increased until 3.502 (peptide B) by aminoacid replacements and also showing the highest positive charge of +7 and +8 respectively between the derived sequences of the natural peptide. In the figure 1, the peptides A and B from cathelicidins LL37, CAP 18 and PMAP36 are showed; the prediction of secondary structure, charge and amphipathic representation in the helical wheel of peptides are also showed.

Table 1. In silico peptide analysis LL37 with AntiBP tool Helical structure underlined

Peptide Sequence	Score	Charge
LLGDKFFRKSKEKIGKEFKRIVQRIKDFLRNLVPRTES		11
GRKSAKKIGKRAKRI B Peptide	3,502	8
LLGDFFRKSKEKIGK	1,065	5
LGDFFRKSKEKIGKE	0,636	5
GDFFRKSKEKIGKEF	1,789	5
DFFRKSKEKIGKEFK	1,412	6
FFRKSKEKIGKEFKR	1,64	6
FRKSKEKIGKEFKRI A Peptide	2,108	7
RKSKEKIGKEFKRIV	0,036	7
KSKEKIGKEFKRIVQ	0,207	5
SKEKIGKEFKRIVQR	-0,197	4
KEKIGKEFKRIVQRI	0,79	6
EKIGKEFKRIVQRIK	-0,437	4
KIGKEFKRIVQRIKD	-0,354	4
IGKEFKRIVQRIKDF	0,673	5
GKEFKRIVQRIKDFL	-0,328	3
KEFKRIVQRIKDFLR	-0,641	4
EFKRIVQRIKDFLRN	-0,058	3
FKRIVQRIKDFLRNL	-0,296	4
KRIVQRIKDFLRNLV	-1,181	4
RIVQRIKDFLRNLVP	0,743	3
IVQRIKDFLRNLVPR	0,519	3
VQRIKDFLRNLVPRT	-0,745	3
QRIKDFLRNLVPRTE	-0,99	2
RIKDFLRNLVPRTES	0,056	2

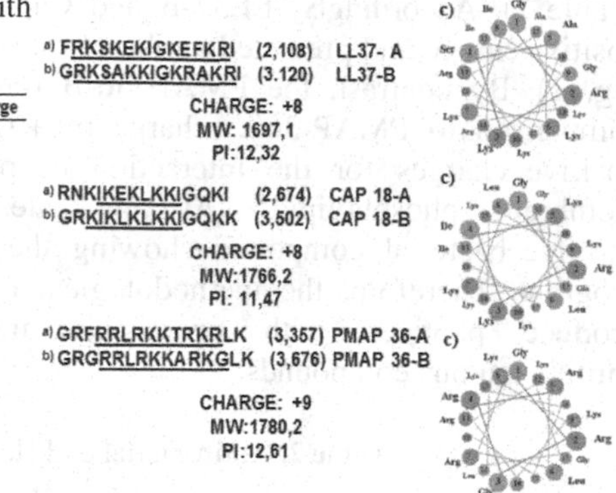

Fig. 1. In silico analysis of peptides a) Reference peptide: Primary structure, secondary (underlined) and antibacterial score. b) Analogue peptide: Primary structure, secondary (underlined) and antibacterial score. c) Helical Wheel of analogue peptide.

3.2 Peptide Synthesis and Characterization

The RP-HPLC analyses indicated high purity of the synthetized peptides that was evidenced by the presence of only one peak in the obtained chromatograms. In addition, the molecular mass of peptides determined by mass-spectrometry analysis corresponded to the expected theoretical masses. The analysis of the structural profile of peptides by DC showed two maximum: one at 190 nm with low molar ellipticity and other at 222 nm with an ellipticity near to zero that suggested an undefined structure with low tendency to helicity for all peptides. Conversely, CAP 18-A showed a maximum of 193 nm (positive π-π^* transition with Θ of 75063.5) and two minimums of 208 nm (π-π^* transition with Θ of -89 031) and 222 nm (n-π^*transition with Θ of -57 182) that suggested a helical structure for this peptide (data not shown). The lack of well-defined helical structure shows that DC could be an inconvenient technique for determining the secondary structure of peptides especially when DC experiments are performed at vacuum conditions that are completely different to the actual peptides environment in aqueous solution.

3.3 Antimicobacterial Activity, Hemolysis on Erythrocytes and Effect in the Plasma Membrane ATPase Activity of the Designed Peptides

In this study, the viability of the environmental, nonpathogenic and fast-growing *M. smegmatis* mc^{2}155 strain was estimated in the presence of peptides using resazurin as indicator of cells viability. We used *M. smegmatis* in the biological assays because this saprophytic species displays similar morphological and metabolic characteristics to the pathogenic mycobacteria *M. tuberculosis*. The MIC displayed by the analogues peptides LL37-B and CAP 18-B was 4 and 10-fold lower respectively than the native peptides; in addition, these peptides also showed a diminished hemolytic activity

(Table 2). Accordingly, LL37-B and Cap 18-B showed a higher helical structure, positive charge and predicted antibacterial score than the native peptides (Table 1 and Fig. 1). By contrast, the PMAP 36-B (charge of +10) showed a higher MIC in comparison to PMAP-36A (charge of +9) strongly suggesting the importance of positive charges for the interaction of peptides with the mycobacterial plasma membrane phospholipids. At any rate, PMAP-36A could be used as an antimycobacterial compound showing the effectiveness of the AntiBP prediction program. Therefore, the methodological approach used in this work was able to produce peptides with promising antimycobacterial activity as potential antituberculous compounds.

Table 2. Antibacterial and Hemolytic activity of peptides

PEPTIDE	MIC (μg/mL)	HC_{50} (μg/mL)	% Hemolysis	Charge
LL37-A	2400	847,7	4,04	+7
LL37-B	600	4735	1,1	+8
CAP 18-A	1500	212,6	24,77	+6
CAP 18-B	150	506,3	4,04	+8
PMAP 36-A	37,5	333,4	4,77	+10
PMAP 36-B	300	280,4	7,34	+9

3.4 Effect of Peptides on the Basal ATPasa Activity of Plasma Membrane Vesicles from *M. smegmatis*

Plasma membrane vesicles of *M. smegmatis* mc^{2}155 showed a basal ATPase activity of 1,46 ± 0,14 nmol Pi.min^{-1}.mg^{-1} corresponding to 100 % activity; when the enzymatic reaction was performed in the presence of the analogues peptides LL37-B, PMAP 36-B and CAP18-B, the ATPase activity of membranes was inhibited 50,60, 18,52 y 4,02 % respectively. Surprisingly, the PMAP 36-B peptide showed inhibition

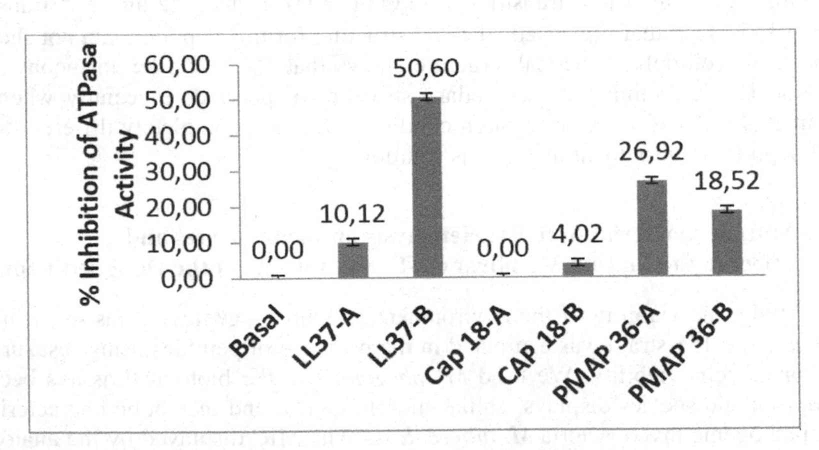

Fig. 2. Inhibition of Basal ATPase activity in vesicules of membrane of *M. smegmatis mc^2* 155 by peptides

of the ATPase activity lower than the showed by PMAP 36-A (Fig. 2); however, both peptides displayed inhibition of ATPase activity higher than the showed for CAP 18 peptides. Thus, the designed peptides significantly affected the mycobacterial ATPase activity of plasma membrane which controls biological processes essential for the cells viability, such as cellular homeostasis and transport of substances across plasma membrane [16].

4 Conclusions

In this work is demonstrated that *in silico* modification of structural and functional characteristics of peptides is an interesting strategy for the rational design of antimicrobial peptides. In this context, the viability assay performed in this work suggested that the predicted positive charge of peptides could be key for the antimycobacterial activity. Although the activity of antimicrobial peptides has not been directly related to the ATPase activity, we observed that the designed peptides inhibit the basal ATPase activity of mycobacterial plasma membrane; however, we can rule out that other factor could be involved in the inhibition of this enzymatic activity. In spite of the high hydrophobicity of the mycobacterial cell envelope, changes of the primary structure of peptides were able to improve their antimycobacterial activity of the native aminoacid sequences displaying low hemolytic activity. Finally, we can postulate that LL37-B could be proposed a potential antituberculous compound.

Acknowledgment. This work was supported by "División de Investigación Bogotá "(DIB) Universidad Nacional de Colombia, grants 13592 and 16060. The authors thank the assistance of Gabriela Delgado and José Julián Perez-Cordero from the Inmunotoxicology Research Group, Universidad Nacional de Colombia for their collaboration in preliminary experiments of hemolysis.

References

1. WHO Report 2012/Global Tuberculosis control.: The Global Plan to Stop TBC, 2006–2015. Geneva, Stop TBC Partnership and World Health Organization, (WHO/HTM/STB/2012.6) (2012)
2. Suchindran, S., Brouwer, E.S., Van Rie, A.: Is HIV infection a risk factor for multi-drug resistant tuberculosis? A systematic review. PloS One 4(5), e5561 (2009)
3. Kant, S., Maurya, A.K., Kushwaha, R.S., Nag, V.L., Prasad, R.: Multi-drug resistant tuberculosis: an iatrogenic problem. Bioscience Trends 4(2), 48–55 (2010)
4. Chou, H.-T., Wen, H.-W., Kuo, T.-Y., Lin, C.-C., Chen, W.-J.: Interaction of cationic antimicrobial peptides with phospholipid vesicles and their antibacterial activity. Peptides 31(10), 1811–1820 (2010)
5. Nguyen, L.T., Haney, E.F., Vogel, H.J.: The expanding scope of antimicrobial peptide structures and their modes of action. Trends in Biotechnology 29(9), 464–472 (2011)
6. Takahashi, D., Shukla, S.K., Prakash, O., Zhang, G.: Structural determinants of host defense peptides for antimicrobial activity and target cell selectivity. Biochimie 92(9), 1236–1241 (2010)

7. Teixeira, V., Feio, M.J., Bastos, M.: Role of lipids in the interaction of antimicrobial peptides with membranes. Progress in Lipid Research 51(2), 149–177 (2012)
8. Ramanathan, B., Davis, E.G., Ross, C.R., Blecha, F.: Cathelicidins: microbicidal activity, mechanisms of action, and roles in innate immunity. Microbes and Infection / Institut Pasteur 4(3), 361–372 (2002)
9. Méndez-Samperio, P.: The human cathelicidin hCAP18/LL-37: a multifunctional peptide involved in mycobacterial infections. Peptides 31(9), 1791–1798 (2010)
10. Sang, Y., Blecha, F.: Porcine host defense peptides: expanding repertoire and functions. Developmental and Comparative Immunology 33(3), 334–343 (2009)
11. Chinpan, C., Brock, R., Luh, F., Chou, P., Larrick, J., Huang, R., Huang, T.: The solution structure of the active domain of CAP18 - a lipopolysaccharide binding protein from rabbit leukocytes. FEBS Letters 370, 46–52 (1995)
12. Santos, P., Gordillo, A., Osses, L., Salazar, L.-M., Soto, C.-Y.: Effect of antimicrobial peptides on ATPase activity and proton pumping in plasma membrane vesicles obtained from mycobacteria. Peptides 36(1), 121–128 (2012)
13. Lata, S., Sharma, B.K., Raghava, G.P.S.: Analysis and prediction of antibacterial peptides. BMC Bioinformatics 8, 263 (2007)
14. Houghten: General method Bismuth citrate in the quantification of inorganic phosphate and its utility in the determination of membrane-bound phosphatases. Analytical Biochemistry 324, 79–83 (2004)
15. Cariani, L., Thomas, L., Brito, J., Castillo, J.R.: Bismuth citrate in the quantification of inorganic phosphate and its utility in the determination of membrane-bound phosphatases. Analytical Biochemistry 324, 79–83 (2004)
16. Chene, P.: ATPases as drug targets: learning from their structure. Nat. Rev. Drug Discovery 1(9), 665–673 (2002)

Harmonizing Protection and Publication of Research Findings in Biosciences and Bioinformatics

Oscar Lizarazo Cortés[1], Natalia Lamprea[2], and Gabriel Nemogá Soto[3]

[1] Universidad Nacional de Colombia, GEBIX, Bogotá D.C, Colombia
oalizarazoc@unal.edu.co,
[2] Independent, Bogotá D.C, Colombia
natalia.lamprea@gmail.com,
[3] University of Winnipeg, Winnipeg, Manitoba, Canada
g.nemoga@uwinnipeg.ca,

Abstract. This paper discusses the criteria in decision-making processes regarding publishing results, particularly of genetic information like nucleotide and amino acid sequences that are usually uploaded into public databases. It also describes property regimes on human and nonhuman genetic information. The main reasons for publishing finding are addressed, including: editorial requisites, replicability, scientific recognition, dissemination of results, legal and contractual obligations from access to genetic resources, and contractual obligations with funders. The discussion is illustrated with brief examples of data sharing policies of scientific institutions. It is contended that research institutions and universities have to balance publishing or postponing the diffusion of genetic data and consider a middle ground option. It is argued that the *release date option* becomes a relevant alternative regarding *data sharing policies.*

Keywords: intellectual property, data sharing policies, genetic information, law, bioinformatics, sequences, publication of results.

1 Introduction.

Genetic Information (GI) is used in a diverse array of basic and applied research, in a multipurpose fashion that ranges from molecular systematics and molecular ecology to studies on protein activity and modifications in genetic expression. The data obtained and their utilizations varies depending upon commercial or non-commercial uses, and its human or non-human origin.

This paper describes the reasons that favor dissemination or a delay in the dissemination of research findings, particularly of GI like nucleotide and amino acid sequences that are usually uploaded into public databases (DB) like GenBank, EBI. Etc. This paper also provides criteria for the decision-making processes that take place in research centers, with the purpose to strengthen intellectual property management (IP). By focusing on property regimes, legal and ethical principles are differentiated for human and non-human GI. . Subsequently, the appropriation via IP on products developed from this information is illustrated. Finally, it concludes with an alternative approach for deciding on the control, dissemination or publication of genetic data.

L.F. Castillo et al. (eds.), *Advances in Computational Biology*,
Advances in Intelligent Systems and Computing 232,
DOI: 10.1007/978-3-319-01568-2_51,

2 Legal and Ethical Background

2.1 Property Regime

Human Genetic Material and Information. The human genome is categorized as the common heritage of humankind (UNESCO Official Declaration, 1997). There is consensus that prior informed consent is required when sampling human populations, although arguments are found in favor and against when dealing with human DNA data. (Bosar, 2006) Within the context of human genetic research bioethical principles of autonomy, human dignity, benefiting and non-maleficence apply, but at this time, no binding international agreement has been established. Finally, there are some guidelines indicate that where possible, the human GI must be published in public databases, this position was enunciated with the Human Genome Project HGP, and is supported by documents such as the Bermuda Rules 1997 (Marshall , 2001) and the Agreement Lauderdale Ford (2003).

Non-Human Genetic Material and Information. Plant, animal and microorganism genetic resources are subject to different property regimes, but in their natural state, they are under the state sovereignty of the country of origin. Access to these resources is subject to state sovereignty and fair and equitable sharing of the benefits arising from its utilization (ABS). The international regime on non-human genetic resources is governed by the Convention on Biological Diversity (CBD) (1992) ratified by 193 countries, and the International Undertaking on Plant Genetic Resources for Food and Agriculture (2003) ratified by 131 countries. In 2010, the parties to the CBD adopted the Nagoya Protocol on Access and Benefit Sharing -ABS (2010) but its ratification is still pending. Additionally, and in accordance with the CBD, approximately 50 countries have established access regimes.

2.2 Application of Intellectual Property Rights

There are national differences in IP rules, but in general, it is more likely to patent innovations related to non-human genetic material and information - particularly speaking that of microorganisms-, than human material.

Human Genetic Material and Information. Most patents systems exclude the protection of inventions related to humans as such. Cloning procedures of humans are excluded or prohibited. The process intended to modify the human genetic identity and the use of human embryos with industrial or commercial purposes, are also proscribed since such intervention would breach public order and moral codes (EU Directive 98/44 article 6). In certain jurisdictions such as the European Union or the CAN (Andean Community) – made up by Colombia, Peru, Ecuador, Bolivia -, the total or partial sequences of a gene is non-patentable, (although this rule is not explicitly restricted to humans) (Andean Decision 486, art. 15; EU Directive 98/44 art. 5, IP French Code L- 611-18). Instead, the technical application of the gene function is patentable in certain cases (Clark et-al, 2000), but patent legislation require the description of the DNA sequences and their function.

Non-Human Material and Information. Patentability of inventions related to microorganisms is widely established (TRIPS Art. 27 3b; The Budapest Treaty). Even in jurisdictions with restrictive IP laws such as the CAN, microorganisms are patentable if they meet the requisites of novelty, inventive level and industrial application and sufficient description is provided (Articles, 14, 29 and Second Transitory Provision[1], Decision 486 of 2000).

In most jurisdictions, patent protection on plants and animals is more restrictive. For example, plants and animals are is not patentable within the CAN (Decision 486, Art. 20), although innovations on genetically modified organisms could be filled for patent protection (Decision 486, Art. 280) , whilst in the European Union and United Kingdom[2] it is possible to patent modified plants and animals. However, patents cannot be apply for a specific variety (EU Directive 98/44 art. 4). Patent protection on plants and animals is extensive in the United States of America if the innovation is "new, non-obvious, and useful". In the following section, the main reasons for disseminating or restricting the diffusion of genetic data are considered.

3 Materials and Methods

Selection of Criteria. We identified the most frequently used arguments by funders and researchers to either justify the publication of results, or to control and postpone the publication in USA and in France (Eisenberg, 2000; Noiville 2006). This was complemented by our expertise in IP in Colombia.

Within the criteria for publishing, we found editorial guidelines- results replicability, scientific prestige, advancement of science, and academic degree requirements. Regarding the criteria for postponing the publication, the following were identified: legal and contractual obligations for accessing genetic resources, contractual obligations with financing entities, conservation of information value and IP protection.

Case Studies. To explain how these apparently contradictory criteria actually work, we selected some case analyses. We choose renowned research centers and financing institutions within the field of bioinformatics and computer biology with visible intellectual property or data sharing policies (DSP). We search for institutions from different countries: United States of America, France, and Germany. The chosen entities work on different fields like biomedicine, human genetics, bio-remediation, and biofuels.

3.1 Pros and Cons of Prompt Publishing or Publication Delay.

These reasons provide inputs for decision making processes by committees involved in the authorizations of data diffusion at research centers. Later, we will analyze how these criteria are incorporated in current the DSP of selected research centers.

---

[1] "Microorganisms shall be patentable until other measures are adopted as a result of the examination provided for in TRIPS article 27 3b).

The commitments assumed by the Member Countries under the Convention on Biological Diversity shall be borne in mind in this regard".

[2] Examination Guidelines for Patent Applications relating to Biotechnological Inventions in the Intellectual Property Office –UK. July 2012.
`http://www.ipo.gov.uk/biotech.pdf`

Editorial Requisite (Replicability). Numerous journals request the data base accession number of the genetic information as an editorial requisite for the publication of the associated research findings. (Gallezot, 2002; Moore, 2002). This is an editorial guideline but also a practice, with few exceptions, within the scientific community. It is not a legal rule; companies certainly are not prone to apply this practice.

An intermediate alternative that enable researchers to fulfill the requested accession number, while exercising some control on data diffusion – at least temporarily – is to provide the link to the data at the corresponding database with a delayed release date (Benson, 2003).

Scientific Recognition. Researchers obtain scientific recognition through publications, contributions to their technical field, advances in the state of art of their area, and wide dissemination of their knowledge. However, researchers also gain recognition and credibility via patents or trade secrets over research findings and technological innovations. A relevant patent portfolio could be instrumental for receiving grants from funding agencies or in achieving strategic alliances with strong industrial and academic partners.

Dissemination of Results. Diffusing of research findings harms future patent applications because universal novelty is lost (Feiler, 1996). At the same time, publication of basic research or lab techniques prevents future low-quality patent applications on obvious advances or basic research tools.

In the short-term, delaying data diffusion may restrict other from using research findings or associated information. In the mid and long-term, those who control the information are in the best position to undertake more in-depth research activities, experiments and assays, and ending up with solutions to technical problems. Eventually, they could develop products and services through monetary investments.

Legal and Contractual Obligations for Accessing Genetic Resources. Because of *Mutual Agreed Terms* (MAT) and *Prior Informed Consent* (PIC) are included in access contract to genetic resources granted by Environmental authorities, like the Environment and Sustainable Development Ministry in Colombia, researchers are forbidden to transfer or exchange genetic resources and derivative products, total or partially, whether for money or in kind. It is not clear if this restriction cover genetic information, but standard practices and laws under development indicate so.

Contractual Obligations with Funding Institutions. It is a generalized practice that funding institutions include scientific articles, presentations at scientific congresses, and patent applications as products of research projects. These obligations force researchers to timely balance dissemination of results and filing for patents. Some patent office's provide a grace period if the diffusions is made by the same inventors before filing the patent application. This option is available only in some jurisdictions. For example, the Andean Community provides one-year grace period (Decision 486 art. 17).

Protect the Value of Information. Research centers and universities are increasingly adopting policies on IP. Thus, research findings could be protected through patents or trade secrets. As it is known, the requisites to obtain a patent are novelty, inventive step and industrial application. The novelty requirement is universal, and it is inexistent when the dissemination of the innovation occurred, whether orally or in written form. This requirement does not concern only to current novelty, but to future "novelty". That is, the results from initial stages in the research are usually not patentable; nevertheless, their early disclosure could affect a future patent application, or the viability of trade secrets grounded on valuable research findings.

3.2 Case Studies, Examples

The following are examples of DSP of some representative research entities and organizations where some of the above criteria are evidenced.

United States of America. National Institutes of Health. NIH. The NIH had a total budget of US $30,860 million dollars for the year 2012. It is composed by 27 Centers and Institutes mainly dedicated to biomedical and human health research. Its current DSP is dated 2003, and it applies to lab research and research with human subjects. Regarding the criteria for authorizing a publication its policy states: *"Data should be made as widely and freely available as possible ..."* On the other hand, it also includes reasons for limiting diffusion of data *"... while safeguarding the privacy of participants, and protecting confidential and proprietary data*[3]*."*

The NIH´s DSP also covers "Data Sharing Methods", "Timeframe for Data Sharing" "Privacy Concerns", and requisites for project financing applications. Under this policy, a funding application shall include a section titled "data sharing plan section", where researchers are invited to evaluate associated aspects beforehand.

United States of America. Department of Energy (DOE). The U.S. Department of Energy has a division for Biological and Environmental Research (BER). It covers research on climate, environmental science and biological systems.[4] The BER includes the Genomic Sciences Program and the Joint Genome Institute (JGI). The JGI sequences about *four trillion genome base pairs per year.* Their research include development of cellulosic biofuels, environmental remediation with microorganisms, among others. The Advisory Committee of the BER drafted the document titled: "Report on the description of current policies and practices for disseminating research results in the fields relevant to the Biological and Environmental Research program" on june, 2011.

Although there are different policies and practices within the BER, the criteria for dissemination of research findings are based on the *"potential values or impacts of the data."* On certain cases, a six-month extendable "delay" is accepted on a case-by-case basis. In this situation, the decision regarding the best timing to publish rely on the

[3] National Institutes of Health.
http://grants.nih.gov/grants/policy/data_sharing/

[4] U.S. Department of Energy's Office of Biological and Environmental Research (BER) http://science.energy.gov/ber/about/

shoulders of the principal researcher, who has the obligation to report all potential inventions to the "BRC Intellectual Property Management Group". This body is the one actually deciding if patents are filed.

France. Centre Nationale de la Recherche Scientifique. CNRS. The National Center of Scientific Research –CNRS of France is a public institution. For the year 2013, it has a budget of 3.415 billion Euros. It is composed by 1,100 research and service units, some of them working in leading research on genomics and bioinformatics. When researchers achieve promising results they shall inform through a form called "Declaration of Invention"[5]. The document warns that "the dissemination in whole or in part of the invention before filing the patent application could entail its rejection. The Center has not adopted a *data sharing policy*, but the standard formats implementing IP policy, include the pros and cons of publishing or withholding results.

Germany. Max-Planck. The guidelines for inventors in this institution clearly incorporate the reasoning for postponing diffusion of data. It says: "*Premature publication of research results represents the greatest danger to the patentability of inventions. Reports, lectures, and similar verbal statements are just as bad for novelty as the publication of abstracts, papers, etc. Due to the fact that such publications – even if they are theoretical by nature - can hint at the innovative idea, they endanger later applications by the inventor, also with a view to the level of inventiveness.*" [6]

4 Discussion

When deciding whether to publish or not GI the individual researcher, the research center or the university often faces a crucial challenge. It is at this juncture that the *release date option,* becomes a relevant alternative for *data sharing and intellectual property policies*. In Europe, 70% of sequences submitted to EBI/EMBL use the *holding period* option. It consists of loading sequences, obtaining an accession number, but keeping the data confidential during a chosen period. In this way, the researcher is able to provide the accession number to fulfill journal requirements, without affecting the eventual protection via patents or trade secrets (Hutter, 2002). It has found that Colombian researchers do not use this option frequently. It shows that whatever results could come, researchers are not envisioning eventual commercial applications.

The chosen examples of worldwide research institutions shows that claims like "to publish or die" or "to obtain patents or die" are unsustainable. Some researchers dogmatically adopt the first claim; the second has been introduced by funding agencies looking for short-term impact. A balanced data sharing policy is viable for publishing and controlling data, as illustrated by the examined research institutions. They file for a substantial number of patents, but at the same time, they publish their research findings in academic journals.

---

[5] Centre Nationale de la Recherche Scientifique
http ://www.dgdr.cnrs.fr/daj/partenariat/textesvalo/textes.htm

[6] Guidelines for Inventors Suggestions for Inventors from the Max-Planck-Gesellschaft, http://www.max-planck-innovation.de/share/guidelines/Guidelines_for_Inventors.pdf

Sometimes it is possible to quickly apply for a patent, then to submit the research finding for publication, and after look for licensees. The different timeframes can be handled without delaying the publication irrationally. An example of this is the 6-month embargo period included in the DOE´s DSP.

In other situations, certain data can be shared (e.g. *raw data*), but other should be keep under control (e.g. industrially modified enzymes with industrial application). This is precisely what the funding entities want to know ahead of time, requesting the timely report of inventions. This kind of control guarantees that decisions and actions are taken for pursuing patents or trade secrets timely. Final decisions are the subject matter of intellectual property committees, technology transfer offices or units alike.

It is true that an excessive use of IP can also hinder technological innovation (Heller, 2008; Eisenberg 1998). Nevertheless, publishing everything could discourage investors for technology innovation, and thus prevent the creation and transference of some research finding into new products and services.

When research is being conducted on human genetic samples, like in the NIH, prior informed consent and privacy issues should be considered. These issues could restrict the availability of the data obtained in the study. In Colombia and other mega diverse countries, access contracts actually limit the transfer of genetic resources to third parties and this limitation could extend to the transfer of GI into databases.

5 Conclusion

Rather than adopting a fixed model like "publishing all data" or "not publishing data", it is convenient to adopt general guidelines and criteria to balance the publication and/or the control of research findings. Optimally, individual researchers and research institutions should adopt an intermediate or mixed model for evaluation on a case by case basis in order to decide on the three following questions What?, When and Why? To publish or not to publish.

Criteria for the prompt publication of data are not totally antagonistic to reasons that favor postponing publication. Actually, examples of DSP indicate that it is quite common to combine dissemination of data, with data confidentiality and control, even in areas like "omics" and bioinformatics.The *release date option* is an alternative for reaching coherence between *data sharing* and *intellectual property policies*. In some cases, however, it would be advisable *not to* upload the data, bearing in mind the potential industrial application or the economic value represented by keeping the data confidential.

References

1. Benson, D.A., Karsch-Mizrachi, I., Lipman, D.J., Ostell, J., Wheeler, D.L.: GenBank. Nucleic Acids Research 31(1), 23–27 (2003)
2. Boussard, H., Smagadi, A.: The principle of Benefit sharing in the utilization of natural plant and human genetic resources: beyond the property and no-property rights paradigms. Journal International de Bioéthique 17(4), 29–53 (2006)

3. Cambon-Thomsen, A.: L'information génétique dans la société de l'information. Revue poli-tique et parlementaire. N°1050 (2009)
4. Chander, A., Sunder, M.: The Romance of the Public Domain. California Law Review 92(5), 1331–1373 (2004)
5. Clark, J.: Patent pools: a solution to the problem of access in biotechnology patents?. United States Patent and Trademark Office (2000), `http://www.uspto.gov/web/offices/pac/dapp/opla/patentpool.pdf`
6. Eisenberg, R.S.: Genomics in the public domain: strategy and policy. Nat. Rev. Genet. 1(1), 70–74 (2000)
7. Eisenberg, R.S.: Can Patents Deter Innovation? The Anticommons in Biomedical Research. Science 280(5364), 98–701 (1998)
8. Feiler, W.S., Auth, D.R.: Publish and perish. Nat. Biotech. 14(11), 1602–1603 (1996)
9. Fort Lauderdale Agreement. Sharing Data from Large-scale Biological Research Projects: A System of Tripartite Responsibility. Report of a meeting organized by the Wellcome Trust (2003), `http://www.genome.gov/Pages/Research/WellcomeReport0303.pdf`
10. Gallezot, G.: La recherche in silico, In: Chartron G (ed).: Les chercheurs et la documentation numérique: nouveaux services et usages. Edition du cercle de la Librairie (2002)
11. Heller, M.: The Gridlock Economy: How Too Much Ownership Wrecks Markets, Stops Innovation, and Costs Lives. Basic Books, New York (2008)
12. Hutter, A.: The Patentability of Biomolecules – Does Online Bioinformatics Compromise Novelty? Comparative and Functional Genomics 3(2), 119–126 (2002)
13. Marshall, E.: Bermuda Rules: Community Spirit, With Teeth. Science 291(5507), 1192–1192 (2001)
14. Moore, P.: Publication with a pinch of privatization. Genome Biology 3(4), spotlight-20020404-02 (2002)
15. Noiville, C., Bellivier, F.: Contrats et vivant: Le droit de la circulation des ressources bi-olo-giques. LGDJ, Paris (2006)

Mathematical Modeling of Lignocellulolytic Enzyme Production from Three Species of White Rot Fungi by Solid-State Fermentation

Sandra Montoya[1], Óscar Julián Sánchez[1,*], and Laura Levin[2]

[1] Institute of Agricultural Biotechnology, Universidad de Caldas, Manizales, Colombia
[2] Department of Biodiversity and Experimental Biology, PRHIDEB-CONICET, FCEN, Universidad de Buenos Aires, Argentina

{osanchez,sandra.montoya}@ucaldas.edu.co,
lale@bg.fcen.uba.ar

Abstract. This research was conducted by growing three species of white-rot fungi (Coriolus versicolor, Lentinus edodes and Pleurotus ostreatus) on twelve formulations of solid substrates using mixtures of different lignocellulosic materials, calcium carbonate salts and copper sulphate (II). The objective of this study was to propose a mathematical model to describe the biomass growth, lignocellulolytic enzymes biosynthesis, production and consumption of reducing sugars, consumption of cellulose and hemicellulose, and lignin degradation. The three species of fungi grew well on all substrate formulations. The response obtained was evaluated by the titles of all enzymatic activities for several combinations fungus – substrate. C. versicolor had the highest capacity to degrade lignin, cellulose and hemicellulose for all combinations, with 65% as the maximum lignin degradation for F1 combination, and 43% cellulose degradation for F9 combination. The mathematical model proposed for C. versicolor consisted of eleven differential equations to describe the behavior of the cultivation system from the experimental data of all the resulting combinations in order to obtain the largest capacity degradation of lignocellulosic substrates by the fungus. In this work, we present the modeling results for combination F9 fungus – substrate combination, which showed the best behavior related to the degradation of lignocellulosic materials used. The results obtained demonstrated that the model proposed represents a powerful tool to design solid-substrate fermentation processes.

Keywords: White-rot fungi, Lignocellulolytic enzymes, Degradation of lignocellulosic materials, Lignocellulosic biomass, Mathematical modeling.

1 Introduction

The mathematical modeling of the kinetics of white rot fungi growth plays a crucial role for the design and scale-up of solid-state fermentation processes using macromycetes.

---

* Corresponding author.

L.F. Castillo et al. (eds.), *Advances in Computational Biology*,
Advances in Intelligent Systems and Computing 232,
DOI: 10.1007/978-3-319-01568-2_52, © Springer International Publishing Switzerland 2014

The most reported models are linear, logistic, and two-phase model (growth acceleration and deceleration) [15], although these models do not include the effect of the concentrations of the components of the culture medium on the growth [2; 10; 17]. The mathematical expressions used to describe the synthesis of different products depend on the type of the metabolite synthesized. In general, the simplest expressions are intended for primary metabolites considering that their production is directly linked to the biomass growth. Currently, there are not available mathematical models describing neither the growth of macromycetes nor the production of different valuable compounds from them. This difficulty arises due to the fact that each one of these fungi is affected by specific intrinsic and extrinsic factors as the physical features and chemical composition of the culture media, origin of the fungal strains, environmental conditions, and the slow growth rate as compared to microfungi.

The objective of this work is to propose and validate a mathematical model to describe the biomass growth of one macromycetes strain as well as the synthesis of lignocellulolytic enzymes by this fungus, the production and consumption of reducing sugars, the consumption of cellulose and hemicellulose, and the lignin degradation under solid-state fermentation conditions.

2 Methodology

Three basidiomycetes species were employed for production of lignocellulolytic enzymes by solid-state fermentation: *Lentinula edodes* CICL54 provided by the National Coffee Research Center, Cenicafé (Chinchiná, Colombia), *Coriolus versicolor* PSUWC430 provided by Pennsylvania State University, USA, and *Pleurotus ostreatus* UCC001 from the internal collection of Universidad de Caldas. The strains were extended on potato-dextrose-agar (PDA) and maintained under refrigeration at 4ºC with periodic transfers. The three species of white-rot fungi were grown on twelve formulations of solid substrates (F1 to F12) using mixtures of different lignocellulosic materials, calcium carbonate salts and copper sulphate (II). The substrates were packed in polypropylene bags and tyndallized. The moisture content was adjusted according to the development requirements of each species. The bags containing the substrate were aseptically inoculated with 4% (wet basis) of spawn. The following variables were measured during the fermentation by taking samples from the bags: fungal biomass, reducing sugars, cellulose, hemicellulose, lignin, laccase activity, manganese peroxidase activity, endoxylanase activity, endoglucanase activity, exoglucanase activity, and β-glucosidase activity.

The extracts for determination of cellulolytic and ligninolytic enzymatic activities were obtained from 1 g substrate in 12 mL of sterile neutral distilled water treated by ultrasound for 5 min and agitation for 10 min with subsequent filtration and centrifugation. The endo-1,4-β-*D*-glucanase activity (EC 3.2.1.4) was determined by using carboxymethyl cellulose (CMC) as described by Montoya [11]. The exo-1,4-β-D-glucanase activity (EC 3.2.1.91) was determined using crystalline cellulose, and the endo-1,4-β-D-xylanase activity (EC 3.2.1.8) was determined on xylan [11]. These reactions were stopped by adding dinitrosalicylic acid (DNS) [7]. The β-glucosidase activity was determined by the reaction with *p*-nitrophenyl β-*D*-glucopyranoside also as described by Montoya [11]. The laccase activity (EC1.10.3.2) was determined by using 2,2'-azino-bis(3-ethylbenzothiazoline-6-sulphonic acid) (ABTS) according to [13]. The manganese peroxidase (MnP) activity (EC 1.11.1.13) was determined by using manganese sulfate with hydrogen peroxide [12].

The basidiomycetes fungi, as those ones used in this work, contain chitin in their cell wall so the fungal biomass in the solid substrates was indirectly estimated by the determination of N-acetyl-*D*-glucosamine (NAGA), the structural unit of chitin, after the hydrolysis with 6N HCl according to the method of Plassard et al. [14]. The content of cellulose, hemicellulose, and lignin in the substrates was determined for each one of the formulations used for cultivation of the three species studied as well as the soluble fraction of the fiber. For this, the results of the determination of neutral detergent, acid, and lignin-acid fiber were used [3]. The concentration of reducing sugars as glucose was quantified by the DNS method [7].

Several mathematical expressions were proposed and tested in order to perform the modeling of fungal growth and enzyme production under solid-state fermentation conditions. The models comprise different systems of ordinary differential equations, which were solved by using the software Matlab® 2010b (MathWorks, USA). For this, the function ode42 based on an explicit Runge-Kutta formula (4,5) using a Dormand-Prince pair [1] was employed as well as the function ode15s based on a variable-order formula that utilizes a numerical differentiation.

3 Results and Discussion

The three species of fungi grew well on all substrate formulations. The response obtained was evaluated by the titles of all enzymatic activities for several combinations fungus – substrate. *C. versicolor* had the highest capacity to degrade lignin, cellulose and hemicellulose for all combinations, with 65% as the maximum lignin degradation for F1 combination, and 43% cellulose degradation for F9 combination.

A mathematical model composed of 11 differential equations (see Table 1) was proposed in order to adjust the experimental data of all the cultivations performed on the different substrate formulations for the fungus *C. versicolor*, since it was the macromycetes with the highest capacity for degrading lignocellulosic components (cellulose, hemicellulose, and lignin) of the three species studied. In this work, the results for the formulation 9 (F9) are presented considering that this cultivation medium showed a better behavior regarding the degradation of the lignocellulosic matrix compared to the other 11 formulations. The results for the rest of formulations are not shown because of space limitation. The adjustment of the experimental data to the mathematical model was evaluated through *F*-test comparing the variance of the residuals between the experimental data and the data calculated by the model, and the variance of the experimental series. This test was applied for all the modeling results and all the experimental data corresponding to the formulations F1 to F12 for *C. versicolor*. The performed test enabled to prove the validity of the model used in terms of the best representation and description of the experimental data obtained.

Table 1. Mathematical model to describe the production of biomass and lignocellulolytic enzymes, and consumption of lignocellulosic matrix

Equation	Description	Parameters
$\frac{dC_b}{dt} = \mu_m . C_b \left(1 - \left(\frac{C_b}{C_{bm}}\right)^n\right)$ (1)	Biomass	μ_m: Specific growth rate (day^{-1}) C_{bm}: Maximum biomass concentration (mg/gss) n: $n < 1$ The organism is relatively sensitive to the auto-inhibition and it occurs for very low values of C_b $n = 1$ logistic equation $n > 1$ The organism is relatively resistant to the auto-inhibition and it occurs only when $C_b \approx C_{bm}$
$\frac{dC_{AR}}{dt} = q_p . \mu_m . \frac{dC_b}{dt} . \left[1 - (n+1) . \left(\frac{C_b}{C_{bm}}\right)^n\right]$ (2)	Reducing sugars	q_p: Production coefficient for reducing sugars ($mg \times mg^{-1} \times day^{-1}$) C_{AR}: Reducing sugars concentration ($mg \times gss^{-1}$)
$\frac{dC_L}{dt} = -k_L . C_{lac} . C_{MnP}$ (3)	Lignin	k_L: Lignin degradation coefficient ($mg \times gss \times day^{-1} \times U^{-2}$) k_{lac}: Laccase production coefficient ($U \times gss \times mg^{-1} \times mg^{-1}$) k_{MnP}: Mn peroxidase production coeff. ($U \times gss \times mg^{-1} \times mg^{-1}$) μ_{lac}: Inhibition coefficient for laccase ($U \times mg^{-1} \times day^{-1}$) μ_{MnP}: Inhibition coefficient for Mn peroxidase ($U \times mg^{-1} \times day^{-1}$) C_{lac}: Laccase activity ($U \times gss^{-1}$) C_{MnP}: Manganese peroxidase activity ($U \times gss^{-1}$) C_L: Lignin concentration ($mg \times gss^{-1}$)
$\frac{dC_{lac}}{dt} = k_{lac} . \frac{dC_b}{dt} . C_L - \mu_{lac} . C_L$ (4)	Laccase	
$\frac{dC_{MnP}}{dt} = k_{MnP} . \frac{dC_b}{dt} . C_L - \mu_{MnP} . C_L$ (5)	Manganese peroxidase (MnP)	
$\frac{dC_{HM}}{dt} = -k_{HM} . C_{ENX}$ (6)	Hemicellulose	k_{HM}: Hemicellulose consumption coefficient ($mg \times day^{-1} \times U^{-1}$) C_{HM}: Hemicellulose concentration ($mg \times gss^{-1}$)
$\frac{dC_{ENX}}{dt} = k_{ENX} . C_{HM} . \frac{dC_b}{dt} - \mu_{ENX} . C_{AR}$ (7)	Endoxylanase	k_{ENX}: Endoxylanase production coefficient ($U \times gss \times mg^{-1} \times mg^{-1}$) μ_{ENX} : Inhibition coefficient for endoxylanase ($U \times mg^{-1} \times day^{-1}$) C_{ENX}: Endoxylanase activity ($U \times gss^{-1}$)
$\frac{dC_C}{dt} = -k_C . C_{ENG} . C_{EXG}$ (8)	Cellulose	k_C: Cellulose consumption coefficient ($mg \times gss \times day^{-1} \times U^{-2}$) C_C: Cellulose concentration ($mg \times gss^{-1}$)
$\frac{dC_{ENG}}{dt} = k_{ENG} . \frac{dC_b}{dt} . C_C - \mu_{ENG} . C_{AR}$ (9)	Endoglucanase	k_{ENG}: Endoglucanase production coeff. ($U \times gss \times mg^{-1} \times mg^{-1}$) μ_{ENG} : Inhibition coefficient for endoglucanase ($U \times mg^{-1} \times day^{-1}$) C_{ENG}: Endoglucanase activity ($U \times gss^{-1}$)
$\frac{dC_{EXG}}{dt} = k_{EXG} . \frac{dC_b}{dt} . C_C - \mu_{EXG} . C_{AR}$ (10)	Exoglucanase	k_{EXG}: Exoglucanase production coeff. ($U \times gss \times mg^{-1} \times mg^{-1}$) μ_{EXG} : Inhibition coefficient for exoglucanase ($U \times mg^{-1} \times day^{-1}$) C_{EXG}: Exoglucanase activity ($U \times gss^{-1}$)
$\frac{dC_{BG}}{dt} = k_{BG} . \frac{dC_b}{dt} - \mu_{BG} . C_{AR}$ (11)	β-glucosidase	k_{BG}: β-glucosidase production coeff. ($U \times mg^{-1}$) μ_{BG} : Inhibition coefficient for β-glucosidase ($U \times mg^{-1} \times day^{-1}$) C_{BG}: β-glucosidase activity ($U \times gss^{-1}$)

Remarks: gss – gram of solid substrate. U – enzyme activity unit.

The proposed equation to describe the biomass growth (equation 1) corresponded to the logistic equation modified by Mitchell et al. [8]. To describe the variation of reducing sugars with the time (equation 2), a constant production factor that affects the biomass growth rate derivative was considered taking into account that the sugars are consumed by the fungus as an energy source for it to grow. The data fit was better than if the factor would affect the biomass concentration itself instead of the biomass growth rate. For lignin degradation and the consumption of hemicellulose and cellulose, the expressions (equations 3, 6, and 8) are depending on the specific activities of the enzymes responsible for the degradation of each one of these substrates. The variation of the specific enzymatic activities (the equivalent of the concentration for enzymes) were considered to be dependent on the corresponding substrate concentration and biomass growth rate with an inhibition factor linked to the reducing sugars concentration (for the two cellulases and the xylanase), and to the lignin concentration (for the two ligninases) as can be seen in equations (4), (5), (7), (9), and (10). The inhibition of ligninolytic enzymes could be explained to the presence of intermediary compounds formed during the fermentation or by high lignin concentrations [16]. Although there is still no certainty of the causes of this inhibition, the experimental data obtained in this work showed a possible inhibition of the ligninases. As such intermediary compounds were not determined during the cultivation, a lignin inhibition was assumed. Finally, the variation (production) of the β-glucosidase was described as a function of the growth rate with an inhibition factor caused by the reducing sugars concentration.

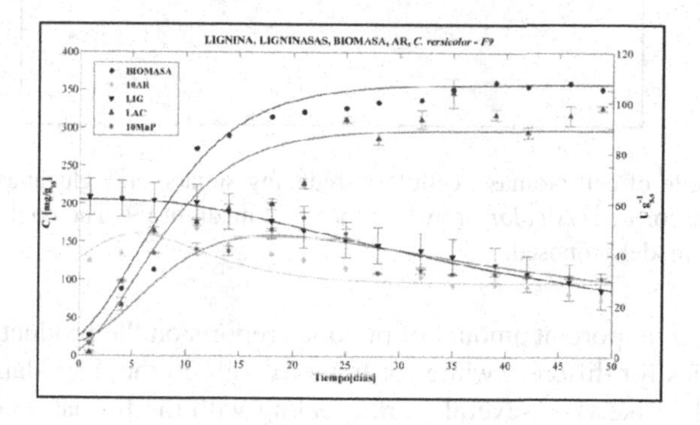

Fig. 1. Time profile of cell biomass, reducing sugars, lignin, laccase, and MnP for *C. versicolor* grown on the solid medium F9. The continuous lines were calculated by the model proposed.

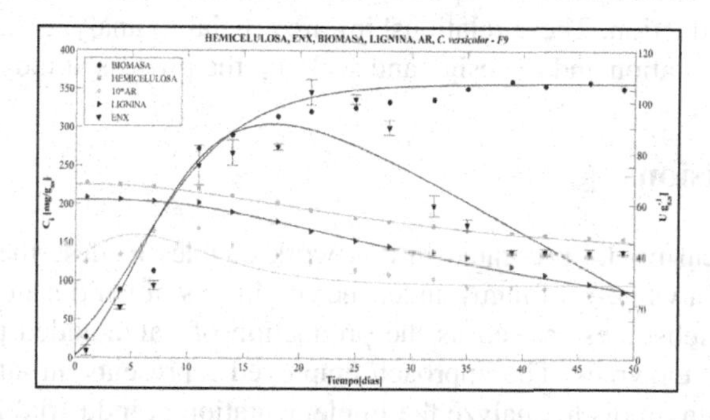

Fig. 2. Time profile of cell biomass, hemicellulose, reducing sugars, lignin, and endoxylanase for *C. versicolor* grown on the solid medium F9. The continuous lines were calculated by the model proposed.

The adjustment of the experimental data to the mathematical model for each one of the measured variables is shown in Fig. 1, Fig. 2, and Fig. 3. The values of the parameters were calculated from the experimental data obtained by non-linear regression using the software Matlab. The adjustment of the model equations describing the behavior of the different variables studies was as follows: biomass – 90%, reducing sugars – 55%, lignin – 90%, laccase – 43%, MnP – 87%, hemicellulose – 86%, endoxylanase – 82%, cellulose – 40%, endoglucanase – 88%, exoglucanase – 78, and β-glucosidase – 96%.

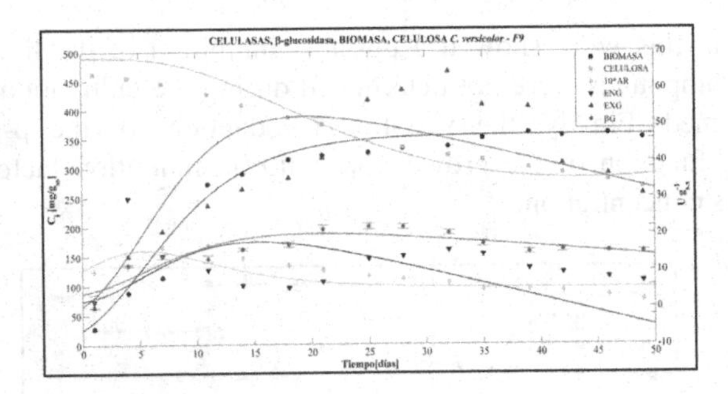

Fig. 3. Time profile of cell biomass, cellulose, reducing sugars, endoglucanase, exoglucanase, and β-glucosidase for *C. versicolor* grown on the solid medium F9. The continuous lines were calculated by the model proposed.

There exists an important amount of previous reports on the production of lignocellulolytic enzymes for different white-rot fungi as well on the degradation of lignocellulosic materials. Likewise, several works dealing with the formation of intermediary products during solid-state fermentation have been published [4-6; 9]. Nevertheless, more information is required to develop the kinetic relationships needed to describe the biomass formation, nutrient uptake, degradation of lignocellulosic components, and enzyme production. These relationships are crucial to analyze the behavior of the solid-state fermentation and to design and scale-up the process at industrial level.

4 Conclusions

The mathematical model presented in this work enables to describe the growth and development of a white-rot fungus under the conditions of solid-state fermentation on lignocellulosic substrates as well as the production of value-added products like the lignocellulolytic enzymes. The approach employed represents an attempt to provide powerful modeling tools to analyze the implementation at industrial level of this kind of processes reducing, at the same time, the experimental efforts required to determine the optimum operating conditions and the data needed for scale-up research.

The development of the kinetic relationships for each one of the measured variables was performed in such a way that the mathematical expressions have some biological sense. The proposed expressions reasonably describe all the variables despite the complexity of the process studied. This complexity is related to the biosynthesis of a large amount of intermediary compounds (besides the enzymes) that are formed during a solid-state cultivation on different mixtures of lignocellulosic materials. Undoubtedly, the follow-up and monitoring of the substances generated by the action of the enzymes studied, or other enzymes not evaluated in this work, would contribute to the acquisition of a valuable and more comprehensive knowledge of the complex phenomena occurring during the growth of the macromycetes on lignocellulosic substrates. And this knowledge has a paramount importance for the implementation of commercial processes utilizing agro-industrial wastes as feedstocks.

References

1. Dormand, J.R., Prince, P.J.: A family of embedded Runge-Kutta formulae. Journal of Computational and Applied Mathematics 6, 19–26 (1980)
2. Ikasari, L., Mitchell, D.A.: Two-phase model of yhe kinetics of growth of *Rhizopus oligosporus* in membrana culture. Biotechnology Bioengineering 68, 619–627 (2000)
3. Leterme, P.: Análisis de alimentos y forrajes. Protocolos de Laboratorio. Universidad Nacional de Colombia Sede Palmira, Palmira (2010)
4. Levin, L.: Biodegradación de materiales lignocelulósicos por Trametes trogii (Aphyllophorales, Basidiomycetes). Tesis Doctoral, Universidad de Buenos Aires, Buenos Aires (1998)
5. Levin, L., Herrmann, C., Papinutti, V.L.: Optimization of lignocellulolytic enzyme production by the white-rot fungus *Trametes trogii* in solid-state fermentation using response surface methodology. Biochemical Engineering Journal 39, 207–214 (2008)
6. Meagher, M.M., Tao, B.Y., Chow, J.M., Reilly, P.J.: Kinetics and subsite mapping of β-D-xylobiose and D-xylose producing *Aspergillus niger* endo-β-1,4-D-xylanase. Carbohydrates Resources 173, 273–283 (1988)
7. Miller, G.L.: Use of Dinitrosalicylic Acid Reagent for Determination of Reducing Sugar. Analytical Chemistry 31(3), 426–428 (1959)
8. Mitchell, D., Stuart, D., Tanner, R.: Solid-State fermentation - Microbial growth kinetics. The Encyclopedia of Bioprocess Technology: Fermentation,Biocatalysis and bioseparation. Wiley, New York (1999)
9. Mitchell, D., Von Meien, O., Krieger, N.: Recent developments in modeling of solid-state fermentation: heat and mass transfer in bioreactors. Biochemical Engineering Journal 13, 137–147 (2002)
10. Mitchell, D.A., Von Meien, O.F., Krieger, N., Dalsenter, F.D.: A review of recent developments in modeling of microbial growth kinetics and intraparticle phenomena in solid-state fermentation. Biochemical Engineering Journal 17, 15–26 (2004)
11. Montoya, S.: Obtención de enzimas lignocelulolíticas y polisacáridos a partir de residuos lignocelulósicos del Departamento de Caldas empleando macromicetos de pudrición blanca por fermentación sumergida y fermentación en estado sólido (Production of lignocellulolytic enzymes and polysaccharides from lignocellulosic wastes from the Department of Caldas using white-rot fungi by submerged fermentation and solid-state fermentation, in Spanish). Ph.D. Thesis, Universidad de Caldas, Manizales, Colombia (2012)
12. Paszczczynski, A., Crawford, R., Huyn, V.: Manganese peroxidase of *Phanerochaete chrysosporium* purification. Methods Enzymology 161, 264–270 (1988)
13. Paszczynski, A., Crawford, R.L.: Degradation of azo compounds by ligninases from *Phanerochaete chrysosporium* Involment of veratryl alcohol. Biochemistry Biophysics Resources Communications 178, 1056–1063 (1991)
14. Plassard, C., Mousain, D., Salsac, L.: Estimation of mycelial growth of *basidiomycetes* by means of chitin determination. Phytochemistry 21, 345–348 (1982)
15. Sangsurasak, P., Nopharatana, M., Mitchell, D.: Mathematical modeling of the growth of filamentous fungi in solid-state fermentation. Journal Science Industrial Resources 55, 333–342 (1996)
16. Tengerdy, R.P., Szakacs, G.: Bioconversion of lignocellulose in solid substrate fermentation. Biochemical Engineering Journal 13, 169–179 (2003)
17. Van de Lagemaat, J., Pyle, D.L.: Modelling the uptake and growth kinetics of *Penicillium glabrum* in a tannic acid-containing solid-state fermentation for tannase production. Process Biochemistry 40, 1773–1782 (2005)

Bioinformatics Tools and Data Mining for Therapeutic Drug Analysis

Juan Manuel Pérez Agudelo[1], Néstor Jaime Castaño Pérez[2], and Jhon Fredy Betancur Pérez[3]

[1] Grupo de Investigación BIOSALUD, Universidad de Caldas, Manizales, Caldas, Colombia
juanmaperez22@gmail.com

[2] Grupo de Investigación en Ciencias Biomédicas de la Universidad de Manizales, Manizales, Caldas, Colombia
njcastanop@gmail.com

[3] Grupo de Investigación en Ciencias Biomédicas de la Universidad de Manizales, Manizales, Caldas, Colombia
jbetancur@umanizales.edu.co

Abstract. Purpose: To identify issues derived from medication usage. Methods: Pharmacological therapeutics analysis generated by a Health Care Institution in Colombia (n= 400); Type of analysis: MATLAB® 7.10.0.499 (R2010a) based data mining over the following variables: "origin", "category", "level", "type of pharmacological interaction", "pharmacogenetic relation" and "chance of adverse reaction to medications". Results: General inadequate medical prescription was found in 3.9% of the cases, while inadequate cardiovascular prescription was found in 37.6% of the cases. 63.6% of the events where catalogued as moderate interactions in the general case when compared with 95.5 % in the cardiovascular case. A significant relation between the type of pharmacokinetic interaction and the chance of having adverse reaction to medication was found. Conclusions: Data mining allows us to get a closer look into under-explored issues related to global analysis such as the interactions derived from pharmacogenetic aspects on the use of medication.

Keywords: Data Mining, drug therapy, drug interactions.

1 Introduction

Responsibility on the use of medications as a technological resource rests on the medical practitioner who, as the first filter on the process of adequate use, establishes in an objective manner the execution of the pharmacotherapeutic act. According to the World Health Organization (WHO) adequate use of medication implies that the patient receives each medication for specific indication, with the right dosage, with specific timing, at the least possible cost (to him and society) and minimizing the chance of adverse drug reactions (ADR). From 1977, a pharmaceutical policy has been proposed for every single country with four basic pillars: Access, Quality, Security and Rational and adequate use, and recently fair cost has been added as a fifth element. In

L.F. Castillo et al. (eds.), *Advances in Computational Biology*,
Advances in Intelligent Systems and Computing 232,
DOI: 10.1007/978-3-319-01568-2_53, © Springer International Publishing Switzerland 2014

Colombia, the Ministry of Social Protection has created the National pharmaceutical policy [1] with the purpose to optimize medication usage, reduce access inequalities and ensure its own quality in the Social Security General System in Health. One of the objectives is to develop and implement strategies targeted to all, including the Health Care professional that improves the correct use of medications. Population transitional epidemiology [2] shows an increase in the growth rate for elder people, which generates changes in the health care system and in the use of pharmacological resources, taking into account that a greater use of medication implies increasing the risk of generating interactions, an inadequate use of medications, hence increasing the chance of adverse medial reactions. Polypharmacy and the adverse effects derived from it represent 5% to 15% of the hospitalization in elder people, more than 45% of the readmissions and 25% are preventable; some of the causes are: inadequate prescription, medication interactions, and inadequate dosage [3]. Good medical prescription practices and optimal treatment results go hand in hand with the provision of proven information exempt from commercial bias, authentic and complete scientific data access, data analysis systems access, pharmacovigilance and on the research on the use of medication in the Health care system [1]; pharmacovigilance defined as "the science and the activities related to the detection, evaluation, understanding and prevention of the adverse effects or any other issue related with medications" [4], should be considered as one the main tools in the process of responsible medication usage.

The National Institute for Drug and Food Surveillance (INVIMA – Colombia) is in charge of handling the reports that contain the results of the therapeutic quality control tests that are performed on all the medications that are used throughout the country. However, this organization doesn't have any clear functions that have to do with the process of investigation of the proper and inadequate use of the pharmacological resources. The Pan American Health Organization (PAHO) references 11 works between 1998 and 2001 that have been developed in Colombia for the adequate of use of drugs. The majority of these works are related to therapeutic formularies and administrative schemes for pharmaceutical services. Calderon [5] preformed an analysis of 13 published studies between 1998 and 2007; he emphasizes that no data mining methods were used in the search for the adverse reactions for the combination of two or more medications. During a medication surveillance project in 2011, Lin [4] presents the analysis and the use of classification technics for the date base of medication consumption in order to find safe signals for the decision making. In an international context, with information technology structure and a strong document storage, the European Sanitary Knowledge Data Base: BOT PLUS [6], is the most essential tool for the development of pharmaceutical attention with in a multidisciplinary health care team. In general, it is an advance pharmacotherapeutic guideline; however, it does not include a multivariate or relational analysis of the different pharmacological data. Data mining technics have been applied in an isolated manner in the health sciences by exploring pathologies such as hypertension in the search for hidden patrons in the clinical data [7]. The data mining in drug surveillance has allowed the creation of new analysis methodologies for the reports of generates ADR (Adverse Drug Reaction) [8], though there are no reports of any predictive types of studies that address the presentation of ADR.

2 Materials and Methods

Files generated by the Health Care Institution were analyzed, from which 400 medical prescriptions were studied. These prescriptions included formulations generated in the consult process for the year 2012 (January – December). The methodology was centered in a global preliminary analysis based in the combination of medication, given that a complete analysis from a holistic pharmacological perspective would consider variables such as: age, economical level, pathological history, adverse medical reactions history, pharmacological compliance, basic biochemical tags (creatinine, liver enzymes) and medication schedule associated with meals. These considerations are out of the scope of this research and are not considered as goals. The following **variables** were considered for analyzing interactions and ADR: **A**: Origin; **B**: Category; **C**: Level; **D**: Type; **E**: Pharmacogenetics; **F**: Potential of ADR.

A. ORIGIN: variable generated with relation to the pathological group that gives origin to the prescription. The following divisions were established: Cardiovascular (CV), where the following pathologies where included: Hypertension, Dyslipidemia, Diabetes, Hypothyroidism or their combinations; General (G), for prescriptions pertaining to other areas.

B. CATEGORY: Related with the adequate follow-up of medical combinations according to the national protocols and guidelines. The following classification was established. 1: Adequate General Combination; 2: Inadequate General Combination; 3: Not Conclusive (More information is required in order to be able to identify prescription as adequate or inadequate).

C. LEVEL: It refers to the chance of generating interactions, regardless of the potential of ADR. The different tools that were chosen for the expert analysis for this variables, established the following level for the interactions: 1: High: High clinical significance. This combination should be avoided. The risk of interaction outpaces the therapeutic benefit; 2: Medium: Medium clinical significance. Combinations that generally should be avoided. Its use is considered in very special cases. It requires clinical and lab monitoring; 3: LOW: Low clinical significance. Minimal risk. Can be considered as an alternative. Can be managed through general clinical monitoring [9].

D. TYPE: Variable defined by the relation of the analyzed prescription interaction with the different pharmacological processes. The following divisions where generated: 1: Pharmacokinetic (FK): Related to absorption, distribution, metabolism (Cytochrome) o excretion. 2: Pharmacodynamic (FD): Related to action mechanism. 3: Mixed: FK and FD.

E. PHARMACOGENETICS (FG): Variable related with the concordance of the interaction with metabolic aspects associated with cytochromes. 0: Absent; 1: Present

F. POTENTIAL OF ADR: Dependent variable, defined as the risk or final probability of issues derived from the use of medications, as a result of interactions of different prescriptions. The following states were defined for this variable: 0= Absent; 1= Moderately possible; 2= Highly possible; 3= Very Highly possible and 4= Certain.

Three different and independent systems for the interaction analysis were used: 1. Interdrugs® Software for Latin America [10]. 2. The interactions check system of the web platform drugs.com [11]. 3. International textbook of adverse reaction and medical interactions "Meyler´s Side Effects of Drugs" [12]. The database was used in a "*in silico*" analysis trough MATLAB® 7.10.0.499 (R2010a). Algorithms for decision trees were generated.

3 Results

The sample was taken of 19733 patients with 25386 different pharmacology combinations. A sample of 400 pharmacological prescriptions was analyzed (the highest occurrence [>= 9]). The first step on the research was classified by origin: General Prescriptions 283 cases y Cardiovascular prescriptions 117 cases. In order to cluster medications, a pseudo-code was made in MATLAB® 7.10.0.499 (R2010a) using Data Mining Analysis, for which the software will group medications according to the specific pseudo-code. Once the pseudo-code was written, the clustering of medications prescribed by a Level 1 Health Care Institution was made. General descriptive statistics of different variables showed 283 general prescriptions (71%) and 117 cardiovascular prescription (29%). The analysis of variable category (1, 2 y 3 see methods) of General Origin and Cardiovascular Origin is shown on Fig. 1 and 2.

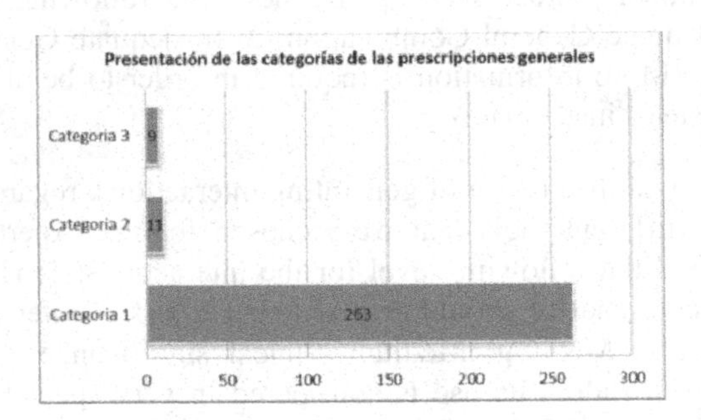

Fig. 1. Characteristics of general prescription

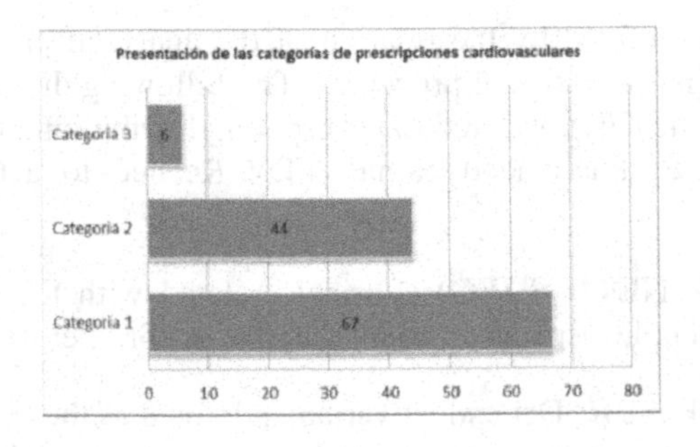

Fig. 2. Characteristics of cardiovascular prescription

After the general descriptive statistics analysis, a decision tree was implemented (Fig. 3) in order to establish a classification method that allows for the implementation of hierarchies on the weight of the different variables for each level of the structure, which then enables the construction of a predictive model related to the dependent variable, that in this specific case, is the potential of ADR.

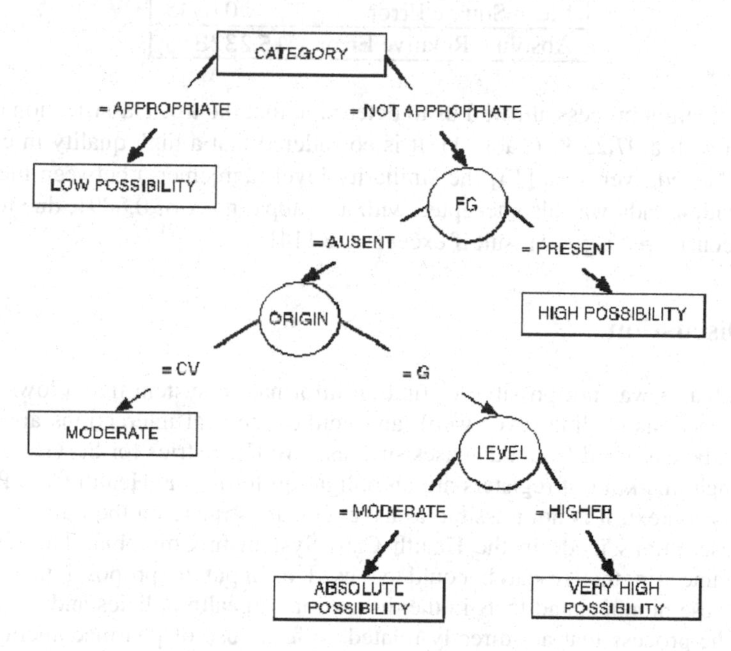

Fig. 3. Decision tree of Adverse Drug Reaction (ADR) derived from pharmacological treatment decision, generated on MATLAB® 7.10.0.499 (R2010a). FG: Pharmacogenetics, CV: Cardiovascular, G: General.

In order to validate the quality of the decision tree, the *cross validation* method was used, for which 10 folds were included, getting the following statistical classifications (Table 1).

Table 1. Classification results using J48

CLASSIFICATION	INSTANCES	%
Correctly Classified	389	97.25 %
Incorrectly Classified	11	2.75 %

On order to determine the similarity grade between the correctly and incorrectly classified data the statistical indicators were calculated (Table 2).

Table 2. Statistical indicators of J48

INDICATOR	VALUE
Kappa	0.8891
Absolute Mean Error	0.0156
Mean Square Error	0.0945
Absolute Relative Error	15.2328 %

The validation process allowed us to determine that the tree classification quality is very high with a 97.25 % (Table 1). It is considered that a high quality in classification is achieved over 90% [13] the similarity level (agreement) between the training and validation dada was also accepted with a Kappa index of 0.8891, due to the fact that it is considered a good result if exceeds 0.8 [14].

4 Discussion

In Colombia, it was not possible to find an information system that allows to report medical reactions on line (over web), and neither medical interactions are analyzed that could be extracted from databases such as RSP (Registries for Service Provision) even though this kind of registers are an obligation for all the Health Care Providers. Given this context, it is not possible to have a clear estimate on the number of inadequate prescriptions made in the Health Care System in Colombia. The results that were obtained in this research could be used as input to proposal that take into account these numbers and facts in the discussion of health policies and more specifically in the process that are directly related with the use of pharmacotherapeutic resources such as: 1. To promote the Access to recent technological innovations in the are specially information systems for medication 2. To implement information system those are adaptable to the needs of pharmaceutical needs. 3. To improve the information centers on medications in different attention levels as a service for practitioners and patients. 4. Carry out research within the institution regarding medication usage using analysis tools such as data mining [15]. It is necessary to take specific actions in the areas of prescription, delivery, public information and sanitary vigilance. Some of the strategies that are proposed in order to optimize pharmacological use are: 1. Search for mechanisms that can enhance medical prescriptions. 2. To improve pharmacovigilance programs, that include identification, analysis and evaluation of issues related to medications of pharmacological treatments in the whole population or in patient groups that are exposed to specific treatments, and the computation of quantitative aspects of the benefit/risk ratio in national territory. Data Mining is also used to spontaneous reporting databases, organized by the WHO. A study carried out by Kim [16] compares this database with data reported in claims for adverse reactions to insurance companies in Korea. Statistical methods were established using Bayesian inference for the indicator definition that would report the existence of safe signs and those indicators were applied to the data base in order to establish relations. The conclusion of the study was that reports in the Korean insurance databases are incipient given the absence of high interest data. Moreover, another study with Data Mining applied to health care sciences was made by Wang [17], where the necessity to use clinical narrative was proposed, taking into account that clinical narrative comprises the symptoms and textual expression that medico enter in the clinical history of the patient, in the feed of automated models of association among symptoms, diagnosis and medications for the formalization and early detection of adverse reactions. The results of this study showed an improvement of the findings from a 43% to 75%.

5 Conclusion

Our investigation witnessed the scarce amount of explorations and investigations that are related with the interaction of medication interactions (ADR), or data mining applied to data bases of prescription drugs on a national level.

The algorithm that was created for the grouping of medical prescriptions allowed identifying the variables of interest at the moment of decision making at a pharmaceutical point of view.

The decision tree permitted a solid, predictive relation to be found between the inadequate use of medicated prescriptions, the polymorphic metabolism of the medications (pharmacogenetics), and the high possibility of generating ADR; therefore, data mining allows us to approach a global analysis of sub-explored topics such as the derived interactions from aspects of pharmacokinetics and in a pharmacogenetics point of view to be more precise. This topic at the moment presents a great future for investigations worldwide.

The tree generated after the data mining process can establish effective working routes related to pharmacotherapeutic decision, which leads to the optimization of the use of the drug, a decrease in ADR and optimization of economic resources in the different programs like cardiovascular and general care in basic and specialized levels of attention.

In countries with growing economies, the application of the analysis *in silico* related with data mining raises new forms of incentive for investigations in the area of pharmacotherapeutics and for the development of solutions to the problems that are related to the inadequate medical prescriptions in all the levels of attention in health care. An important possibility to establish data mining among the various levels of attention in health care opens up, which, permits the creation of clear politics for the adequate use of pharmaceutical resources.

References

1. República de Colombia, Departamento Nacional de Planeación, Consejo Nacional de Política Económica y Social. Ministerio de Salud y Protección Social. Instituto Nacional de Vigilancia de Medicamentos y Alimentos. Documento Lin. Política farmacéutica nacional. Bogotá (2012)
2. Omran, A.R.: The epidemiologic transition: a theory of the epidemiology of population change. Milbank Q 83(4), 731–757 (2005)
3. Fernández, N., Díaz, Z., Pérez, B., Rojas, A.: Polifarmacia en el anciano. Acta Med. 10, 1–2 (2002)

4. Lin, A.R.: Practical Experience in Enterprise Terminology Management. In: 2011 Proceedings of the 14th International Conference on Information Fusion (FUSION), pp. 1–8. IEEE Explore, Chicago (2011)
5. Calderón, O.C.A., Urbina, B.A.: La Farmacovigilancia en los últimos 10 años: actualización de conceptos y clasificaciones. Logros y retos para el futuro en Colombia. MéD.UIS 24, 57–73 (2010)
6. BOT plus: Aplicación utilizada en España para el análisis de interacciones generales de medicamentos, `https://botplusweb.portalfarma.com/`
7. Dávila, F., Sánchez, Y.: Técnicas de minería de datos aplicadas al diagnóstico de entidades clínicas. RCIM 4(2), 174–183 (2012)
8. Fernández de CastroI, O.C., Jiménez, L.G., Gonzales, D.B.E., Avila, P.J.: Aplicación de la minería de datos al Sistema Cubano de Farmacovigilancia. Rev. Cubana Farm. 41, 3 (2007)
9. Drugs.com, `http://www.drugs.com/drug_interactions.html`
10. Interdrugs, `http://www.interdrugs.com.ar`
11. Drugs.com, `http://www.drugs.com/`
12. Aronson, J.K.: Meyler's Side Effects of Drugs: The International Encyclopedia of Adverse Drug Reactions and Interactions, 15th edn., 4192 pages (2006)
13. Lizazo, T.D., Delfor, M.R., Torres, C.V.: Minería de datos en la encuesta permanente de hogares 2009, Universidad Nacional del Litoral. Argentina 1, 19–28 (2011)
14. Warrens, M.J.: Conditional inequalities between Cohen's kappa and weighted kappas. Stat. Method 10, 14–22 (2013)
15. Rodríguez, R.J.E.: Software para clasificación/predicción de datos. Tecnura. año 11. 21 (2007)
16. Kim, J., Ha, J.H.: Signal detection of methylphenidate by comparing a spontaneous reporting database with a claims database. Regul. Toxicol Pharm. 63, 154–160 (2011)
17. Wang, X., Chase, H., Markatou, M., Hripcsak, G., Fri, C.: Selecting information in electronic health records for knowledge acquisition. J. Biomed. Inform. 43, 595–601 (2010)

iTRAQ, The High Throughput Data Analysis of Proteins to Understand Immunologic Expression in Insect

Amalia Muñoz-Gómez[1,2,3], Mauricio Corredor[2,3], Alfonso Benítez-Páez[2,3], and Carlos Peláez[1]

[1] Grupo Interdisciplinario de Estudios Moleculares (GIEM), Instituto de Química, Universidad de Antioquia, Medellín, Colombia
amalia.munoz@laposte.net, directorgiem@gmail.com

[2] Genética y Bioquímica de Microorganismos (GEBIOMIC), Instituto de Biología, Universidad de Antioquia, Medellín, Colombia
mcorredor@matematicas.udea.edu.co

[3] Grupo de Análisis Bioinformático (GABi), Centro de Investigación y Desarrollo en Biotecnología CIDBIO, Bogotá, D.C., Colombia
abenitez@cidbio.org

Abstract. Isobaric tags for relative and absolute quantification of protein expression (iTRAQ®) is a powerful tool which is combined with the accuracy of Mass Spectrometry for protein identification. This tool was approached to detect proteins associated to the innate immune system of *Galleria mellonella* in response to pathogenesis caused by *Fusarium oxysporum*. After experimental approaches, iTRAQ data was used to set up computational analysis based on identification and quantification of peptides and proteins against different protein databases by using ProteinPilot™ and Mascot Distiller search engines. iTRAQ battery was able to identify more than 340 peptides corresponding to 39 putative proteins from *G. mellonella* and close related species. Despite the low level of genomic and proteomic information available for *G. mellonella,* iTRAQ demonstrated to be reliable strategy to determine changes in protein expression as a consequence of the infection process induced in *G. mellonella.* Consequently, it was found differential expression in proteins directly involved in innate immune response such as cecropin-D-like peptide, lysozyme, and hemolin, indicating an active response of *G. mellonella* in early stages of fungal infection.

Keywords: iTRAQ, fungal infection, innate immune system, *Galleria mellonella*, *Fusarium oxysporum,* model insect.

1 Introduction

Insects are frequently attacked and infected by bacterial and fungal organisms. The fungal pathogenesis in model insects like *G. mellonella* [1] has special relevance given implications to disclose evolutionary aspects of ancient innate immune systems. Several studies have shown that *G. mellonella* is able to resist high concentrations of

L.F. Castillo et al. (eds.), *Advances in Computational Biology*,
Advances in Intelligent Systems and Computing 232,
DOI: 10.1007/978-3-319-01568-2_54, © Springer International Publishing Switzerland 2014

Fusarium oxysporum conidia, then reproducing a sub-lethal infection process where activation of the insect innate immune system can be studied by different molecular approaches. On the other hand, *F. oxysporum* is a trans-kingdom pathogen well known for production of harmful secondary metabolites, mycotoxins, responsible for several diseases in humans, livestock animals, insects and even in plants [2].

Regarding the immune response of insect against fungal infections, the insect hemolymph play a critical role becoming in the way to transport cells, proteins and peptides, oxygen, hormones, nutrients, and metabolites associated to the immune system [3,4]. Immunological study of insects comprises an important challenge to understand the molecular mechanisms of innate immune system in whole animal kingdom. Moreover, innate immune system is well known as the primary defense against pathogen or competitors attack because it is wide spread in all three major kingdoms of life [5]. Conversely, adaptive immune system is present only in vertebrates; as a result of gene family duplications and specializations, which allowed developing highly specialized mechanisms to control immune response in time and being selective for pathogens. Notwithstanding, molecules similar to those specialized in the adaptative immune system have been detected in insects given that antimicrobial peptides (AMPs) activity resembles that of cytokines involved to stimulate infection response in hemocytes [6].

Recent contributions in genomic characterization of *G. mellonella* immune system include a transcriptome analysis revealing presence of transcripts encoding a wide repertoire of humoral immunity proteins with potential antimicrobial activity [7]. In order to further analyze the immune system of *G. mellonella* and help in characterization of this model organism, we proposed to analyze its protein expression pattern during fungal infection with sub-lethal doses of *F. oxysporum* conidia. For this aim we designed a proteomic study based on iTRAQ together with a refined bioinformatics approach.

Different techniques for proteomics such as 2D gel-MS [8], LC MS/MS [9], ChIP-on-Chip, or Tandem Affinity Purification (TAP) are important alternatives for identification of protein expression and interaction. Currently, iTRAQ has especial advantages over conventional proteomics techniques because this one identifies and quantifies many proteins from the specific biological environments using label peptides able to be identified by sensitive mass spectrometers. The iTRAQ analysis is further enhanced with use of robust bioinformatic tools [10] and statistic analysis [11], which support the biological evidence discovered. Therefore, using the infection model (*G. mellonella-F. oxysporum*) and iTRAQ approach, we expect to identify and quantify new proteins and peptides produced after activation of innate immune response. Additionally, our experimental workflow will be assisted by an extensive database searching using ProteinPilot™ and Mascot Distiller search engines against non-redundant protein databases with special focusing in the functionally characterized family of proteins belonging to the insecta class.

2 Materials and Methods

2.1 Insect-Pathogen Model

Approximately 3000 *G. mellonella* larvae in the last instar were infected with 1 x10^4 ml^{-1} of *Fusarium* conidia (Foxy-GIEM strain), and incubated by 48 hours at 37 °C. Then, hemolymph was collected and processed free of hemocytes. Hemolymph isolated from surviving larvae (~95%) was concentrated and quantified by Bradford assay. Control larvae were treated conidia-free suspension buffer [12].

2.2 iTRAQ LC-MS/MS

Hemolyph samples were digested with trypsin, cystein blocking, and labeled with isobaric tags in multiplex reaction. Peptide separation was performed in AB Sciex LC-MS/MS 5600 Triple TOFF Mass Spectrometer (Keck Biotechnology Resources Lab., Yale University). The tandem MS/MS spectra were analyzed using ProteinPilot™.

2.3 Bioinformatics and Statistic Analysis

Proteins with 2 or more matching peptides were considered as positive identification. Level of confidence was expressed in ProtScore units. Mascot distiller and Mascot search algorithm were used for database searching. The protein database used was NCBInr_20121109. Protein expression ratios were calculated from the pair-wise comparison of two iTRAQ channels. For each ratio, the iTRAQ peak area was corrected for Observed Bias Correlation. Identified lepidoptera proteins were compared with *G. mellonella* peptides using pairwise alignment with Blastp algorithm [13], and multiple alignments with Muscle [14] and Probcons [15]. Functional annotation was manually performed by mining ModBase, UniProt, Pfam, and KEGG databases. Statistic analysis in ProteinPilot takes in account False Discovery Rate analysis for multiple testing corrections.

3 Results

3.1 Protein Identification

Hemolymph proteins from infected *G. mellonella* with positive identification are shown in Table 1. Exactly 374 peptides were positively resolved which matched to more than 60 different proteins. Among them, 47 proteins were identified with a level of confidence ≥ 99%; 54 proteins were identified at ≥ 95% of confidence. From this set of positively identified proteins, twenty-five corresponded to *G. mellonella*, 4 to *F. oxysporum* (data not show), and 19 to Lepidoptera order. The remaining proteins were identified to belong to other Insecta orders, other invertebrates, bacteria, protozoan and fungi (data not show). Particularly more than 340 peptides were assigned to Lepidoptera proteins (Table 1).

Table 1. List of proteins identified by iTRAQ from the greatest to lowest protein score. Protein coverage corresponds to number of amino acids from peptide matching number of amino acids to protein. Proteins in gray background correspond to *G. mellonella*, in white for related species.

#	Peptides	Protein Score	Percent Coverage	Protein Name
1	89	171.5	73.99	apolipophorin [*Galleria mellonella*]
2	48	137.87	83.9	arylphorin [*Galleria mellonella*]
3	31	55.35	83.09	transferrin precursor [*Galleria mellonella*]
4	22	46.66	72.24	hexamerin [*Galleria mellonella*]
5	20	38.07	88.17	apolipophorin-3; precursor [*Galleria mellonella*]
6	11	34.77	59.1	prophenoloxidase subunit 2 [*Galleria mellonella*]
7	12	34.27	91.53	27 kDa hemolymph protein; precursor [*Galleria mellonella*]
8	21	28.72	82.9	hemolin [*Galleria mellonella*]
9	14	26.98	65.15	prophenoloxidase [*Galleria mellonella*]
10	2	9.91	19.15	apolipophorins [*Danaus plexippus*]
11	8	8.8	70.61	juvenile hormone binding protein [*Galleria mellonella*]
12	6	7.6	81.82	lysozyme; 1,4-beta-N-acetylmuramidase [*Galleria mellonella*]
13	1	7.39	31.84	beta-1,3-glucan recognition protein precursor [*Galleria mellonella*]
14	6	6.72	28.02	hexamerin receptor [*Corcyra cephalonica*]
15	5	6.4	76.42	cationic protein 8 precursor [*Galleria mellonella*]
16	5	6.18	64.71	larval hemolymph protein [*Galleria mellonella*]
17	1	5.85	34.22	apolipophorin-3; precursor [*Hyphantria cunea*]
18	1	5.0	20.75	antennal esterase CXE5 [*Spodoptera exigua*]
19	2	4.89	18.69	apolipophorin precursor protein [*Bombyx mori*]
20	4	4.86	43.53	32 kDa ferritin subunit [Galleria mellonella]
21	3	4.62	59.85	cellular retinoic acid binding protein [*Danaus plexippus*]
22	4	4.55	25.35	imaginal disc growth factor 4 [*Papilio polytes*]
23	3	3.92	22.45	apolipophorin-2; precursor [*Manduca sexta*]
24	1	3.9	40.96	glyceraldehyde-3-phosphate dehydrogenase [*Bombyx mori*]
25	2	3.21	42.57	proline-rich protein [*Galleria mellonella*]
26	3	3.06	55.45	26kDa ferritin subunit [*Galleria mellonella*]
27	1	2.79	64.1	Cecropin-D-like peptide [*Galleria mellonella*]
28	1	2.68	26.32	hexamerin 2 [*Corcyra cephalonica*]
29	2	2.67	28.5	moderately methionine rich storage protein [*Sesamia nonagrioides*]
30	3	2.29	23.19	masquerade-like serine proteinase [*Pieris rapae*]
31	1	2.22	75	Anionic antimicrobial peptide 2 [*Galleria mellonella*]

Table 1. *(continued)*

32	2	2.2	23.99	moderately methionine rich storage protein [*Danaus plexippus*]
33	1	2.08	14.11	serpin 1 [*Danaus plexippus*]
34	1	2.04	40.99	isocitrate dehydrogenase [*Arcte modesta*]
35	1	1.4	41.33	kunitz-type protease inhibitor precursor [*Galleria mellonella*]
36	1	1.14	9.6	seminal fluid protein CSSFP007 [*Chilo suppressalis*]
37	2	1.08	20.67	peptidoglycan recognition protein SA, partial [*Papilio xuthus*]
38	1	0.57	22.71	hemolymph storage protein 1 [*Samia cynthia ricini*]
39	2	0.48	21.79	diapause associated protein 2 [*Choristoneura fumiferana*]

The low level of protein identification reached, after iTRAQ analysis, was directly attributed to the failure of genome information for *G. mellonella*. Notwithstanding, it would be feasible to expect a large amount of proteins identified by similarity with well characterized insects such as *Anopheles gambiae, Aedes aegypti, Apis mellifera,* and *Drosophila melanogaster*. These results would indicate that Lepidoptera species could differ enormously from other insects or that at least *G. mellonella* hemolymph proteins related to innate immune system are totally absent in other insect species. Therefore, further efforts must be conducted to characterize the genetic repertoire of *G. mellonella* and to disclose the level of variation with related insect species.

3.2 Differential and Quantitative Expression

After statistical analysis, we found differential protein expression in several proteins being over-expressed in hemolymph of *G. mellonella* infected with *F. oxysporum* conidia. Among this set of over-expressed proteins we found apolipophorin 3 precursor, cecropin-D-like peptide, and lysozyme. The role in antimicrobial defense of *G. mellonella* was previously demonstrated for last two proteins [16]; whereas function of apolipophorin 3 in the innate immune response is widely studied to act in pathogen recognition [17, 18], then acting like opsin which encapsulates and avoids pathogen dissemination; as well as promoter of phagocytosis and activation of antimicrobial systems [19, 20]. Surprisingly, we also detected down-regulation for proteins such as hemolin, and serpin, well characterized players actively participating in the insect immune response [21, 22]. In case of hemolin proteins, they recently have been associated to viral infections [23; 24], therefore, this would indicate that *G. mellonella* immune system is also able to distinguish viral infections in addition to previously reported [16].

4 Discussion

Proteomics is the study of proteins at large-scale. Hence, the characterization and analysis of proteins expressed in a determined cell, tissue (hemolymph) or organism is called proteome. However, proteome concept has reached new considerations given the differential expression or post-translation pattern of proteins succeeded in a specific time-point. Therefore, cell proteome must be always referred to specific conditions in which those proteins were identified to be present or absent. In our study, hemolymph larva was the main scope. Nowadays, cutting-edge proteomic approaches comprises the usage of iTRAQ, which is considered as a powerful tool for quantitative proteomics then permitting to both identify and measure expression of determined proteins in biological samples.

Proteomic iTRAQ approach was useful to detect and quantify protein expression pattern in hemolymph after an induced sub-lethal infection of *G. mellonella* larva with *F. oxysporum* conidia. Hemolymph protein identification was assisted by using ProteinPilot™ software and NCBI-nr database. Protein identification had low efficiency probably due to poor genomic characterization of *G. mellonella* in biological repositories. Notwithstanding, we are able to identify some dozens of proteins based on robust algorithms with statistical confidence. We were able to to identify 19 new *G. mellonella* peptides associated to humoral immunity. Pairwise comparisons using those peptides informed us about presence of orthologs in other species belonging the insecta class.

Our results are in agreement with previous reports which describe apolipophorin 3 to have a central role in insect immune response, being up-regulated after larval exposure toxins [25]. This function linked to insect immunity has also been reported for *Hyphantria cunea*, *Heliothis virescens*, *Locusta migratoria* and *Anopheles gambiae* species [26, 27].

Despite the number of proteins identified was relatively low, it is very important to highlight that all proteins identified correspond with soluble fraction of hemolimph, therefore we are describing partially the innate humoral immune system of *G. mellonella*. In similar manner, our results corroborate previous analysis showing to apolipophorin 3, cecropin-D-peptide, and lysozyme as proteins involved in the immune response against fungal infection. Interestingly, we present data suggesting that *G. mellonella* humoral innate immune system is able to distinguish fungal/bacterial from viral infections given the expression pattern observed for immunity proteins such a hemolin, which showed down-regulation pattern of expression during *F. oxysporum* infection.

5 Conclusion

Further efforts must be conducted for a deep genomic characterization of *G. mellonella,* then, proteomic approaches like iTRAQ® can be decisive for exploring and disclosing the whole protein network involved in the innate immune response of this model organism.

Acknowledgments. Thanks to Universidad de Antioquia for the budget support.

References

1. Mukherjee, K., Hain, T., Fischer, R., Chakraborty, T., Vilcinskas, A.: Brain infection and activation of neuronal repair mechanisms by the human pathogen Listeria monocytogenes in the lepidopteran model host Galleria mellonella. Virulence 24 4(4) (2013)
2. Navarro-Velasco, G.J., Prados-Rosales, R.C., Ortíz-Urquiza, A.: Quesada-Moraga. E., Di Pietro, A.: Galleria mellonella as model host for the trans-kingdom pathogen Fusarium oxysporum. Fungal Genet. Biol., 1124–1129 (2011)
3. Lehane, M.J.: The Biology of Blood-Sucking in Insects, 2nd edn. Cambridge University Press, New York (2005)
4. Lesch, C., Theopold, U.: Methods to study hemolymph clotting in insects. In: Beckage, N.E. (ed.) Insect Immunology, pp. 1–12. Academy Press, Salt Lake City (2008)
5. Schmid, O., Theopold, U., Beckage, N.E.: Insect and vertebrate inmunity: key similarities versus difference. In: Beckage, N.E. (ed.) Insect Immunology, pp. 1–24. Academic Press, Salt Lake City (2008)
6. Ishii, K., Hamamoto, H., Kamimura, M., Nakamura, Y., Noda, H., Imamura, K., Mita, K., Sekimizu, K.: Insect Cytokine Paralytic Peptide (PP) Induces Cellular and Humoral Immune Responses in the Silkworm Bombyx mori. J. Biol. Chem. 285(37), 28635–28642 (2010)
7. Vogel, H., Altincicek, B., Glöckner, G., Vilcinskas, A.: A comprehensive transcriptome and immune-gene repertoire of the lepidopteran model host Galleria mellonella. BMC Genomics 11(12), 308 (2011)
8. Hu, Y., Wang, G., Chen, G.Y., Fu, X., Yao, S.Q.: Proteome analysis of Saccharomyces cerevisiae under metal stress by two-dimensional differential gel electrophoresis. Electrophoresis 24, 1458–1470 (2003)
9. Aebersold, R., Mann, M.: Mass spectrometry-based proteomics. Nature 22(6928), 198–207 (2003)
10. Muth, T., Keller, D., Puetz, S.M., Martens, L., Sickmann, A., Boehm, A.M.: jTraqX: a free, platform independent tool for isobaric tag quantitation at the protein level. Proteomics 10(6), 1223–1225 (2010)
11. Schwacke, J.H., Hill, E.G., Krug, E.L., Comte-Walters, S., Schey, K.L.: iQuantitator: A tool for protein expression inference using iTRAQ. BMC Bioinformatics 10, 342 (2009)
12. Coleman, J.J., Muhammed, M., Sperkovitz, P.V., Vyas, J.M., Mylonakis, E.: Fusarium pathogenesis investigated using Galleria mellonella as a heterologous host. Fungal Biol. 115, 1279–1289 (2011)
13. Altschul, S.F., Gish, W., Miller, M., Myers, E.W., Lipman, D.J.: Basic Local Alignment Search Tool. J. Mol. Biol. 215, 403–410 (1990)
14. Edgar, R.C.: MUSCLE: multiple sequence alignment with high accuracy and high throughput. Nucleic Acids Res. 32, 1792–1797 (2004), doi:10.1093/nar/gkh340
15. Do, C.B., Mahabhashyam, M.S.P., Brudno, M., Batzoglou, S.: ProbCons: Probabilistic consistency-based multiple sequence alignment. Genome Research 15, 330–340 (2005)
16. Buyukguzel, E., Tunaz, H., Stanley, D., Buyukguzel, K.: Eicosanoids mediate Galleria mellonella cellular immune response to viral infection. J. Insect Physiol. 53, 99–105 (2007)
17. Dunphy, G.B., Oberholzer, U., Whiteway, M., Zakarian, R.J., Boomer, I.: Virulence of Candida albicans mutants toward larval Galleria mellonella (Insecta, Lepidoptera, Galleridae). Can. J. Microbiol. 49(8), 514–524 (2003)
18. Whitten, M.A., Tew, I.F., Lee, B.L., Ratcliffe, N.A.: A Novel Role for an Insect Apolipoprotein. J. Immunol. 172, 2177–2185 (2004)

19. Wiesner, A., Losen, S., Kopácek, P., Weise, P., Gotz, P.: Isolated apolipophorin III from Galleria mellonella stimulates the immune reactions of this insect. J. Insect Physiol. 43, 383–391 (1997)
20. Niere, M., MeiMlitzer, C., Dettlo, M., Weise, C., Ziegler, M., Wiesner, A.: Insect immune activation by recombinant Galleria mellonella apolipophorin III1. Biochim. Biophys. Acta 1433, 16–26 (1999)
21. Lanz-Mendoza, H., Bettencourt, M.F., Faye, I.: Regulation of the Insect Immune Response: The Effect of Hemolin on Cellular Immune Mechanisms. Cell Immunol. 169, 47–54 (1996)
22. Jiang, H., Vilcinskas, A., Kanost, M.R.: Immunity in lepidopteran insects. Adv. Exp. Med. Biol. 708, 181–204 (2010)
23. Terenius, O.: Hemolin—A lepidopteran anti-viral defense factor? Dev. Comp. Immunol. 32, 311–316 (2008)
24. Terenius, O., Pophamb, H.J.R., Shelby, K.S.: Bacterial, but not baculoviral infections stimulate Hemolin expression in noctuid moths. Dev. Comp. Immunol. 33, 1176–1185 (2009)
25. Oppert, B., Dowd, S.E., Bouffard, P., Li, L., Conesa, A., Lorenzen, M.D., Toutges, M., Marshall, J., Huestis, D.L., Fabrick, J., Oppert, C., Jurat-Fuentes, J.L.: Transcriptome profiling of the intoxication response of Tenebrio molitor larvae to Bacillus thuringiensis Cry3Aa protoxin. PLoS ONE 7(4), e34624 (2012)
26. Gupta, L., Noh, J.Y., Jo, Y.H., Oh, S.H., Kumar, S., Noh, M.Y., Lee, Y.S., Cha, S.-J., Seo, S.J., Kim, I., Han, Y.S., Barillas-Mury, C.: Apolipophorin-III mediates antiplasmodial epithelial responses in Anopheles gambiae (G3) mosquitoes. PLoS ONE 5(11), e15410 (2010)
27. Contreras, E., Rausell, C., Real, M.D.: Proteome Response of Tribolium castaneum Larvae to Bacillus thuringiensis Toxin Producing Strains. PLoS ONE 8(1), e55330 (2013)

Author Index